The Lime

Botany, Production and Uses

FSC
www.fsc.org
MIX
Paper from
responsible sources
FSC® C013604

The Lime

Botany, Production and Uses

Edited by

M. Mumtaz Khan
Rashid Al-Yahyai

and

Fahad Al-Said

CABI

CABI is a trading name of CAB International

CABI
Nosworthy Way
Wallingford
Oxfordshire OX10 8DE
UK

Tel: +44 (0)1491 832111
Fax: +44 (0)1491 833508
E-mail: info@cabi.org
Website: www.cabi.org

CABI
745 Atlantic Avenue
8th Floor
Boston, MA 02111
USA

Tel: +1 (617)682 9015
E-mail: cabi-nao@cabi.org

A catalogue record for this book is available from the British Library, London, UK.

Library of Congress Cataloging-in-Publication Data

Names: Khan, M. Mumtaz, editor.
Title: The lime (botany, production and uses) / edited by M. Mumtaz Khan, Rashid Al-Yahyai, Fahad Al-Said.
Description: Boston, MA : CABI, [2017] | Includes bibliographical references and index.
Identifiers: LCCN 2016045652| ISBN 9781780647845 (hbk : alk. paper) | ISBN 9781780647869 (epub)
Subjects: LCSH: Limes.
Classification: LCC SB370.L5 L56 2017 | DDC 634/.337--dc23 LC record available at https://lccn.loc.gov/2016045652

ISBN-13: 978 1 78064 784 5

Commissioning editor: Rachael Russell
Editorial assistant: Emma McCann
Production editor: Tim Kapp

Typeset by SPi, Pondicherry, India
Printed and bound in the UK by CPI Group (UK) Ltd, Croydon, CR0 4YY

Contents

Contributors

Waqar Ahmed, Sector Advisor-Horticulture, United States Agency for International Development – Cultivating New Frontiers in Agriculture (USAID-CNFA), Lahore 54000, Pakistan.

Muhammad Amjad Ali, Department of Plant Pathology and Centre of Agricultural Biochemistry and Biotechnology (CABB), University of Agriculture, Jail Road, Faisalabad 38040, Pakistan.

G. Almaguer-Vargas, Departamento de Fitotecnia, Universidad Autónoma Chapingo, Km. 38.5 Carretera México-Texcoco, Chapingo, Estado de México, C.P. 56230, México.

Ahmed Al-Maskari, Department of Crop Sciences, College of Agricultural and Marine Sciences, Sultan Qaboos University, PO Box 34, Al-khod 123, Sultanate of Oman.

A.M. Al-Sadi, Department of Crop Sciences, College of Agricultural and Marine Sciences, Sultan Qaboos University, PO Box 34, Al-khod 123, Sultanate of Oman.

Fahad Al-Said, Assistant Secretary General for Innovation Development, The Research Council, PO Box 1422, Muscat 130, Sultanate of Oman.

Rashid Al-Yahyai, Department of Crop Sciences, College of Agricultural and Marine Sciences, Sultan Qaboos University, PO Box 34, Al-khod 123, Sultanate of Oman.

Um-e-Ammara, Department of Crop Sciences, College of Agricultural and Marine Sciences, Sultan Qaboos University, PO Box 34, Al-khod 123, Sultanate of Oman.

César Fernandes Aquino, Departamento de Fitotecnia, Universidade Federal de Viçosa, Viçosa, Minas Gerais, 36570-900, Brazil.

N. Cruz-Huerta, Colegio de Posgraduados, Campus Montecillo, Km. 36.5, Carretera México-Texcoco, Montecillo, Estado de México, México.

Dalmo Lopes de Siqueira, Departamento de Fitotecnia, Universidade Federal de Viçosa, Viçosa, Minas Gerais, 36570-900, Brazil.

Philip Donkersley, Department of Entomology, Universidade Federal de Viçosa, Viçosa, Minas Gerais, 36570-900, Brazil.

Simon L. Elliot, Department of Entomology, Universidade Federal de Viçosa, Viçosa, Minas Gerais, 36570-900, Brazil.

Travis J. Esau, Engineering Department, Faculty of Agriculture, Dalhousie University, Halifax, Nova Scotia, B2N 4E3, Canada.

Aitazaz A. Farooque, School of Sustainable Design Engineering, University of Prince Edward Island, Charlottetown, Prince Edward Island, C1A 4P3, Canada.

Francisco García-Sánchez, Department of Plant Nutrition, CEBAS-CSIC, Campus Universitario of Espinardo, Espinardo, PO 30100, Murcia, Spain.
Vicente Gimeno, Department of Agronomy, ISA University, PO Box 166, Av. Pte. Antonio Guzmán Fernández Km. 5 1/2, La Herradura, Santiago, Dominican Republic.
V. González-Hernández, Colegio de Posgraduados, Campus Montecillo, Km. 36.5, Carretera México-Texcoco, Montecillo, Estado de México, México.
M. Asif Hanif, Department of Chemistry, University of Agriculture, Faisalabad 38040, Pakistan.
Rhonda Janke, Department of Crop Sciences, College of Agricultural and Marine Sciences, Sultan Qaboos University, PO Box 34, Al-khod 123, Sultanate of Oman.
Manjunath L. Keremane, National Clonal Germplasm Repository for Citrus and Dates, USDA-ARS, 1060 Martin Luther King Blvd, Riverside, California 92507, USA.
Ahmad Sattar Khan, Institute of Horticultural Sciences, University of Agriculture, Faisalabad 38040, Punjab, Pakistan.
M. Mumtaz Khan, Department of Crop Sciences, College of Agricultural and Marine Sciences, Sultan Qaboos University, PO Box 34, Al-khod 123, Sultanate of Oman.
Yong-Zhong Liu, College of Horticulture and Forestry Sciences, Huazhong Agricultural University/ Key Laboratory of Horticultural Plant Biology, Ministry of Education, 430070 Wuhan, PR China.
S.R. Márquez-Berber, Departamento de Fitotecnia, Universidad Autónoma Chapingo, Km. 38.5, Carretera México-Texcoco, Chapingo, Estado de México, C.P. 56230, México.
Juan José Martínez-Nicolás, Department of Vegetal Production and Microbiology, Escuela Politécnica Superior de Orihuela (Universidad Miguel Hernández), Orihuela, 03312, Alicante, Spain.
A. Nasehi, Department of Crop Sciences, College of Agricultural and Marine Sciences, Sultan Qaboos University, PO Box 34, Al-khod 123, Sultanate of Oman.
Muhammad Azher Nawaz, Department of Horticulture, University College of Agriculture, University of Sargodha, Sargodha 40100, Pakistan and College of Horticulture and Forestry Sciences, Huazhong Agricultural University/Key Laboratory of Horticultural Plant Biology, Ministry of Education, Wuhan 430070, PR China.
Smitha Padmanabhan, Department of Food Science and Nutrition, College of Agricultural and Marine Sciences, Sultan Qaboos University, PO Box 34, Al-khod 123, Sultanate of Oman.
Renan B. Queiroz, Instituto Capixaba de Pesquisa, Assistência Técnica e Extensão Rural, Rodovia BR 101 Norte, Km. 154, PO Box 62, Linhares, Espírito Santo, 29915-140, Brazil.
Chandrika Ramadugu, Department of Botany and Plant Sciences, University of California, Riverside, California 92511, USA.
I. Ramírez-Ramírez, Colegio de Posgraduados, Campus Montecillo, Km. 36.5, Carretera México-Texcoco, Montecillo, Estado de México, México.
Muhammad F. Razi, University of Agriculture, Faisalabad, Sub-campus Burewala, Vehari 61010 Pakistan.
Mikeal L. Roose, Department of Botany and Plant Sciences, University of California, Riverside, California 92511, USA.
Luiz Carlos Chamhum Salomão, Departamento de Fitotecnia, Universidade Federal de Viçosa, Viçosa, Minas Gerais, 36570-900, Brazil.
Arnold W. Schumann, Citrus Research and Education Center, University of Florida, 700 Experiment Station Road, Lake Alfred, Florida 33850-2299, USA.
Rainer W. Scora, Department of Botany and Plant Sciences, University of California, Riverside, California 92511, USA.
Silvia Simón-Grao, Department of Plant Nutrition, CEBAS-CSIC, Campus Universitario of Espinardo, Espinardo, Murcia 30100, Spain.
Zora Singh, Curtin Horticulture Research Laboratory, Department of Environment and Agriculture, School of Science, Faculty of Science and Engineering, Curtin University, GPO Box U 1987, Perth, WA, 6845, Australia.
A.K. Srivastava, Indian Council of Agricultural Research – Central Citrus Research Institute, Nagpur 440033, Maharashtra, India.

Muhammad Usman, Institute of Horticultural Sciences, University of Agriculture, Jail Road, Faisalabad 38040, Pakistan.

Mostafa Waly, Department of Food Science and Nutrition, College of Agricultural and Marine Sciences, Sultan Qaboos University, PO Box 34, Al-khod 123, Sultanate of Oman.

Qamar U. Zaman, Engineering Department, Faculty of Agriculture, Dalhousie University, Halifax, Nova Scotia, B2N 4E3, Canada.

Preface

Lime is among the oldest fruit crops in cultivation and its usage extends back in history from the earliest to modern civilizations. The use of lime has been documented in both folklore and modern practice in ancient and contemporary societies. Earlier, lime use was limited to domestic purposes only; however, its use is now much valued globally due to advances in science and technology and increased understanding of its significance for health and industry. The lime fruit is widely consumed all over the world for culinary and non-culinary purposes. Limes are grown throughout the warm subtropical to tropical regions and have been an important fruit crop all over the citrus-producing areas. The main producers of lime and lemon are India, México, China, Argentina, Brazil and the USA. This fruit is of significant interest to many because of its ornamental value, high vitamin C content, refreshing juice and numerous health applications. Lime cultivation has been afflicted with diverse problems – primarily drought, salinity, and high incidence of pests and diseases. In particular, new emerging diseases like witches' broom disease of lime (WBDL) and citrus greening disease (Huanglongbing, HLB) are problematic. For example, in Oman, more than half of the lime plantations have disappeared due to WBDL, a disease that has spread to neighbouring countries and threatens lime-producing countries around the world.

Lime is ranked third in global citrus production and holds a significant position in the world fresh and processed fruit trade. Nonetheless, little exclusive literature is currently available on lime culture worldwide. The citrus industry has benefited from scientific advancements with particular reference to limes, which are the product of hybridization of basic taxa. The breeding work resulted in broader diversity in this group to enable the selection of the best traits of lime cultivars. Recently seedless lime hybrids including C4-5-27 (a cross of 'Key lime' and tetraploid lemon) with superior fruit and plant characteristics have been released. Similarly, the development of pigmented Mexican lime has been reported through the introduction of transcription factor Ruby (Blood orange) and VvmybA1 (Grapes) in Mexican lime. It has opened up new opportunities for the development of lime cultivars with different colours and nutraceutical values. These new developments are extremely required to establish multi-disease resistant lime cultivars and compatible rootstocks.

The existing literature on *Citrus* has presented the genus in a comprehensive review, or covers major citrus fruits such as oranges, grapefruits, mandarins, etc. This book is the most recent collection of research and scientific literature exclusively dedicated to lime culture and the marketplace. A reader interested in lime production or research will now find ample information related to his/her relevant subject area in a single book. The coverage of each subject area is broad, ranging from basic cultivation practices to post-harvest handling to uses, which makes it the best choice for amateurs and professionals alike. With the latest information presented in simple language, this book will appeal to a range of principal users, e.g. educators, researchers, students, agriculture extension workers, farming communities, industry personnel and professionals/practitioners.

The refereeing help from many colleagues is highly appreciated and acknowledged; however, the editors of this book take full responsibility for any errors to be found. We highly appreciate the staff at CAB International for their continuous support and patience. This endeavour would have been difficult to achieve without the generous support of Sultan Qaboos University through His Majesty's grant on 'Rejuvenating Lime Production in Oman' awarded to the editors of this book, and also additional logistical support provided by the university. We are confident that this work will be a reference book for a broad spectrum of users and look forward to readers' feedback to enhance future volumes of this book.

M. Mumtaz Khan
Rashid Al-Yahyai
Fahad Al-Said

About the Editors

M. Mumtaz Khan is currently serving as an Associate Professor at the Department of Crop Sciences, College of Agricultural and Marine Sciences, Sultan Qaboos University, Oman. He earned his MSc (Hons) in horticulture from the University of Agriculture, Faisalabad, followed by a PhD and post-doctoral training from the University of Sheffield, UK. He has an enduring interest in horticultural crop production systems. His experience in citriculture is highly regarded by his peers and he has published widely on citrus crops. Dr Khan also has a deep interest in organic production systems, impact of climate change and abiotic stresses in horticultural crop plants. The use of marginal water and soil in agricultural crop production is another area of his concern. Teamwork has been the essence of his career as he has demonstrated in many joint research projects/ventures, and he has published in peer reviewed journals and books in collaboration with other academicians. He has taught undergraduate/postgraduate courses in horticulture and has mentored a large number of MSc and PhD candidates.

Rashid Al-Yahyai is the Dean of the College of Agricultural and Marine Sciences, and Associate Professor of Horticulture at the Department of Crop Sciences, Sultan Qaboos University, Oman. He earned his MSc from Cornell University, New York, in 1998 and his PhD in 2004 from the University of Florida. His research work focused on the effects of biotic and abiotic stress factors on the physiology, growth and productivity and post-harvest quality of horticultural crops, with emphasis on heat, drought and salinity. Recent research work includes studies on traditional farming systems, sustainable agriculture, agro-ecology, and the potential impact of changing ecological and climatic variables on fruit production in Oman. Dr Al-Yahyai is the principal investigator of the research project 'Rejuvenating lime production in Oman: Resolving current challenges', from which the idea for this book was born. He has supervised local and international MSc and PhD students working in the field of horticulture and plant production.

Fahad Al-Said is Assistant Secretary General of the Innovation Development Department of the Research Council of Oman. Prior to his current job, he was Associate Professor of Horticultural Sciences at Sultan Qaboos University. He obtained his MSc from the University of Florida in 1994 and his PhD from the University of Reading in 2000. His research interests are in horticultural crop production and post-harvest physiology. He has authored many publications in the area of fruit and vegetable production in arid climates including work on characterizing the physical and chemical constituents of several fruit crops grown in Oman. He has supervised a number of postgraduate students and has been leading principal investigator in several major grants in his area of expertise.

1 Introduction and Overview of Lime

M. Mumtaz Khan[1]*, Rashid Al-Yahyai[1] and Fahad Al-Said[2]
[1]Department of Crop Sciences, College of Agricultural and Marine Sciences, Sultan Qaboos University, Muscat, Oman; [2]Assistant Secretary General for Innovation Development, The Research Council, Muscat, Oman

Lime, lemon and citron belong to the acid citrus fruit group. This group is characterized by an elliptical to round shaped fruit with high citric acid content. Limes can be distinguished from other fruit in the citrus family as they have both acid and sweet varieties. The differences in tree morphology and fruit characteristics may even be sufficient to confer the status of separate species for the acid and sweet types. All citrus fruits have nearly the same structure; however, the elements that comprise these structures vary according to species and variety (Loussert, 1992). The external part of the rind consists of several morphologically different tissues called flavedo because of the presence of flavonoid compounds (Ortiz, 2002). The whole surface of the fruit is covered by polygonal cells to form the isodiametric layer. This layer contains cuticles that are partially enclosed with a waxy substance to prevent excessive loss of water from the fruit. A layer made of collenchyma and parenchyma cells is present under the layer of epidermis, where many oil glands containing essential oil are located at different depths within the parenchyma tissues. The white spongy part of the parenchyma, called the albedo, is located under the layers of collenchyma and parenchyma (Ting and Attaway, 1971). The name albedo is derived from the Latin (albus = white) (Ortiz, 2002). The edible part of the fruit is divided by carpel segments or locules. There are many juice vesicles within the carpels (Ting and Attaway, 1971; Rivera-Cabrera *et al*., 2010) (Fig. 1.1). The number of carpels vary, with acid lime fruit normally containing around 8–11 segments (Loussert, 1992).

The lime plant belongs to the kingdom Plantae; phylum Magnoliophyta; class Magnoliopsida; order Sapindales; family Rutaceae; genus *Citrus*; and species *aurantifolia* (Sethpakdee, 1992). The three main types of lime largely cultivated worldwide are Persian lime, Key lime (Mexican lime) and Makrut lime (Table 1.1). Key limes are a small rounded fruit (*Citrus aurantifolia* (Christ.) Swingle), while the Persian limes bear a larger fruit (*Citrus latifolia* Tanaka) that is triploid and seedless. The Persian lime is the most frequently grown lime variety globally and México is one of the major producing countries. The fruit has an elliptical shape, thin, smooth skin and is juicy with an acidic lime aroma. Less heat is needed for Persian lime fruit maturity than for Key limes, and they are more cold- and frost-tolerant (Reuther *et al*., 1967). The Makrut lime is produced in smaller amounts and is very popular in South-east Asian cookery. The fruit is small sized, the skin is rough and the leaves are aromatic and used in cuisine (Reuther *et al*., 1967).

*E-mail: mumtaz@squ.edu.om

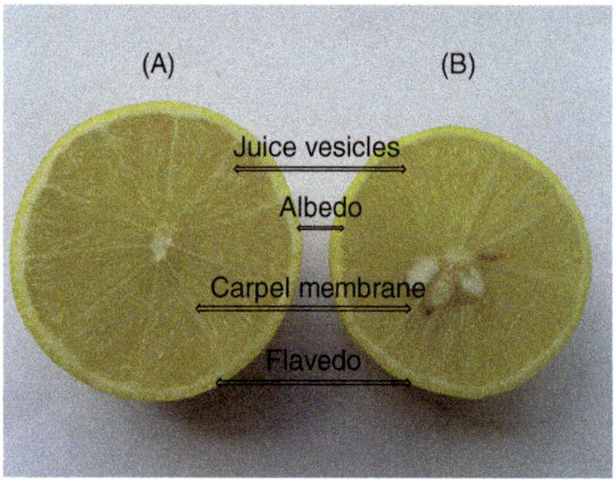

Fig. 1.1. Cross-section of lime fruit. (A) Persian lime, (B) Mexican. (Derived from Rivera-Cabrera *et al.*, 2010.)

Table 1.1. Description of major lime varieties.

Lime type	Other names	Appearance
Key lime (*Citrus* × *auratifolia*)	West Indian, Bartenders, Omani, Mexican	2.5–5 cm. diameter, high acidity, strong aroma, tart and bitter,7–8% citric acid
Persian lime (*Citrus* × *latifolia*)	Shiraz limoo, Tahitian, Bearss (seedless)	5–12.7 cm. diameter, slight nippled end, ripens to yellow but sold green
Makrut lime (*Citrus* × *hystrix*)	Kaffir	5 cm. diameter rough bumpy skin, thick rind, aromatic leaves used in cooking

Source: Reuther *et al.*, 1967

The Key lime (*Citrus aurantifolia* (Christ.) Swingle), is a polyembryonic species, grown globally, generally in warm tropical to subtropical regions such as India, México, the USA, Egypt and the West Indies (Morton, 1987). There are many common names used for Key lime, including West Indian lime, Mexican lime, Egyptian lime, Bilolo and Dayap, etc. The Key lime (*C. aurantifolia* Swingle) is mostly known as tri-hybrid, evolved through an inter-generic cross (a three-way hybrid where three plant species are involved and at least two different genera) of citron (*Citrus medica*), pummelo (*Citrus grandis*) and a *Microcitrus* species (*Citrus micrantha*).

The Key lime plant is spiny, less vigorous and less robust compared with Persian lime trees, and requires more heat for fruit development. It produces smaller fruit when it is grown in the Mediterranean climate due to the sporadic cold snaps and lower temperature regimes (Reuther *et al.*, 1967). On the other hand, it grows vigorously in tropical environments and produces higher yields.

History, Origin and Distribution

It has been suggested that the lime perhaps evolved from a tri-hybrid cross amongst *C. medica*, *C. grandis* and a *Microcitrus* species (Barrett and Rhodes, 1976). However, another study reported that mandarin was a parent of lime, while another parent could be citron, pummelo or *Papeda* (Handa *et al.*, 1986). Recent studies have strongly argued that Key lime was a hybrid of *Papeda* and *Citron*. The molecular evidence of these studies offered more conclusive information compared with all previous studies reported relating to the origin of lime (Li *et al.*, 2010). Like Key lime, 'Tahiti' lime is probably a tri-hybrid inter-generic cross of citron (*C. medica*), pummelo (*C. grandis*) and a *Microcitrus* species, *C. micrantha* (Moore, 2001). However, unlike Key lime, 'Tahiti' lime is a triploid.

The ancestral place of lime origin is also controversial, similar to that of other citrus species, but many researchers and historians believe that lime originated from South-east Asia around 4000 BC and its native home is the Indo-Malayan region (Nicolosi *et al.*, 2000). It is believed that the Europeans were not familiar with this fruit before the crusades, and it was Arab travellers who carried it to North Africa and the Near East. After the crusades it was taken to Palestine and then finally spread to Mediterranean Europe (Cooper and Chapot, 1977; Eckert and Eaks, 1989). During the middle of the thirteenth century, it was thought to be cultivated in some parts of Italy, France, Spain and Portugal. It was introduced in México during the time of Spanish colonization. It is strongly believed that both Portuguese and Spanish voyagers took it to the Americas during the early sixteenth century (Ziegler and Wolfe, 1961). Once introduced as an exotic plant, it has shown wide adaptation to diverse climates and was largely naturalized in México, the Caribbean, tropical areas of South America, Central America and the Florida Keys. During 1839 its cultivation started to expand in Florida and it was developed as a common home yard fruit. By 1883 it began to be produced on a small scale commercially in south-central Florida.

A hurricane in 1906 along with soil nutrient depletion resulted in the pineapple culture being abandoned. At that time, people began to plant limes as a substitute crop on the west coast of Florida. Lime fruit pickles became a prime snack choice among schoolchildren in Boston. A small-scale lime pickle business was active from 1913 to 1923 but it crumbled in 1926 due to the devastating impact on lime groves of another hurricane. Afterwards, the lime was largely grown as a common yard and garden plant in the Keys and the southern part of the Florida mainland.

In the Middle East, lime is a major traditional crop that has been used for a variety of culinary and medicinal purposes. Both fresh and dried limes are used for juice and as a condiment for food flavouring. Historically, sun-dried limes were a major export commodity of Oman, and are frequently used in sauces and as a flavouring agent in other Arab cuisines. Oman is a leading lime-producing country in the region. However, lime production has been drastically reduced in recent decades due to the spread of witches' broom disease of lime (WBDL), which has reduced the cultivated area to less than half since 1990 (Al-Yahyai *et al.*, 2012). Since its discovery in the 1970s, the disease has spread to other countries in the Middle East, including the United Arab Emirates (UAE), Saudi Arabia and Iran, threatening many major lime-producing countries.

Global Lime Production

Mexican lime is largely cultivated in tropical and subtropical areas of the globe. It is grown commercially, semi-commercially and in gardens. Lime is cultivated on a range of continents all over the globe, particularly in Asia and the Americas (Fig. 1.2a). Lime and lemon are mainly grown in India, México, China, Argentina, Brazil, the USA, Turkey, Spain, Iran and Italy (FAOSTAT, 2016). The percentage share of the top ten lime and lemon-producing countries is presented in Fig. 1.2b.

The relationship between area planted and production of the major lime-producing countries becomes clearer regarding total area under crop and the total harvest during 2013 upon examination of Fig. 1.3. Recognizing the importance of lime and lemon for both domestic and industrial uses, their cultivation has been increased considerably throughout the globe since 2000 (Fig. 1.4).

It is noteworthy to observe the tremendous increase in yields, which may be attributed to better plant material and the advancement in technologically driven crop management. The

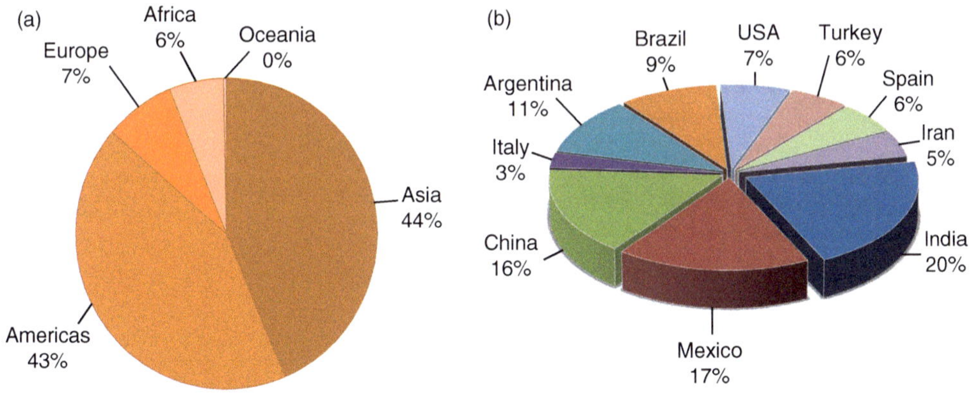

Fig. 1.2. The percentage of lime and lemon production on different continents (a) and top ten global producers in 2013 (b). (Source: FAOSTAT, 2016.)

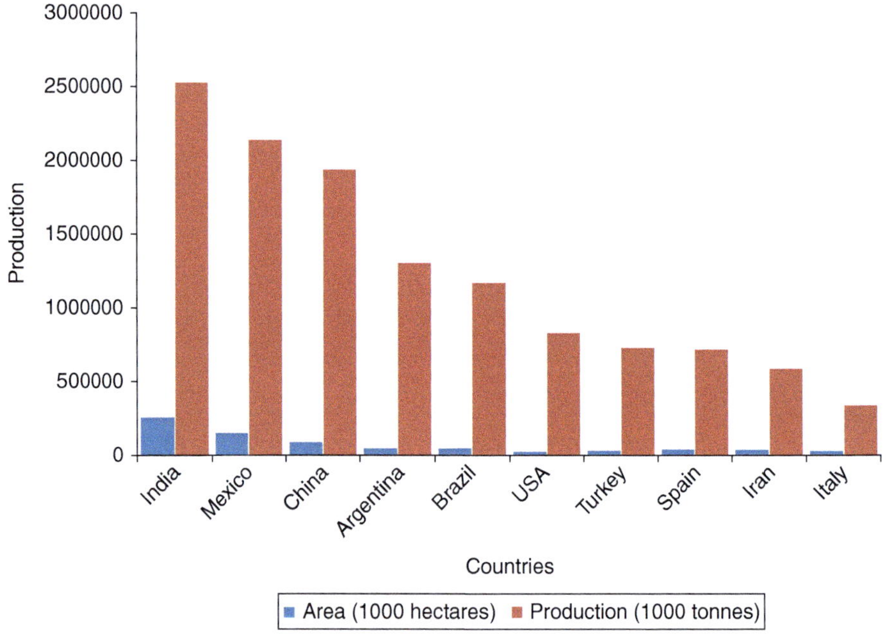

Fig. 1.3. Lime and lemon production and area harvested in top ten producing countries during 2013. (Source: FAOSTAT, 2016.)

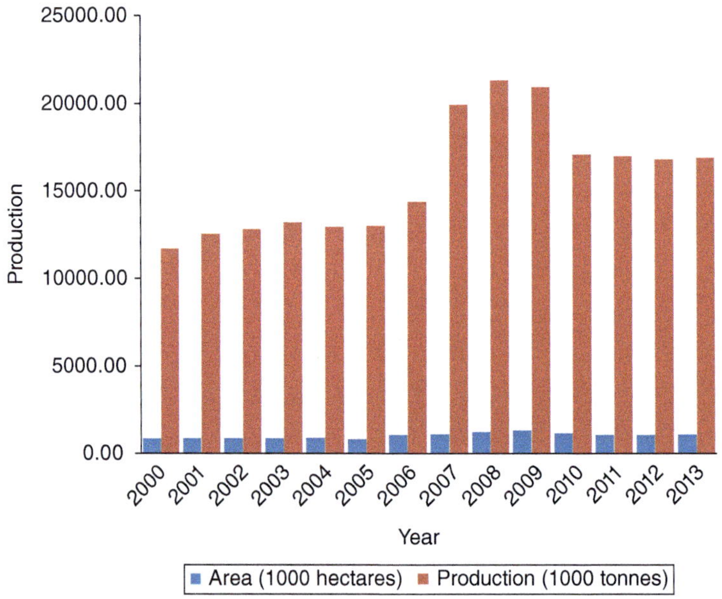

Fig. 1.4. Lime and lemon production and area harvested 2000–2013. (Source: FAOSTAT, 2016.)

fruit yield was 13 million tonnes during 2005 and has recently been maintained at approximately 18.9 million tonnes from 2010 to 2013 (FAOSTAT, 2016). Lime-producing countries harvest fruit for both domestic consumption and export. It is interesting to note that the top lime-producing country (India) has negligible lime exports, while Spain and México had the highest quantity of lime exported during 2013 (Fig. 1.5a). Europe, the USA and Russia were the largest importers of lime and lemon (Fig. 1.5b).

Plant Description

In general, the lime plant is largely a small shrub-like tree, approximately 5 m in height. It is an evergreen, ever-bearing tree that is densely and irregularly branched and possesses short, stiff spines (thorns). The leaves are alternate; elliptical to oblong-ovate (4–8 cm × 2–5 cm) in shape and have a crenulated margin. The flower diameter is about 2.5 cm and flower colour is yellow to white with a little purple tinge on the margins. The fruit are globose to ovoid berries of about 3–6 cm in diameter and sometimes have apical

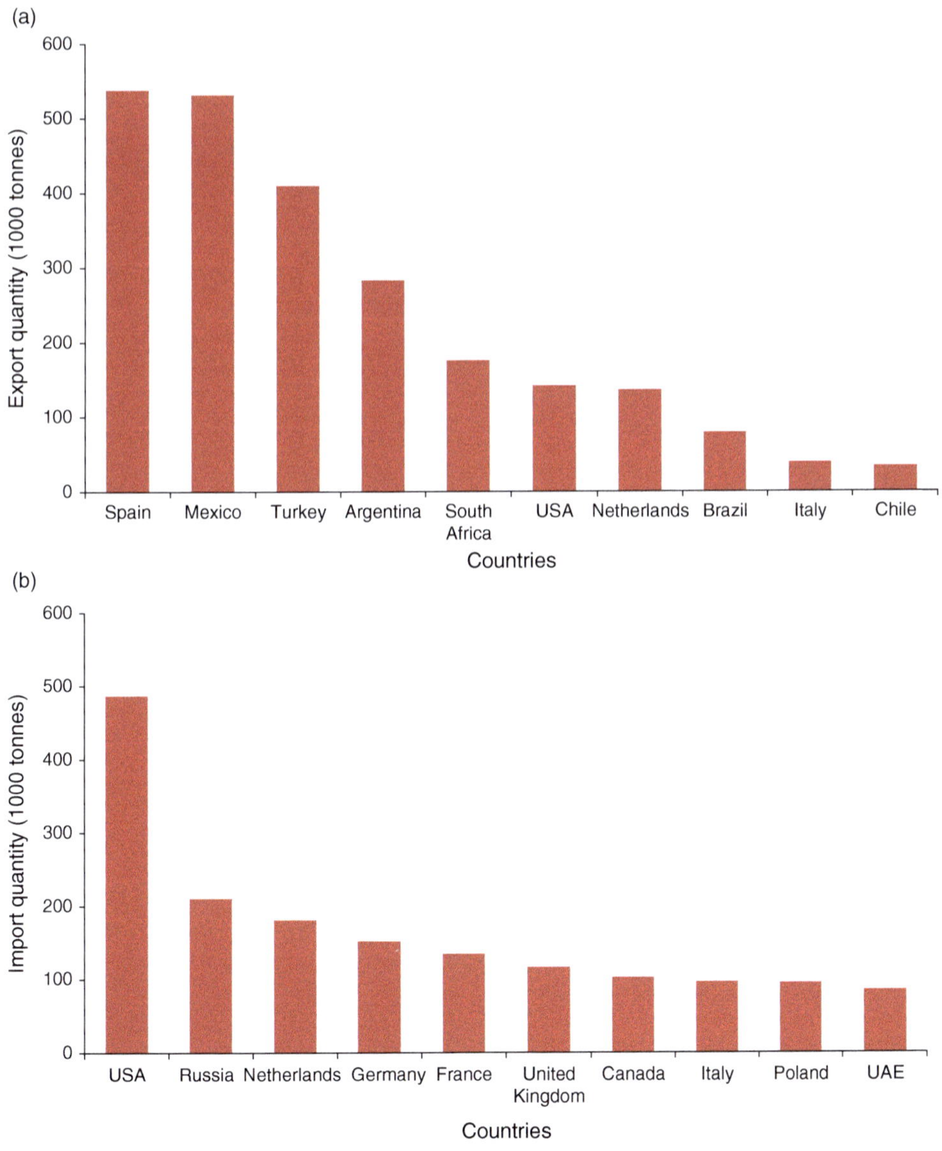

Fig. 1.5. Lime and lemon top ten (a) exporters and (b) importers of the world during 2013. (Source: FAOSTAT, 2016.)

papilla. Limes are usually picked when green for commercial production; however, if kept on the tree longer, the fruit turns yellow at maturity. The lime tree keeps bearing fruit and flowers throughout the year, but blooms most profusely during May–September in the northern hemisphere. The fruit peels are very thin with dense glandular segments with yellow-green pulp vesicles. The fruit juice is acidic and fragrant, as sour as lemon juice but more aromatic. The *C. aurantifolia* limes enjoy high standing due to this exceptional aroma compared with other lime types. The seeds are small, plump, ovoid, pale and smooth with a white embryo.

Propagation

Key lime seeds are poly-embryonic and generally multiplied sexually (seeds). To maintain identical clones in limes asexual propagation is practised. Mature or hardwood cuttings are used for propagation but generally are not able to carry strong root systems. Root sprouts are also used in some areas. Sprouting is stimulated by digging round the parent plant to separate the roots completely or partially. Ground and air layering are also very popular methods of propagation in many regions. For example, in Oman, ground layering is practised, while in Indonesia and Florida, air layering is more common. Using indole butyric acid (IBA) to aid root development, air layering can result in 100% success when propagating the 'Kaghzi' lime in India (Morton, 1987). The improvement of rootstock is highly desirable for obtaining certain benefits for lime cultivars, e.g. biotic and abiotic stress tolerance, enhanced fruit quality, earliness and a better canopy with a robust root system. For example, clones budded onto Rough lemon/sour orange can provide more resistance to strong winds or hurricanes.

Cultivars

There are relatively small differences among wild or cultivated varieties of the Key lime excluding a few thornless cultivars.

'Everglade'

Also known as Philippine Islands 218, this is a seedling of Key lime that was fertilized by grapefruit/pummelo pollen. However, the fruit does not show any dominant feature of grapefruit or pummelo. This fruit was first introduced in Trinidad in 1922, but it showed little or no distinctive characteristics compared with cultivated Key lime when grown in the Citrus Experiment Station collections in Riverside, California. It is similar to lime: the fruit shape is elliptical, having a small nipple at the fruit apex; fruit size is 4–5 cm wide, 4.5–5.4 cm in height; the fruit rind becomes light yellow on ripening; the oil glands in fruit peel are marginally dipped; it has very small size glands about 1.5 mm in length; pulp colour is light greenish; it has eight to ten segments with softer covering walls; scented aroma; and high juice content. The texture and quality are excellent; the aroma lively acidic; the number of seeds may vary from two to ten, with an average number of approximately five. The flowers are perfect and large clusters of fruit are borne on tree branches. The tree is extremely sensitive to lime anthracnose 'wither tip' infection caused by the *Colletotrichum* species of fungus.

'Kaghzi'

This is an acid lime cultivar that is very popular and cultivated on a commercial scale in India. It has several subtypes with varied tree and fruit size, shape and colour. In general, fruit are small to medium; pulp is greenish with a strong aroma; the juice vesicles are heavily adhered to the skin; the skin is green to yellow, thin or papery and shiny. These limes are largely grown for processing purposes.

'Palmetto'

This is a Key lime seedling selection cross-pollinated by the 'Sicily' lemon. Dr H. J. Webber reported this first in the United States Department Yearbook in 1905. Its form is globular to elliptical having a minor nipple on the fruit at the apex. Fruit size is small; peel colour is pale yellow on ripening; the pulp colour is light green to yellow; and juice vesicles are juicy and enriched with a lively flavour. It has a small number of seeds (three to six).

'Yung'

This is a spineless/thornless Key lime and was introduced from México and into California in 1882 by George Yung.

A thornless bud sport was reported in the Dominican Republic in 1892, and a similar line was introduced from Trinidad to the USA in 1910. Many spineless bud sports were recorded in lime plantations near Weslaco, Texas, after a 1925 freeze. Lime plant seeds found growing in the Yuma desert, Arizona, were introduced to southern Florida in 1967. After germinating about 50 seedlings, there were eight that were virtually spineless. The bud stock from these was selected and grafted onto Rough lemon for dissemination to growers.

During 1925 in Trinidad a hybridization programme was started to develop wither tip (anthracnose) resistant genotypes. Promising selections were made and hybrids with desired traits were given the name 'T-l'. Its fruit was slightly bigger compared with the typical Key lime. The fruit was not juicy at the green stage. However, back crosses were carried out to achieve typical Key lime traits.

Pests and Diseases

Lime is prone to a large number of insects/pests and diseases similar to other citrus species, leading to mild to huge losses. These losses are chiefly dependent on the type of pest, disease, environmental conditions and resistance of the plants. Leaf miners, leafhoppers, scales, mites and psyllids, etc. are considered economically damaging insects of lime. The lime is susceptible to diverse bacterial, fungal and viral diseases (Morton, 1987). The lime anthracnose (wither tip), canker, nematodes and certain *Fusarium* species also result in negative impacts on tree health and productivity. More recently WBDL and Huanglongbing (HLB) or greening have been considered new potential threats to lime plantations in many regions of the world. It has been reported that an important pest elsewhere is the snow scale (*Unaspis citri*), particularly during protracted droughts. It can cause dieback of branches once heavily infested, and it pierces the bark, which encourages other insects and fungi to invade. Ants are also a frequent visitor, moving from plant to plant. Insects cause direct and indirect (vector) threats to the lime industry. Post-harvest decay-related organisms are another dilemma for sustained lime businesses. A chapter in this book is dedicated to lime diseases and pests, and includes a great deal of information regarding infestation control measures.

Uses of Lime

Fruit uses

Lime fruit peel and leaves have been used for culinary and non-culinary purposes around the globe. In addition, lime is known for its juice extract, which is used as a cleansing agent and in cooking (Morton, 1987; Bocco *et al.*, 1998). The use of lime in human history has been documented for centuries, though its usage was mainly limited to folk custom. In general citrus fruits including lime have been reported to have high anti-oxidant properties, e.g. anti-cancer, anti-inflammatory, and anti-fungal and blood clot inhibition (Guimarães *et al.*, 2010; Karoui and Marzouk, 2013). The lime essential oils are used in pharmaceutical forms, fragrances and perfumes, and food flavouring (Dongmo *et al.*, 2013). Lime twigs and leaves are used in the perfume industry and its twigs and the leaves are used to extract petitgrain cedrat oil.

Lime carries a strong aromatic and acidic tang, which makes it valuable in drinks, curries, rice dishes, cakes, desserts, pickles, salads, sauces, jams and jellies (Hardy *et al.*, 2010). Dried lime is abundantly available in Middle Eastern markets, since it is commonly used in Arabian cuisines (Fig. 1.6). Limes contain vitamins, flavonoids, especially ascorbic acid (vitamin C) and citric acid, and provide a potential alternative to synthetic antioxidants (Morton *et al.*, 1994). Although the lime industry is not well developed, increasing health consciousness and industrial usage have increased demand. Limes with special features, for example, finger

Fig. 1.6. Dried lime being sold in Middle East countries. (Source: Courtesy of S.A. Siddiqui.)

lime and coloured flesh lime have high demand in restaurants (Hardy *et al.*, 2010; Siebert *et al.*, 2010).

Tree uses

Lime seedlings are frequently used as a rootstock for other citrus cultivars. Limes and lemons are generally considered as salt tolerant with vigorous growth and are used as rootstock for other citrus cultivars on a limited scale in Iran, India and other Asian countries (Bitters, 1986). However, lime germplasm is a potential source for crop improvement programmes for the production of new plant materials (scions and rootstocks) with better fruit and plant characteristics (Kahn *et al.*, 2001a, b; Krueger and Navarro, 2007; Nawaz *et al.*, 2007a, b). Lime trees are used in landscaping and for ornamental use in front yards and backyards. Their flowers and leaves have a specific scent, and the small plants, lush green foliage and small yellow coloured fruit look very attractive. Lime trees carry flower blossoms and young to fully ripened fruit at the same time, which further augments its aesthetic look. A detailed agronomy of 'Tahiti' lime cultivation as a home landscape tree has been explained (Crane and Osborne, 2013). Recently, the Citrus Research and Education Centre in Florida released transgenic Mexican lime plants exhibiting unique pigmented leaf, flower and fruit pulp. These pigmented transgenic materials carry great value for ornamental horticulture and for human health (Dutt *et al.*, 2016).

Challenges and Future Perspectives

Although the documented history of limes is more than 700 years old, the parentage and origin of limes is not clear. Advancement of biotechnological and molecular tools can aid in resolving these issues. A recent report revealed the phylogenetic origin of lemons and limes by using cytoplasmic and nuclear markers (Curk *et al.*, 2016). These studies are helpful in understanding the history of limes and will open up new avenues for research and development work on limes.

Lime plants are facing a number of problems across the world, including excessive attacks from insect pests (citrus psyllids, leaf miners, mealy bugs, citrus canker, bacterial diseases, fungal disease, viral and viroid-related diseases and some physiological problems). The emerging threats to the lime industry are climate change impact and new diseases like WBDL and HLB. These natural calamities have drastically reduced lime plantations in many regions of the globe. Substantial work on rootstock intervention on other closely related citrus fruits (e.g. oranges, mandarins and grapefruit) has led to remarkable achievements. However, very little work has been reported on rootstock scion interaction on lime, and it is hard to find standard rootstocks being used for limes. The role of rootstocks in other fruits has been well documented; lime needs further attention from plant breeders to develop resistant genotypes with the best traits to help in promoting lime cultivation across the world. Lime genetic resources are eroding and face serious threats and losses. These valuable resources need to be preserved and broadened for future crop improvement programmes by the breeders and geneticists.

Despite lime cultivation having faced numerous setbacks in the past, there is massive potential as the demand from both the domestic and industrial consumer grows. It has been reported that the consumption and trading of limes between countries has been on an upward trend in recent years (Plattner, 2014). Keeping in view the customer and industry demands, new lime cultivars/hybrids are expected to emerge to meet the demands of different countries across the world. Recently, seedless lime hybrids including C4-5-27 (a cross of 'Key lime' and tetraploid lemon) with superior fruit and plant characteristics have been released (Grosser *et al.*, 2015). Similarly, the development of pigmented Key lime has been described through the introduction of transcription factor Ruby (blood orange) and VvmybA1 (grapes). This has opened up new opportunities for the development of lime cultivars with different colours for ornamental and nutraceutical values (Dutt *et al.*, 2016). These conventional and emerging biotechnological tools are creating interesting new lime genotypes, which will include biotic/abiotic resistance as well as possessing health/ornamental values.

About this Book

A survey of the current literature indicates the limited availability of authoritative reference material on limes. Current books on citrus cover mainly oranges, grapefruits and mandarins. There is little exclusive literature available on lime cultivation worldwide, despite lime being ranked third in global citrus production and having a significant share of the citrus trade. Citrus species including lime are grown throughout the globe and particularly in the northern to southern hemisphere and Mediterranean regions. Detailed research-based information presented by dedicated academicians with in-depth knowledge will greatly enhance the understanding of lime cultivation. The diversity of the subjects covered in an array of chapters brings an opportunity to citrus students, amateurs, agriculture extension workers, researchers, growers and fruit industry experts to comprehend the advances and deficits in lime culture. A reader interested in lime can find much information related to his/her subject in a single book.

The book covers an assorted collection of subjects. For example, in Chapter 1, lime importance and global production are emphasized; in Chapter 2, its comprehensive systematic classification and distribution is covered; in Chapter 3, lime breeding, genetics and biodiversity are discussed; and in Chapter 4, lime tree growth, development and reproductive physiology are elaborated. In Chapter 5, propagation techniques, nursery production and certification are explained; in Chapter 6, schematic planning for lime orchard establishment is discussed; in Chapter 7, crop water requirements, irrigation systems and fertigation are highlighted; and in Chapter 8, cultural practices, e.g. tree pruning, thinning, weeding, planting density and intercropping, are covered. In Chapter 9, precision agriculture in lime is elaborated; in Chapter 10, insect pests and diseases and their control measures are covered; in Chapter 11, innovative production technologies, e.g. the use of plant growth regulators, remote sensing and organic production systems, are highlighted. In Chapter 12, harvesting and post-harvest management of lime fruit are emphasized; and in Chapter 13, uses (folk to modern) and future dynamics are discussed in depth.

References

Al-Yahyai, R., Khan, I., Al-Said, F., Al-Sadi, A., Al-Wahaibi, A. and Deadman, M. (2012) Status of *Citrus aurantifolia* infected with witches' broom disease of lime in Oman. *Acta Horticulturae* 928, 375–381.

Barrett, H.C. and Rhodes, A.M. (1976) A numerical taxonomic study of affinity relationships in cultivated citrus and its close relatives. *Systematic Botany* 1, 105–136.

Bitters, W.P. (1986) Citrus rootstocks: their characters and reactions. University of California Riverside Science Library. Available at: http://www.citrusvariety.ucr.edu/links/documents/Bitters.pdf (accessed 19 July 2016).

Bocco, A., Cuvelier, M.E., Richard, H. and Berset, C. (1998) Antioxidant activity and phenolic composition of citrus peel and seed extracts. *Journal of Agricultural and Food Chemistry* 46, 2113–2129.

Cooper, W.C. and Chapot, H. (1977) Fruit production with special emphasis on fruit for processing. In: Nagy, S., Shaw, P.E. and Veldhuis, M.K. (eds) *Citrus Science and Technology*, 1st edn, Vol 2. AVI Publishing Co., Westport, Connecticut, pp. 1–127.

Crane, J.H. and Osborne, J.L. (2013) *Growing 'Tahiti' Limes in the Home Landscape*. US Department of Agriculture, University of Florida, IFAS extension HS8. University Cooperative Extension Program and Boards of County Commissioners Cooperating. Nick T. Place, dean for UF/IFAS Extension.

Curk, F., Ollitrault, F., Garcia-Lor, A., Luro, F., Navarro, L. and Ollitrault, P. (2016) Phylogenetic origin of limes and lemons revealed by cytoplasmic and nuclear markers. *Annals of Botany* 117, 565–583.
Dongmo, P.M.J., Tchoumbougnang, F., Boyom, F.F., Sonwa, E.T., Zollo, P.H.A. and Menut, C. (2013) Antiradical, antioxidant activities and anti-inflammatory potential of the essential oils of the varieties of citrus limon *and citrus aurantifolia* growing in Cameroon. *Journal of Asian Scientific Research* 3(10), 1046–1057.
Dutt, M., Stanton, D. and Grosser, J.W. (2016) Ornacitrus: development of genetically modified Anthocyanin-expressing citrus with both ornamental and fresh fruit potential. *Journal of the American Society for Horticultural Science* 141(1), 54–61.
Eckert, J.W. and Eaks, I.L. (1989) Postharvest disorders and diseases of citrus fruits. In: Reuter, W., Calavan, E.C. and Carman, G.E. (eds) *The Citrus Industry*. Division of Agriculture and Natural Resources, University of California Press, Berkeley, California, pp. 179–260.
FAOSTAT (2016) The FAO Statistical Database online. Available at: http://faostat.fao.org/site/567/default.aspx#ancor (accessed: 19 July 2016).
Grosser, J.W., Viloria, Z. and Dutt, M. (2015) New acid citrus selections for Florida. Citrus Industry 30–33. Available at: http://www.crec.ifas.ufl.edu/extension/trade_journals/2015/2015_July_acid.pdf (accessed 18 July 2016).
Guimarães, R., Barros, L., Barreira, J.C., Sousa, M.J., Carvalho, A.M. and Ferreira, I.C. (2010) Targeting excessive free radicals with peels and juices of citrus fruits: grapefruit, lemon, lime and orange. *Food and Chemical Toxicology* 48(1), 99–106.
Handa, T., Ishizawa, Y. and Oogaki, C. (1986) Phylogenetic study of fraction I protein in the genus *Citrus* and its close related genera. *The Japanese Journal of Genetics* 61, 15–24.
Hardy, S., Wilk, P., Viola, J. and Rennie, S. (2010) Growing Australian native finger limes. *Primefact* 979, 1–11.
Kahn, T.L., Krueger, R.R., Gumpf, D.J., Roose, M.L., Arpaia, M.L. *et al.* (2001a) Genetic Resources Conservation Program. Division of Agriculture and Natural Resources, University of California, USA. pp. 1–58. Available at: http://www.grcp.ucdavis.edu (accessed 19 July 2016).
Khan, I.A., Jaskani, M.J., Khan, M.M. and Fatima, B. (2001b) Breeding for triploid seedless citrus fruits: a biotechnology approach. In: Al-Shayji, Y.A., Sidhu, J.S., Saleem, M. and Guerinik, K. (eds) *Biotechnology Applications for the Arid Regions*. Kuwait Institute for Scientific Research, Kuwait, pp. 229–234 (ISBN 99906-41-16-1).
Karoui, I.J. and Marzouk, B. (2013) Characterization of bioactive compounds in Tunisian bitter orange (*Citrus aurantium* L.) peel and juice and determination of their antioxidant activities. *BioMed Research International* 2013, pp. 1–12.
Krueger, R.R. and Navarro, L. (2007) Citrus germplasm resources. In: Khan, I.A. (ed.) *Citrus Genetics, Breeding and Biotechnology*. CAB International, Wallingford, UK, pp. 45–140.
Li, X., Xie, R., Lu, Z. and Zhou, Z. (2010) The origin of cultivated citrus as inferred from internal transcribed spacer and chloroplast DNA sequence and amplified fragment length polymorphism fingerprints. *Journal of the American Society for Horticultural Science* 135(4), 341–350.
Loussert, R. (1992) *Los Agrios*, 1st edn. Mundi-Prensa, Madrid, p. 319.
Moore, G. (2001) Oranges and lemons: clues to the taxonomy of citrus from molecular markers. *Trends Genetics* 17, 536–540.
Morton, J.F. (1987) Mexican lime. In: Morton, J.F. (ed.) *Fruits of Warm Climates*. Creative Resource Systems, Winterville, North Carolina, pp. 168–172.
Morton, J.F., Miami, F.L., Morel, I., Lescoat, G., Cillard, P. and Cillard, J. (1994) Role of flavonoids and iron chelation in antioxidant action. *Methods in Enzymology* 234, 437–443.
Nawaz, M.A., Khan, M.M., Jaskani, M.J. and Iftikhar, Y. (2007a) Present status of citrus gene pool in Pakistan. Proceedings of the International Symposium on Prospects of Horticulture Industry in Pakistan. University of Agriculture, Faisalabad. pp. 70–74.
Nawaz, M.A., Khan, M.M., Jaskani, M.J., Iftikhar, Y. and Ahmed, W. (2007b) *Citrus Germplasm Status in Pakistan and its Role in Citrus Nurseries. Proceedings of Workshop on Production of Disease Free Citrus Nursery Plants (Production, Certification and Challenges).* University of Agriculture, Faisalabad, pp. 91–98.
Nicolosi, E., Deng, Z.N., Gentile, A., Malfa, S.L., Ciotinella, G., and Tribulato, E. (2000) Citrus phylogeny and genetic origin of important species as investigated by molecular markers. *Theoretical and Applied Genetics* 100, 1155–1166.
Ortiz, J.M. (2002) Botany: taxonomy, morphology and physiology of fruits, leaves and flowers. In: Dugo, G. and Di Giacomo, A. (eds) *Citrus: The Genus Citrus*, 1st edn. CRC Press, Taylor and Francis Group, Boca Raton, Florida, pp. 114–147.

Plattner, K. (2014) *Fresh-Market Limes. Fruit and Tree Nuts Outlook: Economic Insight*. Economic Research Service, United States Department of Agriculture, FTS-357SA, pp. 1–9.
Reuther, W., Webber, H.J. and Batchelor, L.D. (1967) *The Citrus Industry Volume 1: History, World Distribution, Botany and Varieties*. University of California, Board of Regents, Division of Agricultural Sciences. Available at: http://websites.lib.ucr.edu/agnic/webber/Vol1/Vol1TP.html (accessed 19 July 2016).
Rivera-Cabrera, F., Ponce-Valadez, M., Díaz de León Sánchez, F., Villegas-Monter, A. and Pérez-Flores, L.J. (2010) Acid limes. A review. In: Sivakumar, D. (ed.) *New Trends in Postharvest Management of Fresh Produce II*. Fresh Produce 4 (Special Issue 1). Global Science Books, Japan, pp. 116–122.
Sethpakdee, R. (1992) *Citrus aurantifolia* (Christm. and Panzer) Swingle. In: Coronel, R.E. and Verheij, E.W. (eds) *Plant Resources of South-East Asia. Edible Fruits and Nuts*. Prosea Foundation, Bogor, Indonesia, 2, pp. 126–128.
Siebert, T., Krueger, R., Kahn, T., Bash, J. and Vidalakis, G. (2010) Descriptions of new varieties recently distributed from the citrus clonal protection program. *Citrograph* March/April 2010, 20–26.
Ting, S.V. and Attaway, J.A. (1971) Citrus fruit. In: Hulme, A.C. (ed.) *The Biochemistry of Fruit and their Products*, Vol I. Academic Press, London, pp. 107–369.
Ziegler, L.W. and Wolfe, H.S. (1961) *Citrus Growing in Florida*. University of Florida Press, Gainesville, Florida, pp. 51–53.

2 Systematic Classification, Distribution and Botany

Chandrika Ramadugu[1]*, Muhammad F. Razi[2], Manjunath L. Keremane[3], Rainer W. Scora[1] and Mikeal L. Roose[1]
[1]*Department of Botany and Plant Sciences, University of California, Riverside, California, USA;* [2]*University of Agriculture, Faisalabad, Sub-campus Burewala, Pakistan;* [3]*National Clonal Germplasm Repository for Citrus and Dates, USDA-ARS, Riverside, California, USA*

Limes are well known for their distinctive flavour, intense acidity, pleasant aroma and high Vitamin C content. The unique flavour imparted by limes is considered indispensable in many cuisines and in certain alcoholic beverages. The aroma of the zest of the lime fruit, the acidity and the intense flavour of sour limes are highly valued in the culinary industry. In many cultures limes are considered to be important because of the medicinal properties associated with their antioxidant activity, flavonoid content and polyphenolic compounds. Limes, lemons and citrons constitute the acid members of the citrus group. Limes probably originated in South Asia, mainly north-eastern parts of India and surrounding regions of Myanmar (Burma), and northern Malaysia (Bhattacharya and Dutta, 1956; Nicolosi, 2007). They have been cultivated for many centuries in tropical and subtropical regions (Hodgson, 1967). The common name 'lime' is used to refer to many different citrus hybrids known to have diverse origins. The heterogeneous group includes citrus plants with similar fruit flavour used primarily as essential ingredients in certain alcoholic and non-alcoholic beverages, and as flavour-enhancing condiments. Several of the accessions classified as limes are not related to each other. In this sense, it is not a natural or monophyletic group.

Botany

Lime trees are small to medium sized, evergreen, usually with axillary thorns and rounded or angled stems. Most limes are sensitive to cold temperatures and are drought tolerant. The pattern of arrangement of the leaves on the stem (phyllotaxy) is typically alternate. The petioles have narrow wings and are visibly articulated with broad, oblong shaped leaf blades. The tender foliage is usually purple in acid limes and green in the majority of non-acid forms. The trees are generally ever-flowering and ever-bearing, as is typical of the other acid citrus fruit trees like citrons and lemons. In certain lime cultivars, flowering is seasonal as is observed in mandarins, pummelos and other types of citrus. The flowers are pentamerous, fragrant and usually white in colour, sometimes with pink or purple buds. The fruit is a berry, described as a hesperidium, which is typical of the family Rutaceae. The size of the fruit varies from very small (about 2–5 cm in diameter in acid limes) to medium

*E-mail: chandram@ucr.edu

sized (7–10 cm in sweet limes). The fruit shape is usually round or slightly oblong; the Australian finger lime has characteristic finger-shaped berries. The outer surface of the fruit can be smooth or very rough, glandular and bumpy. Most limes have a mammilla, which is a projection at the stylar end of the fruit, a characteristic they share with lemons and certain citrons. Occasionally, the mammilla may be absent in certain lime varieties. The fruit has an outer rind consisting of a colourful, greenish, yellowish or bright orange coloured flavedo layer and an internal, white albedo layer. The rind is extremely thin in Kaghzi lime; about 0.5–1 cm thick in sweet limes and Rangpur lime. The inside of the fruit consists of a central spongy columella surrounded by six to ten distinct segments separated by membranes. The columella is usually solid, not as prominent as in citrons and does not disintegrate as in mandarins. The segments are filled with juice vesicles, which are thin and lanceolate, fusiform, compressed laterally or globose. The juice vesicles are borne on short stalks and are generally greenish in colour in the common lime cultivars. Certain varieties have yellow, bright orange (as in Rangpur lime) or brightly coloured juice vesicles (red, pink, bright green or burgundy as in certain types of finger limes and related hybrids). The fruit flavour ranges from intensely sour (as in sour limes) to mild (in non-acid forms; sometimes considered insipid). Sweet lime is a misnomer for the non-acid types since the perceived sweet flavour is due to lack of sourness. The number of seeds per fruit can vary from zero (Persian lime) to 15 (Mexican lime). Most limes are highly polyembryonic; the seeds give rise to many nucellar seedlings that are derived from the vegetative cells and are identical to the maternal parent. The Australian limes are predominantly monoembryonic. Certain limes are also used as rootstocks for other citrus cultivars. When limes are used as rootstocks, the fruit of the scion grafted on these rootstocks are larger, and have lower acidity and Brix levels (determined by the total soluble solids, mainly sugars) (Krezdorn, 1979). The entomophilous flowers are pollinated in nature by bees. Lime plants grown indoors are generally pollinated by hand.

Limes and lemons have many common characters and distinguishing between the two groups can be challenging since there are many intermediate forms. In Spanish and Portuguese, the word used to designate both limes and lemons is 'limon'. Such nomenclature issues add to the confusion that exists in distinguishing among the acid citrus fruits. Lemons are generally described as large, oval, yellow fruits, while limes are traditionally small, green and round. However, immature lemons are green and many limes are yellow when mature. Availability of newer cultivars of limes has made this differentiation even more difficult. Commercially cultivated 'Persian' lime can be as large as a lemon and colour distinctions are not always applicable. The perceived flavour of the fruit is a result of a combination of different acids, sugars, flavonoids, limonoids and many volatile components. The composition and ratio of the different compounds impact the perception of flavour associated with each cultivar. Lime juice accents the flavour of food because of its acidic taste and presence of certain phytochemicals. The pH of lime juice is often around 2.0; the acidity imparts a sour taste to foods, considered to be a very important quality for the acid citrus fruits. Acidity is due to the presence of citric and malic acids. In the culinary industry, the distinct flavours of limes and lemons are well recognized. Lemons are not preferred as substitutes for limes, especially in the beverage industry.

Limes and lemons have been associated with health benefits. The citrus fruits have flavonoids with antioxidant and radical scavenging activities, and are associated with a reduced risk of certain chronic human diseases. Consumption of citrus fruits rich in bio-active compounds is associated with lowering the risk of oesophageal, colorectal and stomach cancers, obesity, improvement of blood lipid profiles and prevention of stroke (Roowi and Crozier, 2011). The anti-cancerous effects are known to be due to the presence of kaempferol, a natural flavonol, known to stop cell division in many types of cancers (Luo *et al.*, 2010). The carotenoids, limonoids, coumarins, essential oils and other compounds present in lime fruit have numerous beneficial effects. The bitter taste in many lime fruits is due to the presence of neohesperidoside conjugates and certain limonoids such as limonin (Albach and Redman, 1969; Roowi and Crozier, 2011). Glycosides (compounds with sugar groups) and flavanone aglycones (compounds lacking the sugar moieties) affect the taste of citrus fruits and juice.

Systematic Classification of Limes

The term 'biological species' refers to a group of organisms that can interbreed and that is reproductively isolated from other such groups (Mayr, 1940). Species classified under the genus *Citrus* are an exception to this general rule since the different taxonomic species in the group are known to interbreed with each other. Mandarins, pummelos, citrons, papedas and kumquats are genetically distinct, yet can crossbreed forming interspecific hybrids. Members of the genus *Citrus* are also known to be sexually compatible with most other closely related genera belonging to the subtribe citrinae (a group consisting of the 'true citrus fruit trees'). Intergeneric hybrids of *Citrus, Fortunella* (kumquat), *Poncirus* (trifoliate), *Microcitrus* (Australian finger lime) and *Eremocitrus* (Australian desert lime), five of the six genera in the subtribe citrinae, have been reported in the literature (Barrett, 1977; Iwamasa *et al.*, 1988). Numerous morphological forms that are generated by natural hybridization of plants belonging to different citrus groups are now cultivated extensively. In addition, many biotypes of probable hybrid origin are grown as improved root stock cultivars, novel scion varieties and ornamentals. Many of these have been propagated successfully, despite probable sexual imperfection, because of apomixis, which involves development of embryos derived directly from maternal cells of the ovary rather than from the fusion of male and female gametes. Clonal mode of propagation ensures survival of many aberrant types. Somatic variants known as 'bud sports' arise sporadically in all citrus including lime. Various types of lime hybrids derived from citron, lemon, papeda and finger lime have been cultivated. Because of natural and deliberate interbreeding of different citrus cultivars, selection of desirable characters observed in bud sports and propagation of horticulturally interesting forms, there are a large number of lime varieties.

Citrus belongs to the family Rutaceae, subfamily Aurantioideae, tribe Citreae and subtribe Citrinae according to the classification proposed by Swingle and Reece (1967). Many different classifications of plants belonging to this group exist; in this article, we will use the most widely accepted system proposed by Swingle and Reece (1967), allude to the species specific names proposed by Tanaka (1977), while mentioning other alternate names proposed by different taxonomists, for clarity. In the tribe Citrinae, Swingle and Reece recognize three groups based on the pulp vesicle morphology and fruit characteristics: (i) primitive citrus fruit trees (consisting of the genera *Severinia, Pleiospermium, Burkillanthus, Limnocitrus* and *Hesperethusa*); (ii) near-citrus fruit trees (including genera *Citropsis* and *Atalantia*); and (iii) true citrus fruit trees (comprising genera *Fortunella, Eremocitrus, Poncirus, Clymenia, Microcitrus* and *Citrus*). Some of the most widely accepted classifications of the genus *Citrus* are those of Swingle, recognizing 16 species (Swingle, 1943; Swingle and Reece, 1967), and Tanaka, recognizing 162 species (Tanaka, 1977).

Early taxonomy of limes

Early taxonomists classified *Citrus* based entirely on morphological attributes. This is reflected in the works of Volkamer (a two-volume treatise on citrus entitled *Nürnbergische hesperides* published in 1708), Linnaeus (published *Species plantarum* in 1753), Gallesio (*Traité du Citrus* published in 1811) and Poiteau and Risso (*Histoire naturelle des orangers*, a volume on citrus classification published in 1818). In all of these taxonomies, limes, lemons and citrons were placed under one species: *Citrus medica* Linn. (Bhattacharya and Dutta, 1956). In 1832, in his publication *Flora Indica*, Roxburgh recognized five *Citrus* species and classified all limes under *Citrus acida*. Hooker (*Flora of British India*, published in 1872) and Brandis (*Forest Flora of Northwest and Central India*, 1874 and 1906) included limes, lemons and citrons under *Citrus medica* var. acida (Bhattacharya and Dutta, 1956). In 1910 Lushington endorsed Roxburgh's circumscription and clustered limes under *C. acida*. Swingle separated limes from lemons and classified a majority of the cultivated hybrid limes under a separate single species; *Citrus aurantifolia* (Christm.) Swingle (Swingle, 1913). Tanaka recognized about 16 species names for *Citrus* taxa belonging to the lime group and assigned species rank to many of the limes that are obviously hybrids. These are sometimes referred to as secondary species (Curk *et al.*, 2015). The classification proposed by Tyôzaburô Tanaka was supported by his successor, Yuichirô Tanaka. Although Tanaka's classification has been described as anomalous and criticized by other taxonomists as excessive splitting of the species, recognition of many species names for limes emphasizes the wide horticultural variability observed in the group (Bhattacharya and Dutta, 1956; Barrett and Rhodes, 1976). Despite taxonomical controversies, Tanaka's classification is considered very useful since it is much more comprehensive than many other systems including Swingle's (Hodgson, 1967). Tanaka divided the limes (section limonellus) into small fruited (subsection eulimonellus) and large fruited (subsection megacarpa). Subsection eulimonellus includes three species: *C. aurantifolia* Swing. (Mexican lime), *Citrus limettioides* Tanaka (Palestine sweet lime) and *Citrus latifolia* Tanaka (Persian lime). Subsection megacarpa includes 13 species: *Citrus javanica* Blume, *Citrus papaya* Hassk., *Citrus ovata* Hassk., *Citrus obversa* Hassk., *Citrus pseudolimonum* Wester, *Citrus webberi* Wester (Kalpi lime), *Citrus montana* Tanaka, *Citrus excelsa* Wester, *Citrus davaoensis* Tanaka (Davao lime), *Citrus longispina* Wester (Winged lime), *Citrus hyalopulpa* Tanaka, *Citrus bergamia* Risso & Poit. (Bergamot orange) and *Citrus pennivesiculata* Tanaka (Gajanimma) (Swingle and Reece, 1967).

Biochemical and molecular studies

During the early twentieth century biochemical tools became available to taxonomists. Swingle, Kefford and Horowitz were the first to recognize the presence of specific flavonoid substances in different citrus groups (Albach and Redman, 1969). Horowitz observed that the type of rutinose and non-hesperidose derivatives present in the fruit can be utilized for the taxonomy of citrus (Albach and Redman, 1969). Colorimetric reactions have been used to distinguish between different types of citrus. Powdered bark treated with Almen reagent turned the extracts deep pink (in sweet orange), light pink (in grapefruit), very light pink (in Rough lemon) or brown (in sour orange) (Halma and Haas, 1929). Scora *et al.* utilized gas chromatographic profiles of distilled essential oil from citrus cultivars to identify taxa (Scora *et al.*, 1966; Scora, 1988).

Flavanone composition of citrus has been utilized to identify *Citrus* species and to deduce parentage of certain citrus hybrids (Albach and Redman, 1969). Quantitative analysis of the different limonoids indicated a correlation between the form of limonoids and the type of citrus. The ratio of nomilin to limonin and the relative proportions of deacetylnomilin and obacunone were reported to be characteristic of different citrus species (Rouseff and Nagy, 1982). Sweet oranges, mandarins, lemons and citrons have hesperidin as the major flavonoid and hence were clustered together; pummelo and grapefruit have naringin as the principal flavonoid and formed a distinct group. Natsudaidai has a mixture of hesperidin and naringin, indicating the possibility of a mandarin and a pummelo in its parentage (Albach and Redman, 1969). Swingle incorporated such biochemical information—in addition to the morphological characters—into his taxonomical system and identified 16 major citrus species. Current availability of molecular data consisting of DNA sequences, microsatellite marker information, single nucleotide polymorphism profiles and genomic data has contributed towards a better understanding of the relationship between various accessions classified as limes (Barrett and Rhodes, 1976; Nicolosi *et al.*, 2000; Bayer *et al.*, 2009; Garcia-Lor *et al.*, 2013; Ramadugu *et al.*, 2013; Curk *et al.*, 2015; Curk *et al.*, 2016).

Sour and sweet limes

True limes form a diverse group with two natural clusters: acid or sour limes, and acidless or sweet limes. Acid limes can be either small fruited or large fruited types that also vary in their ability to thrive under different climatic conditions, resistance or susceptibility to various diseases, etc. (Hodgson, 1967). These two groups were considered by certain taxonomists to be distinct enough to merit creation of new species. In addition to the acidity/sweetness of the juice vesicles, there are many distinct differences between the two groups. Sour lime trees are generally bushy with thin branches consisting of small, dark green leaves with a prominent winged petiole. All parts of a sour lime plant have oil glands and exude a characteristic aroma when crushed. The trees flower all year long and the small flowers are borne in racemes. The fruit are small (2–5 cm diameter) with a thin rind, a solid columella and very sour, greenish juice vesicles. The seeds are small with greenish white cotyledons. In contrast, the sweet lime trees are mostly medium sized, larger than the sour lime trees and have thicker branches. The leaves of sweet lime are yellowish green, medium or large without the winged petiole present in sour limes. The distinct aroma of sour limes is missing in the sweet lime group. Sweet lime trees flower annually and have large flowers borne in cymes. Fruits are large (7–10 cm in diameter) with a thick rind, a semi-hollow columella and white, sweet (non-sour) pulp vesicles. Seeds are large with white cotyledons (Bhattacharya and Dutta, 1956). Sour lime (*C. aurantifolia*) and sweet lime (*C. limettioides*) can be distinguished by a colorimetric reaction of the bark tissue using Almen reagent ($FeCl_3$, $(NH_4)_2MoO_4$; Albach and Redman, 1969). In 1832, Roxburgh classified sweet lime under *C. acida*, while Hooker clustered sweet limes under *C. medica* var. limetta in 1872. Swingle did not recognize sweet lime as a separate species but included it under *C. aurantifolia* (Swingle, 1913). Webber classified sweet lime as a variety of the sour lime, *C. aurantifolia* Swingle. (Webber, 1943). Bhattacharya and Dutta endorsed the classification of sweet lime proposed by Risso and recognized *C. limetta* (Risso) Lushington (Bhattacharya and Dutta, 1956). Tanaka classified sweet limes under *C. limettioides* Tan.

Major groups of cultivated limes

Five main groups are recognized here:

1. Small fruited acid limes (*C. aurantifolia* (Christ.) Swingle): Mexican lime (Fig. 2.1) and its clonal derivatives like West Indian lime, Kaghzi lime and Key lime are the most common cultivars. Other small fruited acid limes are cultivated in certain regions only; examples are, Egyptian lime (Fig. 2.2), Abhayapuri lime (Fig. 2.3), Everglade lime and India lime (Fig. 2.4).
2. Large fruited acid limes (*C. latifolia* (Yu. Tanaka) Tanaka): Persian lime (Fig. 2.5) and its clonal selections like Tahiti lime or Bearss lime are well known in this category.
3. Sweet limes (*C. limettioides* Tan.): common cultivars are Palestine sweet lime or Indian sweet lime (Fig. 2.6), Soh synteng, Columbia, Lemonade lemon hybrid (lemon lime hybrid) and Mary Ellen sweet lime (Fig. 2.7).
4. Australian lime group consisting of finger limes (Fig. 2.8) belonging to the genus *Microcitrus* and the Australian desert lime, *Eremocitrus* (Fig. 2.9). According to Mabberley, the Australian limes are grouped under the genus *Citrus* (Mabberley, 1997; Mabberley, 1998). The Australian limes are not considered to be ancestors or derivatives of the three lime groups described above.
5. Lime hybrids: numerous lime hybrids are cultivated in various citrus growing regions. Perrine lemonime is a popular hybrid of lemon × lime (lemonimes). Limequats are a result of the cross between lime and kumquat. Examples are, Eustis, Lakeland and Tavares (Fig. 2.10). Certain lime hybrids are grown exclusively in some regions. Addanimma hybrid is only grown in India.

Hodgson recognized four of the five groups mentioned above in his published work on horticultural varieties of citrus (Hodgson, 1967). At that time Australian limes were not popular in the global market. This has changed since many Australian lime cultivars are now grown in many parts of the world.

Parentage of Limes

Cultivated citrus is presumed to have been derived through hybridization of four basic types of citrus: mandarin (*Citrus reticulata*), pummelo (*Citrus maxima*), citron (*C. medica*) and papeda (*Citrus micrantha*) (Scora, 1975; Nicolosi *et al.*, 2000; Moore, 2001; Barkley *et al.*, 2006; Ollitrault *et al.*, 2012; Garcia-Lor *et al.*, 2013; Curk *et al.*, 2016). The major techniques used to understand the parentage of citrus cultivars include: (i) analysis of chromosome structure (Guerra *et al.*, 1997; Roose *et al.*, 1998; Guerra *et al.*, 2000; Yamamoto et al., 2007); (ii) comparison of the types of flavonoids and limonoids present in the hybrids and the parental types (Peterson *et al.*, 2006); (iii) composition of volatile oils (Scora and Kumamoto, 1983); (iv) profile of isoenzymes (Esen and Scora, 1977; Torres *et al.*, 1978; Hirai *et al.*, 1986; Novelli *et al.*, 2000); (v) chloroplast sequence analysis (Nicolosi *et al.*, 2000; Araújo *et al.*, 2003; Abkenar *et al.*, 2004; Bayer *et al.*, 2009; Penjor *et al.*, 2010; Curk *et al.*, 2016); (vi) restriction fragment length polymorphism (RFLP) and random amplified polymorphic DNA analysis (RAPD)

(Federici *et al.*, 1998; Nicolosi *et al.*, 2000; Abkenar and Isshiki, 2003); (vii) simple sequence repeat information (SSR or microsatellite; Gulsen and Roose, 2001a; Barkley *et al.*, 2006; Garcia-Lor *et al.*, 2013); (viii) single nucleotide polymorphism information (SNP; Ollitrault *et al.*, 2012; Garcia-Lor *et al.*, 2013; Ramadugu *et al.*, 2013; Curk *et al.*, 2015; Curk *et al.*, 2016); and many variants of these methods. In hybrids, the level of admixture and presence of specific markers or traits are indicative of the parentage (Pritchard *et al.*, 2000). Because of repeated hybridization events, a hybrid cultivar can become introgressed with heterologous genome fragments. Any of these common methods of analysis, when used alone, will not be adequate to definitively assess the nature of admixture in a hybrid. Consideration of all available data may be necessary to infer the parentage. Often, conflicting results are obtained when methods with different algorithms and assumptions are used in such analyses. The number and type of accessions and marker loci included in the dataset are known to influence the quality of data generated. Information about the place of origin of natural hybrids, inclusion of all possible parental types and other sexually compatible taxa from this geographical location may be essential for such a study. Often, certain putative parental types may not be available. Certain taxa included as representatives of different groups like mandarins, pummelos, citrons, papedas, etc. may have admixtures and this can confound the results of the analysis. Because of the complexities involved, determining the correct parentage of hybrids based on indirect data can be challenging. Confirmatory evidence regarding the exact parentage of any hybrid can be generated by two approaches, both equally challenging. The first approach consists of recreating a hybrid by controlled breeding of putative parental types. Since essentially all citrus is heterozygous, progeny from hybridization have high levels of variability, so it is highly unlikely that a cultivar that is created in nature can be precisely recreated by such breeding efforts. In addition, many cultivated hybrids probably result from several generations of crosses sometimes involving many parents, further complicating such studies. The second approach to confirm parentage of a hybrid is to compare the complete nucleotide sequences of all accessions that may have a role in the creation of the hybrid. It may be possible to determine – piece by piece – whether certain parts of the genome originated from a particular parental type. Recently, using complete genomic information, the genetic admixture in sweet orange was determined (Wu *et al.*, 2014). Generating genomic information for a large number of accessions may not always be feasible due to the expense, effort and time involved. Furthermore, the exact individuals that hybridized to produce a modern cultivar may no longer exist. It may be necessary to approximate ancestry using information from existing accessions.

Comparison of chloroplast sequences is a commonly used technique to study kinship between taxa. Organelles are generally maternally inherited in citrus, although certain exceptions are reported (Bernet *et al.*, 2010). Until recently, phylogenetic inferences deduced by comparison of sequence information were made by analysis of a few nuclear or chloroplast genes. Complete chloroplast genome sequences of lime accessions are now available for many lime cultivars including: Mexican lime, Rangpur lime, Australian finger lime, Australian round lime, Australian desert lime and Omani lime (a cultivar derived from Mexican lime) (Su *et al.*, 2014; Carbonell-Caballero *et al.*, 2015). In the near future it may be possible to analyse whole chloroplast genome sequences for all the accessions involved in a study, but organelle genome data may provide information about the maternal lineage only.

Analysis of nuclear genes is considered valuable to deduce the ancestry of citrus hybrids and to understand their interrelationships. SNPs, indels (insertions and deletions) and SSRs from both coding and intergenic regions have been studied to delineate the relationships between citrus accessions and to determine the probable ancestry of certain hybrids (Garcia-Lor *et al.*, 2013; Ramadugu *et al.*, 2013; Curk *et al.*, 2015). Nuclear internal transcribed spacer regions (ITS) of ribosomal DNA have also been studied to understand relationships between citrus cultivars (Li *et al.*, 2010; Hynniewta *et al.*, 2014). Genomes of many economically significant citrus cultivars are now being sequenced. When sequence information becomes available for an adequate number of citrus accessions, a better understanding of the nature of admixture in different lime hybrids will be possible. Although whole nuclear genomes can be sequenced, these data may not provide information on the two haplotypes and therefore on the number of generations of hybridization in the ancestry.

Some citrus cultivars are derived from limes. For example, Bergamot (*Citrus bergamia* Risso), a sour orange hybrid was presumed to be a hybrid of *C. aurantium* and *C. limetta* by RFLP analysis (Federici *et al.*, 2000). The distinctive aroma of bergamot oil also occurs in the limettas that are grown in the Mediterranean basin (Swingle and Reece, 1967).

Distribution

Lime has been known since ancient times. Since lemons and limes were not distinguished by earlier botanists, it is likely that the reference to lemons in Chinese literature and in Sanskrit writings could include limes (Bhattacharya and Dutta, 1956). Limes are believed to have originated in north-eastern India, neighbouring Myanmar (Burma) and northern Malaysia (Hodgson, 1967). In Assam, which is located in the north-eastern part of India, many wild forms of limes are known (Bhattacharya and Dutta, 1956).

Most citrus-growing regions are located between 40° north–south latitude. Tropical and subtropical regions where the minimum temperature is greater than −6°C are suitable for citrus. Depending on the type of cultivar, arid, semi-arid and humid environments may be desirable for lime cultivation. The major lime production regions are: Argentina, Brazil, China, Egypt, India, Iran, Iraq and México. Compared to lemons, limes are more sensitive to cold and hence their cultivation is restricted to the warmer parts of the world citrus belt. Small fruited acid limes like Mexican lime are very cold sensitive and grown close to the equator. There is a very large lime plantation in Colima, México with about two million seedling trees of Mexican lime. Certain large fruited acid lime cultivars like Persian lime are more cold tolerant than Mexican lime. Sour limes perform well in warm climates and high humidity, and can be successfully cultivated in semi-tropical and tropical regions.

Mexican lime (*C. aurantifolia* (Christm.) Swingle)

The Mexican lime is one of the popular small fruited sour limes, well known for its intense flavour. It is probably the longest known cultivated lime variety. Some of the popular names of the clonal selections of Mexican lime are: West Indian lime, Kaghzi lime (name in India and Pakistan; derived from the word *kagaz*, referring to its thin paper like rind), Key lime (due to its extensive cultivation in the Florida Keys during the 1840s to 1920s), Bartender's lime (owing to its widespread use in alcoholic drinks), Limão Galego (in Brazil), Limûn baladi (in Egypt and Sudan), Limette or Limettier acide (France), Ndimu (East Africa), Dalayap or Dayap (Philippines) and Lima ácida (Spain). The Mexican lime is cultivated commercially in México, India, Egypt, the West Indies, the Dominican Republic and some other tropical regions of the old world. The Peruvian lime, Limon Peruano, widely cultivated in Peru is presumed to be very closely related to the Mexican lime. Many landraces of acid lime (*C. aurantifolia*) are known near the place of origin (Indian subcontinent), and they have considerable genetic diversity (Shreshtha *et al.*, 2012; Munankarmi *et al.*, 2014).

The Mexican lime presumably originated in the Indo-Malayan regions (Hodgson, 1967). The sour limes were probably one of the first citrus fruits to be carried from the east by the crusaders. Arabs carried the sour lime to North Africa and surrounding regions. It was then transported from Palestine to Mediterranean Europe. By the mid-thirteenth century the small fruited acid lime was well known in Italy and France. Spanish and Portuguese explorers probably transported this cultivar to the Americas during the sixteenth century. The Mexican lime was naturalized throughout the Caribbean, eastern México, tropical South America, Central America and the Florida Keys (Crane, 2010). This cultivar is very well adapted to the dry and wet tropical conditions in the southern and central Pacific regions of México (Robles-González *et al.*, 2008). Cultivation of Mexican lime in Florida was made possible by the significant efforts of Dr Henry Perrine, a horticulturist. Perrine brought the Mexican lime from Yucatan province to the Florida Keys in 1838 at the request of President John Quincy Adams (Robinson,

1942). In the 1920s, the lime industry was flourishing in south Florida and this has been ascribed to its wide culinary consumption, its use in fruit juices, carbonated drinks and alcoholic beverages. The hurricane of 1926 destroyed most of the lime groves in the Florida Keys (Jackson, 1991). According to John Gifford, the lime industry flourished in Florida since lime was an essential ingredient in the development of the Gin Rickey cocktail. Prohibition of alcoholic beverages (in addition to unfavourable climatic factors) resulted in the near demise of the commercial Mexican lime industry in Florida (Gifford, 1972). Lime was subsequently replanted in southern Florida but frequent hurricanes and citrus canker disease reduced the Mexican lime acreage significantly by the year 2000.

The Mexican lime performs well in hot, tropical and semi-tropical regions; it is very sensitive to frost and hence most of the cultivation is restricted to warmer parts of the world. It can also withstand drought conditions better than most citrus cultivars. The tree is vigorous, 2–4 m high with spreading branches, prominent sharp spines, alternate, aromatic, elliptic or oblong-ovate leaves that are articulate with the petiole. The foliage is light purplish in colour when young, and the leaves have rounded dentate margins and narrowly winged petioles. The axillary flowers are about 0.5 cm across, faintly fragrant, white in colour with a purple tinge. The fruit are either single or borne in clusters at the tips of the branches, small (2–5 cm in diameter), round, obovate or elliptical, with an apical papilla or a nipple-like projection at the stylar end. Immature fruit are pale green and turn pale yellow when ripe, although limes are typically harvested while green. The outer rind is thin with a rough or smooth exterior. The Kaghzi lime of India is one of the smallest commercial citrus fruits and has the thinnest rind. The pulp in the segments is pale green, aromatic, juicy, very acidic with an intense flavour. It is a typical lime with many small seeds, a very high level of polyembryony and often raised as seedlings that are identical to the mother plant (Davies and Albrigo, 1994). Mexican lime is also budded on rootstocks like sour orange or Rough lemon that can tolerate high winds during hurricanes. Multiplication by air layering is popular in Indonesia and India. Seedlings typically fruit after 3–6 years and can reach full production by 8–10 years (Morton, 1987). Popular selections of Mexican lime are Everglade, Kaghzi, Palmetto and Yung. An autotetraploid of Key lime, known as the Giant Key lime, has fruit that are twice the size of normal Key lime but this cultivar is not commercially grown. Minor genetic differences have been documented in clonal selections of Mexican lime (Robles-González *et al.*, 2008).

Due to its characteristic intense flavour, Mexican lime is the preferred lime in several cuisines. The rind is used in making jam, jelly and marmalade. The juice is an ingredient in cold drinks, mainly in the preparation of limeade and for flavouring alcoholic beverages, especially margaritas. In the Indian subcontinent, the whole fruit is used in preparation of lime pickle, a popular preserved lime dish used as a condiment. In México and Peru, the fresh fruit is widely used to garnish foods and the extracted lime oil is exported to various parts of the world. Distilled lime oil is used in a large number of lime flavoured products. The peel oil is used in the perfume industry. In certain parts of the world, Mexican lime fruit are preserved by boiling in salt water, dried and consumed as Black lime or Omani lime. The black lime has a slightly bitter and fermented flavour and is used for flavouring in Persian and Iraqi cuisine. The Mexican lime is used in many cultures for medicinal purposes to relieve indigestion, as a vermifuge (to expel parasitic worms from intestines), as an astringent, an antiseptic, a diuretic in patients with liver ailments, for heart palpitations, in curing arthritis and as a disinfectant (Morton, 1987).

Mexican lime was considered to be a hybrid derived from citron, pummelo and Australian finger lime (Barrett and Rhodes, 1976). Admixture with papeda was suggested based on chromosomal analysis that indicated a 'G' chromosome type, also present in *Citrus hystrix* (Guerra *et al.*, 1997). Isoenzyme analysis indicated that Mexican lime is a hybrid between citron and papeda (Torres *et al.*, 1978). Based on the composition of volatile oils, Scora and Kumamoto (1983) identified *Microcitrus* as one of the parents of Mexican lime. According to Nicolosi *et al.* (2000), Mexican lime has a citron as the male parent and either *C. micrantha* or something very closely related to it as the female parent. SSR data indicate Mexican lime to be 50% citron and 50% papeda (Barkley *et al.*, 2006; Ramadugu *et al.*, 2013). Analysis of complete chloroplast sequences revealed that *C. micrantha* is the maternal parent of Mexican lime (Carbonell-Caballero *et al.*, 2015). Studies based on nuclear gene sequences, mitochondrial and chloroplast markers indicate that the Mexican lime group consists of hybrids with papeda (*C. micrantha*) and citron (*C. medica*) parentage (Curk *et al.*, 2015; Curk *et al.*, 2016).

Persian lime (*C. latifolia* (Yu. Tanaka) Tanaka)

Persian lime (Fig. 2.5) and Mexican lime are the two most widely grown limes in the world. The Persian lime and its clonal selections like Tahiti lime and Bearss lime are cultivated in many parts of the world. In Brazil, Persian lime is referred to as Limão Tahiti. Although the Persian lime is less tart and lacks the distinctive flavour of Mexican lime, the fruit have a thicker rind and are easy to transport and store; because of this fruit character, the Persian lime is now more popular than Mexican lime. Commercial cultivation of Tahiti lime or Persian lime is confined to México, Cuba, Guatemala, Honduras, El Salvador, Egypt, Israel and Brazil.

Persian lime probably originated in South-east Asia, specifically the region that includes north-eastern India, northern Myanmar, south-west China and the Malay Archipelago (Moore, 2001). It was probably transported to Persia (now Iran) from South-east Asia, similar to the citron; however, by the mid-twentieth century, this cultivar was not grown in Iran. It was introduced into Australia around 1824 (Bowman, 1955) from Brazil. Persian lime was transported to the Polynesian islands and was named Tahiti lime because of its association with the Tahitian citrus industry. It was imported to the USA from Tahiti in the 1850s and subsequently, a seedless form was first recognized in a home landscape in California in 1875 (Ziegler and Wolfe, 1961;

Fig. 2.1. Mexican lime from Citrus Variety Collection (CVC), Riverside, CA, USA; CRC (citrus research center) number 1710. *Citrus aurantifolia* (Christm.) Swingle.

Fig. 2.2. Egyptian lime from CVC; CRC number 2883. *Citrus aurantifolia* (Christm.) Swingle.

Crane and Osborne, 1991). In the 1880s, the cultivar Tahiti lime was commercially grown in south central Florida. The large fruited acid limes are susceptible to wood pocket, a genetic disease (Roistacher and Velezquez-Monreal, 2012). At present, due to environmental conditions and citrus canker problems, Tahiti lime is not commercially grown in Florida. Bearss lime is a cultivar that was developed as a selection from a seedling of Tahiti lime, by T.J. Bearss, a nurseryman in California (Hodgson, 1967). Persian lime and many other cultivars derived from this are triploids (Bacchi, 1940; Curk *et al.*, 2016) and generally seedless. Idemor is another cultivar that is a limb sport obtained from Tahiti lime. Sakhesli lime from the island of Djerba (Tunisia) is a clonal selection of the Persian lime.

The Persian lime is a large fruited lime cultivar. The trees are more vigorous than Mexican lime trees. The drooping branches have medium sized leaves that are broadly lanceolate with winged petioles. The trees flower throughout the year, more in spring. The tender shoots and flowers have a faint purple coloration. Persian lime is more cold tolerant than Mexican lime and is similar to the lemons in this attribute. The fruit can reach up to 7 cm in diameter and have small nipple like projections at the distal end. The fruit has about ten segments with a small, solid, central columella, pale greenish yellow coloured flesh that is juicy, very acidic and with true lime flavour. However, the pungency and characteristic aroma of the Mexican lime are lacking in the Persian lime. As is typical of limes, the flowers and fruit are borne throughout the year but the main harvest season in the USA is July to September. The Persian lime does not require cross-pollination to set fruit (Crane and Osborne, 1991). The seeds, which are few, are known to have a high level of monoembryony unlike other limes, which are predominantly polyembryonic. Propagation by air layering (marcottage) is common (Schaffer and Moon, 1990). Rough lemon and Alemow (*Citrus macrophylla*) are used as common rootstocks for Persian lime cultivation. The Persian lime has more cold hardiness

Fig. 2.3. Abhayapuri lime from CVC; CRC number 3762. *Citrus aurantifolia* (Christm.) Swingle.

than Mexican lime since it can tolerate temperatures as cold as −2°C. Severe frost damage occurs below −4°C (Crane and Osborne, 1991). Its lower heat requirement and cold hardiness (compared with the Mexican lime), seedlessness, absence of thorns and longer shelf life had a major role in expanding its commercial cultivation in the USA. Because of its large size and less pronounced flavour, it was initially not preferred in the USA. However, the Persian lime is currently widely accepted in the markets in the USA.

A physiological disorder commonly observed in Persian lime is the stylar end rot caused by the rupture of juice sacs of a turgid fruit. The stylar end rot eventually causes decay of the fruit affecting market value (Davenport *et al.*, 1976; Browning *et al.*, 1995). The Persian lime is widely used for culinary purposes, for preparation of canned lime juice, frozen limeade, powdered lime juice and consumed as fresh fruit. About 40% of the Persian lime grown is used for preparation of lime juice concentrate. Peel oil is extracted from the culled fruit after juice extraction. In many regions, the Persian lime has now replaced Mexican lime both for culinary use and as a source of peel oil.

Persian lime was presumed to be a hybrid of a small acid lime (probably the Mexican lime) and either a citron or a lemon (Reece and Childs, 1962). Based on cytoplasmic and nuclear sequence information, it is now proposed that sour orange is the maternal ancestor of Persian lime, and four ancestral taxa (*C. maxima*, *C. medica*, *C. micrantha* and *C. reticulata*) may have contributed to the development of this cultivar. Furthermore, it is hypothesized that Persian lime might have resulted from fertilization of a haploid lemon ovule by a diploid gamete from Mexican lime (Curk *et al.*, 2015; Curk *et al.*, 2016).

Palestine sweet lime: *Citrus limettioides* Tan.

The Palestine sweet lime (Fig. 2.6), also known as Indian sweet lime is native to north-eastern

Fig. 2.4. India lime from CVC; CRC number 2450. *Citrus aurantifolia* (Christm.) Swingle.

India, where it was called Mitha nimboo (Bhattacharya and Dutta, 1956; Nicolosi, 2007). Other common names for this variety are: Limon helou or Succari (Egypt), Lima dulce (Spain), Limettier doux (France) and Laymŭn-helo (Syria and Palestine). It is widely grown in India, Pakistan, Iran and Egypt. This variety has been cultivated in India for a long time and there are variants of this cultivar occurring naturally in this region. Soh synteng grown in the Assam region of India is very similar to the Palestine sweet lime and differs in having purple coloured flowers and intense acid pulp vesicles unlike the sweet lime; it is considered to be the acid form of the Indian sweet lime (Bhattacharya and Dutta, 1956; Hodgson, 1967). Commercial production of Palestine sweet lime is limited to Egypt, western Asia and Latin America. Columbia is a clonal selection of Palestine sweet lime.

Palestine sweet lime is an acidless variety. The fruit is larger in size than most limes (5–7 cm in diameter), oblong, the areolar area often protrudes into a low, flat nipple surrounded by a shallow circular furrow. The flowers are generally pure white, although many variations with fruit shape, flavour and tree productivity are known in the wild state in India. The fruit is medium to large in size, subglobose to oblong, with pale yellow flesh, tender and juicy with some seeds. The fruit has a thin rind, greenish to orange yellow at maturity, with a distinctive rind oil aroma. The fruit has ten very juicy segments, sometimes considered insipid. Although described as a sweet lime, the pulp is not sweet – it is just low in acid; it has a subtle flavour compared with other limes. The citric acid content is low but Vitamin C content is high; the mild flavour is valued in certain cultures and the fruit pulp is used in the preparation of beverages with delicate citrus flavour. Like typical limes, the seeds of Palestine sweet lime are polyembryonic. It is generally self-compatible; cross pollination with sweet orange or grapefruit has been reported to increase fruit size. Hence, interplanting with other types of citrus is practised (Morton, 1987).

Fig. 2.5. Persian lime from CVC; CRC number 4170. *Citrus latifolia* (Yu. Tanaka) Tanaka.

The Palestine sweet lime is highly prized in certain cultures for its special medicinal values in treating fevers and for providing relief to patients with liver problems. In many Arabic countries, it is used to prevent colds and flu. A study based on essential oils extracted from the mature fruit of Palestine sweet lime demonstrated antibacterial and antifungal activity against many microbes. The essential oils can be used in acne control, treatment of typhoid fever, food poisoning, sepsis, bladder and prostate infections. It is also used as a safe fumigant for preserving food products (Vasudeva and Sharma, 2012). Palestine sweet lime has been ascribed to provide relief for many ailments including jaundice, stomach disorders and morning sickness, and is believed to soften the effects of chemotherapy. In Egypt it is used to prepare delicately flavoured beverages. This variety is also used as a rootstock in India, Israel and Palestine. When used as a rootstock, the acidity of the fruit of the scion is significantly lower (Swingle and Reece, 1967; Morton, 1987).

Palestine sweet lime was reported to be a hybrid of Mexican lime and a sweet lemon or a sweet citron (Webber, 1943). Barrett and Rhodes considered Palestine sweet lime to be the result of a cross of Mexican lime and sweet orange (Barrett and Rhodes, 1976). Cluster analysis of Palestine sweet lime based on microsatellite makers indicates that the genome has an admixture with 60% citron and 28% papeda (Barkley *et al.*, 2006). SNP analysis of six nuclear genes indicated that Palestine sweet lime is probably a citron–mandarin hybrid (Ramadugu *et al.*, 2013). In a study based on seed limonoid profile, Rouseff and Nagy studied Columbia sweet lime, a clonal selection of the Palestine sweet lime and compared the nomilin–limonin ratios of putative orange ancestors. The Jaffa orange (*Citrus sinensis*) was observed to have a limonoid pattern very similar to the Colombia sweet lime. Using this biochemical method, they speculated that Jaffa orange may be a putative parent (Rouseff and Nagy, 1982). Curk *et al.* (2016) rule out sour lime as the

Fig. 2.6. Palestine sweet lime from CVC; CRC number 1482. *Citrus limettioides* Tan.

maternal parent of Palestine sweet lime since the chloroplast sequences of Palestine sweet lime are very similar to pummelo.

Australian lime

There are three main types of indigenous limes grown in Australia. Australian finger lime (*Microcitrus australasica* Swingle), Australian round lime (*Microcitrus australis* (Planch.) Swingle) and Australian desert lime (*Eremocitrus glauca* (Lindl.) Swing.). The Australian genera, *Microcitrus* and *Eremocitrus* were originally grouped under the genus *Citrus* by earlier taxonomists F. M. Mueller (1825–1896) and F. M. Bailey (1827–1915). Swingle assigned separate genera, *Microcitrus* and *Eremocitrus*, since there were significant differences between the Australian citrus types and *Citrus* (Swingle, 1914; 1915). The basis for identifying a new genus for *Microcitrus* was the presence of free filaments in the flower. In the genus *Citrus*, the presence of connate or free stamens remains a variable character and hence other taxonomists did not consider these differences as sufficient to justify creation of an independent genus. Similarly, the morphological characters that distinguish *Eremocitrus* from other types of *Citrus* were considered as adaptations necessary for its growth and survival in the semi-arid climate. These characters were also not considered to be taxonomically significant. Hence some taxonomists have now reclassified the Australian limes under *Citrus australasica* F. Mueller, *Citrus australis* (Mudie) Planchon and *Citrus glauca* (Lindley) Burkill (Mabberley, 1998). Similar to *Citrus*, the Australian limes have evergreen foliage. The fruit are used for preparing marmalades, sauces and specialty dishes in the place of origin. In Australia many other species of *Microcitrus* are known to exist. The Australian citrus types can hybridize with many citrus species and may have potential in developing drought resistant, salinity tolerant, and disease-resistant cultivars.

Fig. 2.7. Mary Ellen sweet lime from CVC; CRC number 4053. *Citrus limettioides* Tan.

Australian finger lime

The finger lime (*Microcitrus australasica* Swingle) is an understorey shrub or small tree native to Australia and cultivated mainly in this region (Fig. 2.8). Recently it has been cultivated in the USA as a specialty crop. The trees are small, thorny and evergreen. Many registered cultivars are known in Australia and prized for the colourful pulp ranging from white, yellow, green, pink and red. The bright red variant, *M. australasica* var. sanguinea, is found wild in Australia and can be propagated from seed; other variants are probably budsports that have been cultivated by horticulturists (Native Australian Citrus, 2014). To maintain the trueness to type of these selections, propagations are made using budwood that is grafted on trifoliate rootstock or Troyer citrange. The leaves are usually small (2–4 mm × 3–5 mm), borne on very short petioles and always wingless. Short internodes and small axillary spines are typical. Depending on the age and the maturity of the plant, the size of the leaves and spines may be larger. Leaves have dentate margins and are often emarginate. The tender shoots are purple in colour. The flowers are small (1 cm long), generally white with pink tips, not very fragrant, axillary and pentamerous with 20–25 free stamens. In certain regions, finger limes flower throughout the year similar to typical limes. Often male flowers predominate and only a percentage of the flowers are perfect with both male and female parts. Fruit is finger shaped, described as cylindric fusiform, often slightly curved, about 1.5–2.5 cm × 6–10 cm. Fruit of commercially grown cultivars have coloured rind and juice vesicles ranging from light green to dark green, black, red, purple and pink. The fruit do not have a nipple that is present in most limes but have a blunt protuberance at the stylar end. The pulp vesicles are numerous, subglobose, tart and usually free. The compressed juice vesicles typically burst out when the fruit rind is cut. Described as 'citrus caviar', the pulp

Fig. 2.8. Australian finger lime from CVC; CRC number 3670. *Microcitrus australasica* Swingle.

vesicles have a delicate flavour reminiscent of true limes. The fruit have many monoembryonic seeds that are small and flattened on one side. The pulp of finger limes is used to garnish seafood and to impart unique texture and flavour. The pulp is used in the preparation of sauces, jams, jellies and marmalade.

Australian round lime

The Australian round lime (*Microcitrus australis* (Planch.) Swingle) is native to north-eastern Australia. Locally known as 'Dooja', the round lime tree can reach 30–60 feet. The stems are angled with long internodes. Leaves are much larger than the finger lime reaching up to 6 cm in length, oval, entire, glabrous, glandular, with blunt tips and lacking a winged petiole present in most citrus types. Slender acute spines that are about 1 cm long are present. Flowering season is restricted to spring unlike Australian finger lime and other more typical limes. The flowers are axillary, 1–1.5 cm long, pentamerous, delicately fragrant, generally white with pink tips. They have about 16–20 stamens with free filaments. Fruit are up to 5 cm in diameter, green, globose, with a rough rind. The pulp vesicles are very firm, pale green, fusiform, often angular with blunt tips and with characteristic yellow oil droplets in the centre. The vesicles are often twisted due to the pressure created by the compression of juice vesicles inside the fruit. The acidic pulp is edible when raw and is used for making jams and marmalade in Australia.

Sydney hybrid

The Sydney hybrid is a result of a cross between the Australian finger lime and the Australian round lime, and is grown mainly in Australia (Swingle, 1943). Classified as *Microcitrus* × *virgata* H.Hume or *Citrus* × *virgata* Mabb, this accession is more vigorous than either parent and has leaves resembling the Australian finger lime. The fruit is elongated, 3 × 5 cm with an abruptly rounded apex. The fruit resembles the finger lime parent except for the blunt projection at the stylar end. The Sydney hybrid is known to be extremely drought resistant.

Australian desert lime

The Australian desert lime (*Eremocitrus glauca* (Lindl.) Swing.) is the only xerophytic plant in the orange subfamily, Aurantioideae (Fig. 2.9). Native to Australia, the plant can survive desert conditions (up to 45°C), hot dry winds, drought and also freezing temperatures down to −24°C. The Australian desert lime is a large shrub or a small tree that can grow in soils with high salt concentrations. The plants have a characteristic appearance with green branches bearing greyish green, leathery foliage. The leaves are oblong, obtuse at the apex, about 2 cm long, described as paraheliotropic with the leaf edge facing the sun light; both abaxial and adaxial surfaces appear similar. The seedlings often have very narrow post-cotyledonary leaves known as cataphylls. The fragrant small axillary flowers are about 6 mm long, tetra- or pentamerous. The flowers are seasonal unlike the ever-blooming limes. The fruit are pale green, glandular, sub-globose, about 1.5 cm long, juicy and acidic. In California, the fruit bear one or two monoembryonic seeds. The Australian desert lime is graft compatible with citrus and sexually compatible with certain types of citrus. Fruit is edible and used to prepare sauces in Australia.

Rangpur lime

Originally from the Indian subcontinent, the Rangpur lime is a popular rootstock for many types of citrus. It is classified as *Citrus* × *limonia* Osbeck and is commonly known as Mandarin lime. The fruit is very acidic and can be used as a substitute for other commercial limes. Other common names include Canton lemon (south China), Hime lemon (Japan), Cravo lemon (Brazil), Marmalade lime (India) and Tahitian orange (USA). The fruit is suitable for preparing marmalade. Many named cultivars of Rangpur lime exist and differ mainly in the fruit characteristics. The Rangpur lime cultivars are valued as rootstock varieties in the Orient and also in South America, especially because of its tolerance to Citrus tristeza virus and drought tolerance. By 1995, about 85% of the sweet orange trees in Brazil were on Rangpur lime rootstocks (Bové and Ayres, 2007). In 1999, citrus sudden death disease was observed in sweet orange on Rangpur lime rootstock resulting in death of millions of trees (Müller *et al.*, 2002). Because of this problem, other disease-tolerant rootstocks are gaining importance in Brazil (Bové and Ayres, 2007).

The fruit is mandarin-like, depressed at the stylar end, globose to round with a furrowed collar. A nipple that is not very prominent is surrounded by a shallow furrow. The fruit is loose skinned, with a slightly rough exterior and a bright orange or yellow rind. The mature fruit has a hollow columella (similar to mandarins) and eight to ten segments. The flesh is orange, very juicy and strongly acidic with many seeds that are polyembryonic. Nucellar seedlings genetically identical to the seed parent are produced from the seeds. After maturity, the

Fig. 2.9. Australian desert lime from CVC; CRC number 3463. *Eremocitrus glauca* (Lindl.) Swingle.

fruit persists on the tree for a long period of time. The tree is vigorous, medium sized (can grow up to 6 m tall) with small thorns, mandarin-like leaves and purple tender growth. The small flowers are also like those of the mandarin but with purple coloration in the corolla (similar to many limes). Rangpur lime is more cold tolerant than typical limes and lemons. The highly acidic fruit, polyembryonic seeds and presence of mammilla are lime-like but other features are mandarin-like. Rangpur lime was introduced from the western part of India to Florida in the late nineteenth century by Reasoner Brothers of Oneco (Webber, 1943).

According to Webber (1943), the Rangpur lime may be a hybrid with lime as one parent and either mandarin or sour mandarin as the other parent. According to Barrett and Rhodes (1976), Rangpur lime is *C. reticulata* genotype introgressed with citron. Singh and Schroeder (1962) speculated that Rangpur lime may be a result of a mandarin and Rough lemon cross. Using RFLP of chloroplast DNA, Gulsen and Roose (2001b) demonstrated that mandarin was the maternal parent of Rangpur lime. Based on SSR data, most of the genome of Rangpur lime was similar to citron with a small percentage of mandarin; the Rangpur lime clusters with citrons indicating a definite citron parentage (Barkley *et al.*, 2006). Federici *et al.* (1998) using RFLP and RAPD markers determined that Rangpur lime clustered with limes within the citron group. Cytogenetic studies indicated distinct similarities between karyotypes of Rangpur lime and the presumed parents, citron and Cleopatra mandarin (Caravalho *et al.*, 2005). According to an analysis conducted with complete chloroplast genome sequence, mandarin was presumed to be the maternal parent of Rangpur lime (Carbonell-Caballero *et al.*, 2015). Rangpur lime shares the chlorotype of Cleopatra mandarin confirming the maternal parentage (Curk *et al.*, 2016). The most detailed study of ancestry to date included analysis of nuclear genes, mitochondrial and chloroplast markers, and indicates that Rangpur lime is a hybrid of acid mandarin and citron (Curk *et al.*, 2015, 2016).

Calamondin (Musk lime)

Calamondin is not a true lime; it is a natural, intergeneric hybrid of sour mandarin and a kumquat (Swingle, 1943). Various botanical names have been used for Calamondin including *Citrus mitis* Blanco, *Citrus microcarpa* Bunge and *Citrus madurensis* Lour. Currently it has been given the hybrid name, × *Citrofortunella mitis* (Blanco) J.W. Ingram & H. E. Moore (Morton, 1987). Common names include Calamondin orange, Panama orange, Golden lime and Scarlet lime. In the Philippines where it is widely cultivated, Calamondin is known by several vernacular names such as Kalamansî, Limonsito or Agridulce. In Malaysia it is known as Limau kesturi (musk lime) and Limau chuit. In Thailand, it is known as Ma-nao-wan.

A native of China, Calamondin was transported to Indonesia and the Philippines (Morton, 1987). It is a cold-resistant medium sized tree that can reach up to 7.6 m in height, has thorny branches that grow close to the ground, broad oval leaves that are toothed at the tips and narrowly winged petioles. The flowers are self-fertile; the fruit are round, about 4 cm wide with a thin orange rind. The pulp is sweet, bright orange and bears polyembryonic seeds. Since Calamondin is a preferred host for the Mediterranean fruit fly, it is usually not planted in regions where the fruit fly problem persists.

The fruit are used instead of regular limes to flavour foods and to prepare beverages. In Asian countries, Calamondin juice is used for seasoning fish and poultry, to prepare marmalade, cakes, coulis and lemonade (kalamansi-ade). The fruit has antioxidants similar to many other citrus fruits and the pulp is rich in ascorbic acid (111 mg per 100 g fruit). Medicinally, it is used for insect bites to relieve itching and irritations, as acne medication, as a natural anti-inflammatory substance, for constipation, etc. (Morton, 1987). In combination with pepper, Calamondin fruit are used in Malaysia to expel phlegm, treating coughs, colds and sore throats. The fruit are also used to provide relief from nausea, and to lower hypertension (Olowa *et al.*, 2012). The leaf oil is used as a carminative to prevent gas formation in the gastrointestinal tract. It is used as a bleaching agent and stain remover in the Philippines. In certain parts of the world, it is prized for its recognized value as an anti-microbial agent, anti-depressant and hepatoprotective agent (Franchesca *et al.*, 2010). Musk lime seed oil is rich in unsaturated fatty acids (Manaf *et al.*, 2007). An alkaloid from the seeds is used as an ingredient in certain commercial expectorants. In the USA, a variegated form of Calamondin, known as Peters, is grown as an ornamental plant.

Kaffir lime

Citrus hystrix, popularly known as Kaffir lime is an important ingredient of Pacific Rim cuisine. Other common names for this cultivar include Kieffer lime, Makrut and Magrood. The cultivar is native to Indonesia; it is not a true lime but belongs to the papeda group. The tree is grown for the edible leaves that are used in Thai, Malaysian and Indonesian cooking. The leaves are large, glossy and have a prominent winged petiole characteristic of papedas. The fruit is round, deep green with a bumpy rind and a slightly elongated neck. The fruit juice is very tangy, sour and used in medicine. The aromatic oils from the leaves have an intense unique flavour. The glossy leaves are also very fragrant. Unlike true limes, Kaffir lime can be grown in cold regions.

Other lime hybrids

Limetta

Citrus limetta Risso. also known as Sweet lime, or Sweet limetta originated in the Mediterranean basin. These hybrids are assumed to be from a cross between three citrus species: (*C. maxima* × *C. reticulata*) × *C. medica* based on molecular analysis (Curk *et al.*, 2016). Three types are common: Marrakech (acid form), Millsweet lemon limetta (low acid form) and Pomona sweet lemon (acid-less form). The limetta cultivars are not commercially significant.

Lemonimes (lemon × lime hybrids)

Perrine lemonime is the most well known cultivar in the group and is a result of a deliberate cross between *C. aurantifolia* (Mexican lime) and *C. limon* Genoa. (Hodgson, 1967). Perrine lemonime is more cold tolerant than Mexican lime and was grown in Florida for some time. It was replaced with Persian lime, which is even more cold tolerant.

Limequat

Described by Swingle (Swingle, 1913), limequats are bigeneric hybrids of *Citrus* and *Fortunella*. The widely used botanical name is × *Citrofortunella floridana* J. Ingram and H. Moore; it is also referred to as *Citrus* × *floridana* (J. Ingram and H. Moore) Mabb. The first limequats likely originated in China but various cultivars are now popular around the world. The fruit is available during autumn and winter. Eustis, Lakeland and Tavares are three known distinct varieties. Eustis and Lakeland are sister hybrids of West Indian lime (*C. aurantifolia*) and round kumquat (*Fortunella japonica*). Tavares is a hybrid of the West Indian lime and an oblong kumquat (*Fortunella margarita*) (Fig. 2.10). The fruit of all three cultivars is small and oval like a kumquat, and the rind colour is greenish yellow instead of orange. Like the kumquat, the fruit has a sweet rind, acidic pulp and the entire fruit is edible. Eaten whole, the fruit has a characteristic bittersweet flavour. The fruit rind is fragrant, thin with an acidic fleshy interior. It is similar to Mexican lime in size, form and composition of the fruit juice and may be useful as a substitute. Limequats are also used in cooking, in preparing syrups and preserves. They are more tolerant of cold weather than Mexican lime but less tolerant compared to the kumquat parent.

Blood lime and other Australian limes

Blood lime is a hybrid developed in Australia and has a blood red rind, flesh and juice. It is the result of a cross between the Australian finger lime (*M. australasica* var. sanguinea) and *Citrus* (assumed to be either Ellendale mandarin or Rangpur lime). Also known as Australian Blood or Australian Red Centre lime, it is smaller and sweeter than most limes. Australia has several native citrus relatives that are adapted to the dry climate. In addition to the Australian limes described above, there are many citrus types with unique textures, flavours and colours cultivated in Australia. Dr Steve Sykes of Commonwealth Scientific and Industrial Research Organisation (CSIRO) has developed three high-yielding citrus varieties with larger than normal fruit, consistent yields and the unique flavour and texture of native Australian limes (Sykes, 2002). The varieties have proved suitable for production using standard horticultural practices and are now available for domestic gardeners. The fruit is used in preparing preserves, juices and certain types of beverages. Another lime developed in Australia is Sunrise lime, a trigeneric finger lime hybrid of *Microcitrus* × Calamondin (Calamondin is a hybrid of *Fortunella* × *Citrus*) (Sykes, 2002). Many lime hybrids that share the qualities and culinary uses of lime are popular.

There are several other lime cultivars that are grown in different parts of the world. Innumerable variants of named cultivars exist in the place of origin. The lime germplasm available in the world citrus belt can be very useful if there is better documentation of the variations observed and protection of the wild type germplasm. Certain lime hybrids show resistance to disease when both parents do not. Also, some lime hybrids are suitable for growth in cooler climates where true limes cannot be cultivated. More research in this area may be helpful to utilize the diverse germplasm of limes.

Limes are a very important citrus fruit because of their cultural, culinary and medicinal uses in many countries. Although limes are not consumed as fresh fruit, their tartness and unique flavour are highly valued in the culinary industry. They are considered essential in

Fig. 2.10. Tavares limequat from CVC; CRC number 4048. × *Citrofortunella swinglei* J.W. Ingram & H.E. Moore.

preparation of tropical punches as well as certain alcoholic beverages. Lime oil is used widely because of its association with many health benefits and in the cosmetic industry. Genetically, they are a diverse group of cultivars and are generally hybrids.

Acknowledgements

The photographs of the lime trees and fruit included in the article were based on the trees maintained in the Citrus Variety Collection located in Riverside, CA, USA.

References

Abkenar, A.A. and Isshiki, S. (2003) Molecular characterization and genetic diversity among Japanese acid citrus (*Citrus* spp.) based on RAPD markers. *Journal of Horticultural Science and Biotechnology* 78, 108–112.

Abkenar, A.A., Isshiki, S. and Tashiro, Y. (2004) Phylogenetic relationships in the 'true citrus fruit trees' revealed by PCR-RFLP analysis of cpDNA. *Scientia Horticulturae* 102, 233–242. doi: 10.1016/j.scienta.2004.01.003.
Albach, R.F. and Redman, G.H. (1969) Composition and inheritance of flavanones in citrus fruit. *Phytochemistry* 8, 127–143.
Araújo, E.F., Queiroz, L.P. and Machado, M.A. (2003) What is *citrus*? Taxonomic implications from a study of cpDNA evolution in the tribe Citreae (Rutaceae subfamily Aurantioideae). *Organisms Diversity and Evolution* 3, 55–62. doi: 10.1078/1439-6092-00058.
Bacchi, O. (1940) Observaçóes citólogicas em citrus. I. Número de cromosômios de algumas espécies e variedades. *Jornal de Agronomia, Piracicaba* 3, 249–258.
Barkley, N.A., Roose, M.L., Krueger, R.R. and Federici, C.T. (2006) Assessing genetic diversity and population structure in a citrus germplasm collection utilizing simple sequence repeat markers (SSRs). *Theoretical and Applied Genetics* 112, 1519–1531. doi: 10.1007/s00122-006-0255-9.
Barrett, H.C. (1977) Intergeneric hybridization of citrus and other genera in citrus cultivar improvement. *Proceedings of the International Society of Citriculture* 2, 586–589.
Barrett, H.C. and Rhodes, A.M. (1976) A numerical taxonomic study of affinity relationships in cultivated *Citrus* and its close relatives. *Systematic Botany* 1, 105–136. doi: 10.2307/2418763.
Bayer, R.J., Mabberley, D.J., Morton, C., Miller, C.H., Sharma, I.K., Pfeil, B.E., Rich, S., Hitchcock, R. and Sykes, S. (2009) A molecular phylogeny of the orange subfamily (Rutaceae: Aurantioideae) using nine cpDNA sequences. *American Journal of Botany* 96, 668–685. doi: 10.3732/ajb.0800341.
Bernet, G.P., Fernandez-Ribacoba, J., Carbonell, E.A. and Asins, M.J. (2010) Comparative genome-wide segregation analysis and map construction using a reciprocal cross design to facilitate citrus germplasm utilization. *Molecular Breeding* 25, 659–673.
Bhattacharya, S.C. and Dutta, S. (1956) *Classification of Citrus Fruits of Assam*. Indian Council of Agricultural Research, Scientific Monograph 20, New Delhi, p. 110.
Bové, J.M. and Ayres, A.J. (2007) Etiology of three recent diseases of citrus in São Paulo State: sudden death, variegated chlorosis and huanglongbing. *IUBMB Life* 59, 346–354.
Bowman, F.T. (1955) A history of citrus growing in Australia. 1788–1900. *Citrus News* 31, 102–103.
Browning, H.W., McGovern, R.J., Jackson, L.K, Calvert, D.V. and Wardowski, W.F. (1995) *Florida Citrus Diagnostic Guide*. Florida Science Source, Inc., Lake Alfred, Florida, pp.1–2441.
Caravalho, R., Soares Filho, W.S., Brasileiro-Vidal, A.C. and Guerra, M. (2005) The relationships among lemons, limes and citron: a chromosomal comparison. *Cytogenetic and Genome Research* 109, 276–282.
Carbonell-Caballero, J., Alonso, R., Ibañez, V., Terol, J., Talon, M. and Dopazo, J. (2015) A phylogenetic analysis of 34 chloroplast genomes elucidates the relationships between wild and domestic species within the genus *Citrus*. *Molecular Biology and Evolution* 32, 2015–2035. doi: 10.1093/molbev/msv082.
Crane, J.H. (2010) Key lime growing in the Florida home landscape. Document FC47, *UF/IFAS extension*. Available at: http://edis.ifas.ufl.edu (accessed 12 October 2016).
Crane, J.H. and Osborne, J.L. (1991) *Growing 'Tahiti' Limes in the Home Landscape*. Florida Cooperative Extension Service, IFAS, University of Florida, Gainesville, Florida, pp.1–9.
Curk, F., Garcia-Lor, A., Snoussi-Trifa, H., Froelicher, Y., Ancillo, G., Navarro, L. and Ollitrault, P. (2015) New insights on limes and lemons origin from nuclear and cytoplasmic markers genotyping and targeted nuclear gene sequencing. *Acta Horticulturae* (ISHS) 1065, 135–146. Available at: http://www.actahort.org/books/1065/1065_13.htm (accessed 12 October 2016).
Curk, F., Ollitrault, F., Garcia-Lor, A., Luro, F., Navarro, L. and Ollitrault, P. (2016) Phylogenetic origin of limes and lemons revealed by cytoplasmic and nuclear markers. *Annals of Botany* 117, 565–583.
Davenport, T.L., Campbell, C.W. and Orth, P.G. (1976) Stylar-end breakdown in 'Tahiti' lime: some causes and cures. *Proceedings of Florida State Horticultural Society* 89, 245–248.
Davies, F.S. and Albrigo, L.G. (1994) *Citrus*. CAB International, Wallingford, UK, p. 41.
Esen, A. and Scora, R.W. (1977) Amylase polymorphism in citrus and some related genera. *American Journal of Botany* 64, 305–309.
Federici, C.T., Fang, D.Q., Scora, R.W. and Roose, M.L. (1998) Phylogenetic relationships within the genus *Citrus* (Rutaceae) and related genera as revealed by RFLP and RAPD analysis. *Theoretical and Applied Genetics* 96, 812–822. doi: 10.1007/s001220050807.
Federici, C.T., Roose, M.L. and Scora, R.W. (2000) RFLP analysis of the origin of *Citrus bergmania*, *Citrus jambhiri*, and *Citrus limonia*. *Acta Horticulturae* (ISHS) 535, 55–64.

Franchesca, C.M., Gutierrez, M., Leano, D.R. and Solidum, J.N. (2010) Evaluation of the hepatoprotective activity of *Citrus microcarpa* Bunge (Family Rutaceae) fruit peel against acetaminophen-induced liver damage in male BFAD-Sprague Dawley rats. *International Journal of Chemical and Environmental Engineering* 1, 127–132.

Garcia-Lor, A., Curk, F., Snoussi-Trifa, H., Morillon, R., Ancillo, G., Luro, F., Navarro, L. and Ollitrault, P. (2013) A nuclear phylogenetic analysis: SNPs, indels and SSRs deliver new insights into the relationships in the 'true citrus fruit trees' group (Citrinae, Rutaceae) and the origin of cultivated species. *Annals of Botany* 111, 1–19.

Gifford, J.C. (1972) *On Preserving Tropical Florida*. Elizabeth Ogren Rothra (compiled), University of Miami Press, Coral Gables, Florida.

Guerra, M., Pedrosa, A., Silva, A.E.B., Cornélio, M.T.M., Santos, K.G.B. and Soares Filho, W.S. (1997) Chromosome number and secondary constriction variation in 51 accessions of a Citrus germplasm bank. *Brazilian Journal of Genetics* 20, 489–496.

Guerra, M., Santos, K.G.B., Silva, A.E.B. and Ehrendorfer, F. (2000) Heterochromatin banding patterns in Rutaceae-Aurantioideae – A case of parallel chromosomal evolution. *American Journal of Botany* 87, 735–747.

Gulsen, O. and Roose, M.L. (2001a) Lemons: diversity and relationships with selected citrus genotypes as measured with nuclear genome markers. *Journal of the American Society for Horticultural Science* 126, 309–317.

Gulsen, O. and Roose, M.L. (2001b) Chloroplast and nuclear genome analysis of the parentage of lemons. *Journal of the American Society for Horticultural Science* 126, 210–215.

Halma, F.F. and Haas, A.R.C. (1929) Identification of certain species of citrus by colorimetric tests. *Plant Physiology* 4, 265–268.

Hirai, M., Kozaki, I. and Kajiura, I. (1986) Isozyme analysis and phylogenic relationship of citrus. *Japanese Journal of Breeding* 36, 377–389. doi: 10.1270/jsbbs1951.36.377.

Hodgson, R.W. (1967) Horticultural varieties of *Citrus*. In Reuther, W., Webber, H.J. and Batchelor, L.D (eds) *The Citrus Industry*, Vol I. History, World Distribution, Botany and Varieties. University of California Press, Riverside, California, pp. 431–588.

Hynniewta, M., Malik, S.K. and Rao, S.R. (2014) Genetic diversity and phylogenetic analysis of *Citrus* (L) from north-east India as revealed by meiosis, and molecular analysis of internal transcribed spacer region of rDNA. *Meta Gene* 2, 237–251.

Iwamasa, M., Nito, N. and Ling, J.T. (1988) Intra- and intergeneric hybridization in the orange subfamily, Aurantioideae. In: Goren, R. and Mendel, K. (eds) *Proceedings of the Sixth International Citrus Congress*. Tel Aviv, Israel, pp. 123–130.

Jackson, L. (1991) *Citrus Growing in Florida*. University of Florida Press, Gainesville, Florida.

Krezdorn, A.H. (1979) Selecting citrus rootstocks for fruit size and quality. *Florida Grower and Rancher* 72, 16–17.

Li, X., Xie, R., Lu, Z. and Zhou, A. (2010) The origin of cultivated citrus as inferred from internal transcribed spacer and chloroplast DNA sequence and amplified fragment length polymorphism fingerprints. *Journal of the American Society for Horticultural Science* 135, 341–350.

Luo, H., Daddysman, M.K., Rankin, G.O., Jiang, B.-H. and Chen, Y.C. (2010) Kaempferol enhances cisplatin's effect on ovarian cancer cells through promoting apoptosis caused by down regulation of cMyc. *Cancer Cell International* 10, 16. doi: 10.1186/1475-2867-10-16.

Mabberley, D.J. (1997) A classification for edible *Citrus* (Rutaceae). *Telopea* 7, 167–172.

Mabberley, D.J. (1998) Australian Citreae with notes on other Aurantioideae (Rutaceae). *Telopea* 7, 333–344.

Manaf, Y.N.A., Osman, A., Lai, O.M., Lomng, K. and Ghazali, H.M. (2007) Characterization of musk lime (*Citrus microcarpa*) seed oil. *Journal of the Science of Food and Agriculture* 88, 676–683.

Mayr, E. (1940) Speciation phenomena in birds. *The American Naturalist* 74, 249–278.

Moore, G.A. (2001) Oranges and lemons: clues to the taxonomy of citrus from molecular markers. *Trends in Genetics* 17, 536–540.

Morton, J.F. (1987) Mexican lime. In: Morton, J.F. (ed.) *Fruits of Warm Climates*. Florida Flair Books, Boynton Beach, Florida, pp. 168–172.

Müller, G.W., De Negri, J.D., Vildoso, C.I.A., Mattos, D. Jr, Pompeu, J. Jr, Sobrinho, J.T., Machado, M.A., Carvahlo, S.A. and Girotto, L.F. (2002) Citrus sudden death: a new citrus disease in Brazil. *Proc. Fifteenth IOCV Conference*. Duran-Villa, N., Milne R.G. and da Graca J.V. (eds). IOCV, University of California, Riverside, California, pp.405–407.

Munankarmi, N.N., Shrestha, R.L., Rana, N., Shrestha, J.K.C., Shrestha, S., Koirala, R. and Shrestha, S. (2014) Genetic diversity assessment of acid lime (*Citrus aurantifolia* Swingle) landraces of eastern Nepal using RAPD markers. *International Journal of Applied Sciences and Biotechnology* 2, 315–327.
Native Australian Citrus (2014) Available at: http://citruspages.free.fr/australian.html (accessed 12 October 2016).
Nicolosi, E. (2007) Origin and taxonomy. In: Khan, I.A. (ed.) *Citrus Genetics, Breeding and Biotechnology*. CAB International, Wallingford, UK, pp.19–44.
Nicolosi, E., Deng, Z.N., Gentile, A., La Malfa, S., Continella, G. and Tribulato, E. (2000) Citrus phylogeny and genetic origin of important species as investigated by molecular markers. *Theoretical and Applied Genetics* 100, 1155–1166. doi: 10.1007/s001220051419.
Novelli, V.M., Machado, M.A. and Lopes, C.R. (2000) Isoenzymatic polymorphism in *Citrus* spp. and *Poncirus trifoliata* (L.) Raf. (Rutaceae). *Genetics and Molecular Biology* 23, 163–168.
Ollitrault, P., Terol, J., Garcia-Lor, A., Berard, A., Chauveau, A., Froelicher, Y., Belzile, C., Morillon, R., Navarro, L., Brunel, D. and Talon, M. (2012) SNP mining in *C. clementina* BAC end sequences; transferability in the *Citrus* genus (Rutaceae), phylogenetic inferences and perspectives for genetic mapping. *BMC Genomics* 13, 13.
Olowa, L.F., Torres, M.A.J., Aranico, E.C. and Demavo, C.G. (2012) Medicinal plants used by the Higaonon tribe of Togongon, Lligan city, Mindanao, Philippines. *Advances in Environmental Biology* 6, 1442–1449.
Penjor, T., Anai, T., Nagano, Y., Matsumoto, R. and Yamamoto, M. (2010) Phylogenetic relationships of *Citrus* and its relatives based on *rbcL* gene sequences. *Tree Genetics and Genomes* 6, 931–939. doi: 10.1007/s11295-010-0302-1.
Peterson, J.J., Beecher, G.B., Bhagwat, S.A., Dwyer, J.T., Gebhardt, S.E., Haytowitz, D.B. and Holden, J.M. (2006) Flavonones in grapefruit, lemons, and limes: a compilation and review of the data from the analytical literature. *Journal of Food Composition and Analysis* 19, S74–S80.
Pritchard, J.K., Stephens, M. and Donnelly, P. (2000) Inference of population structure using multilocus genotype data. *Genetics* 155, 945–959.
Ramadugu, C., Pfeil, B.E., Keremane, M.L., Lee, R.F., Maureira-Butler, I.J. and Roose, M.L. (2013) A six nuclear gene phylogeny of *Citrus* (Rutaceae) taking into account hybridization and lineage sorting. *PLOS One* 8, e68410. doi: 10.1371/journal.pone.0068410.
Reece, P.C. and Childs, J.F.L. (1962) Character differences among seedlings of the Persian lime. *Proceedings of the Florida State Horticultural Society* 75, 110–116.
Robinson, R.T. (1942) Henry Perrine, pioneer horticulturist of Florida. *Tequesta*, 2, 16–24.
Robles-González, M.M., Medina-Urrutia, M.V., Velázquez-Monreal, J.J. and Simpson, J. (2008) Field performance and molecular profiles of Mexican lime selections. *Euphytica* 161, 401–411.
Roistacher, C.N. and Velazquez-Monreal, J.J. (2012) The wood pocket genetic disease of large fruited lime trees. Available at: http://ecoport.org/ep?SearchType=earticleView&earticleId=731&page=-2 (accessed 12 October 2016).
Roose, M.L., Schwarzacher, T. and Heslop-Harrison, J.S. (1998) The chromosomes and *Citrus* and *Poncirus* species and hybrids: identification of characteristic chromosomes and physical mapping of rDNA loci using in situ hybridization and fluorochrome banding. *Journal of Heredity* 89, 83–86.
Roowi, S. and Crozier, A. (2011) Flavonoids in tropical citrus species. *Journal of Agricultural and Food Chemistry* 59, 12217–12225.
Rouseff, R.L. and Nagy, S. (1982) Distribution of limonoids in citrus seeds. *Phytochemistry* 21, 85–90.
Schaffer, B. and Moon, P.A. (1990) Influence of rootstock on flood tolerance of 'Tahiti' lime trees. *Proceedings of the Florida State Horticultural Society* 103, 318–321.
Scora, R.W. (1975) On the history and origin of *Citrus*. *Bulletin of the Torrey Botanical Club* 102, 369–375.
Scora, R.W. (1988) Biochemistry, taxonomy and evolution of modern cultivated Citrus. In: Goren, R. and Mendel, K. (eds) *Proceedings of the Sixth International Citrus Congress*. Balaban Publishers, Philadelphia, Pennsylvania, pp. 277–289.
Scora, R.W. and Kumamoto, J. (1983) Chemotaxonomy of the genus *Citrus*. In: Waterman, P.G. and Grundon, M.F. (eds) *Chemistry and Chemical Taxonomy of the Rutales*. Academic Press, London, pp. 343–351.
Scora, R.W., England, A.B. and Bitters, W.P. (1966) The essential oils of *Poncirus trifoliata* (L.) Raf. and its selections in relation to classification. *Phytochemistry* 5, 1139–1146.
Shreshtha, R.L., Dhakar, D.D., Gautum, D.M., Paudyal, K.P. and Shrestha, S. (2012) Genetic diversity assessment of acid lime (*Citrus aurantiifolia* Swingle) landraces in Nepal, using SSR markers. *American Journal of Plant Sciences* 3, 1674–1681.

Singh, R. and Schroeder, C.A. (1962) Taxonomic and physiological relationships of the so-called mandarin-lime group of citrus. *Proceedings of American Society for Horticultural Sciences* 80, 291–295.
Su, H.-J., Hogenhout, S.A., Al-Sadi, A.M. and Kuo, C.-H. (2014) Complete chloroplast genome sequence of Omani lime (*Citrus aurantiifolia*) and comparative analysis within the Rosids. *PLOS One* 9, e113049.
Swingle, W.T. (1913) The botanical name of the lime, *Citrus aurantifolia*. *Journal of the Washington Academy of Sciences* 3, 99–102.
Swingle, W.T. (1914) *Eremocitrus*, a new genus of hardy drought-resistant citrus fruits from Australia. *Journal of Agricultural Research* 2, 85–100.
Swingle, W.T. (1915) *Microcitrus*, a new genus of Australian citrus fruits. *Journal of the Washington Academy of Sciences* 5, 569–578.
Swingle, W.T. (1943) The botany of citrus and its wild relatives of the orange subfamily. In: Webber, H.J. and Batchelor, L.D. (eds.) *The Citrus Industry*, Vol. 1. Univ. of California, Berkeley, California, pp. 129–474.
Swingle, W.T. and Reece, P.C. (1967) The botany of citrus and its wild relatives. In Reuther, W., Webber, H.J. and Batchelor, L.D. (eds) *The Citrus Industry*, Vol. 1. History, World Distribution, Botany and Varieties. University of California Press, Riverside, California, pp.190–422.
Sykes, S.R. (2002) 'Australian Outback', 'Australian Blood' and 'Australian Sunrise'. *Plant Varieties Journal* 15, 18–21.
Tanaka, T. (1977) Fundamental discussion of citrus classification. *Studia Citrologia* 14, 1–6.
Torres, A.M., Soost, R.K. and Diedenhofen, U. (1978) Leaf isozymes as genetic markers in citrus. *American Journal of Botany* 65, 869–881.
Vasudeva, N. and Sharma, T. (2012) Chemical composition and antimicrobial activity of essential oil of *Citrus limettioides* Tanaka. *Journal of Pharmaceutical Technology and Drug Research* 1, 2.
Webber, H.J. (1943) Cultivated varieties of citrus. In: Webber, H.J. and Batchelor, D.L. (eds) *The Citrus Industry*, vol. 1. University of California, Berkeley, California, pp. 475–668.
Wu, G.A., Prochnik, S., Jenkins, J., Salse, J., Hellsten, U., *et al.* (2014) Sequencing of diverse mandarin, pummelo and orange genomes reveals complex history of admixture during citrus domestication. *Nature Biotechnology* 32, 656–662.
Yamamoto, M., Abkenar, A.A., Matsumoto, R., Nesumi, H., Yoshida, T., Kuniga, T., Kubo, T. and Tominaga, S. (2007) CMA banding patterns of chromosomes in major *Citrus* species. *Journal of the Japanese Society for Horticultural Science* 76, 36–40.
Ziegler, L.W. and Wolfe, H.S. (1961) *Citrus Growing in Florida*. University of Florida Press, Gainesville, Florida, pp. 51–53.

3 Advances in Lime Breeding and Genetics

Muhammad Amjad Ali[1]* and Muhammad Azher Nawaz[2,3]
[1]*Department of Plant Pathology and Centre of Agricultural Biochemistry and Biotechnology (CABB), University of Agriculture, Faisalabad, Pakistan;* [2]*Department of Horticulture, University College of Agriculture, University of Sargodha, Sargodha, Pakistan;* [3]*College of Horticulture and Forestry Sciences, Huazhong Agricultural University/Key Laboratory of Horticultural Plant Biology, Ministry of Education, Wuhan, China*

Introduction

The genus *Citrus* belongs to the family *Rutaceae* with great diversity in its species and hybrids, which are largely different from each other on the basis of variations in shape, size and colour of fruit and leaves, and canopy size and plant structure. The lime appears to be highly diverse due to polyploidization in several cultivars. All lime accessions are highly heterozygous, with interspecific admixture of two, three or sometimes four ancestral 'taxa' genomes (Curk *et al.*, 2016). The magnificence of breeding is to exploit this huge intra- and interspecific variation by selecting more adaptive and acclimatized cultivars for a particular locality.

Plant breeding has been practised for a long time in the history of human beings. Nevertheless, the twist in the history occurred in the 1860s, when an Austrian monk Gregor Mendel, using his hybridization experiments in peas, coined the basic laws of inheritance. This development was not only a landmark in classical genetics and breeding, but also resulted in the foundation of today's modern genetics. This advancement renovated classical breeding into the modern science of molecular breeding and genetics. Moreover, Mendel's work paved the way for several improvements in crop plants, which in turn led to the advent of the so-called 'Green Revolution'. The citrus industry was also benefited by these advancements with particular reference to limes, which were the product of hybridization of basic taxa. This hybridization resulted in more diversity in this citrus group, which was exploited by the breeders to select high yielding and good quality lime cultivars (Grosser *et al.*, 1989). This chapter gives an account of germplasm and genetic diversity, and advancement in lime breeding and genetics. It also discusses the use of various classical and modern strategies used for improvements in lime scions and rootstocks.

Germplasm Resources and Biodiversity

The term 'germplasm' refers to the living tissues that contain complete genetic information and can be used to reproduce plants with identical characteristics, and can be used to link the generations of a specific plant species or cultivar.

*E-mail: amjad.ali@uaf.edu.pk, amjad.ali2001@gmail.com

The lime's germplasm is represented by seed, pollen, budwood and, in certain cases, leaves and other vegetative tissues can be used for the purpose. In fact germplasm is biological information that has passed down for generation after generation in the form of an unbroken chain (Wilkes, 1988). Once this chain is broken due to natural disasters or human interference, specific genetic resources are lost forever. This situation necessitates the conservation of germplasm for the use of current and future generations (Holden and Williams, 1984; Brown *et al.*, 1989; Holden *et al.*, 1993). Germplasm is used for the conservation of exciting plant types, and the creation of new plant types (rootstocks and scions) through conventional and advanced biotechnological crop improvement approaches. Keeping in view the importance of plant germplasm, several citrus germplasm protection programmes have been launched and operated in different citrus-producing countries.

Lime cultivars

All the reported lime species belongs to the genus *Citrus*. Lime is an important group of citrus fruits. It is known as lime in English and French, and lima in Italian and Spanish. In Arab countries, limes and lemons are grouped together and called limun (limoon, limoun), and nimbu or limbu, respectively. Limes are believed to have originated in north-east India, nearby areas of Burma or northern Malaysia (Reuther *et al.*, 1967). Currently limes are grown in a number of countries including the USA, Brazil, China, India, Australia, Pakistan, Iran, Israel, Japan and the Gulf countries. It is difficult to get exact information about limes and in some references limes and lemons are dealt with collectively. According to a report 56 accessions of limes are in Brazil, 17 in Argentina, five in Pakistan, three in Cuba, two in Chile and one in Uruguay (Anderson, 2000; Nawaz *et al.*, 2007a, b). Some details about different lime species, cultivars and hybrids are given below.

West Indian/Key lime/Mexican lime/Kaghzi nimbu

This is the most popular cultivar among limes. It is believed to have originated in India. In India it is called Kaghzi nimbu; it is known as Limunbaladi in Egypt, Doc in Morocco, Gallego lime in Brazil and Limon corriente in Latin American countries. In North America it is also known as Key lime (Reuther *et al.*, 1967). The known age of *Citrus aurantifolia* Swingle is believed to be about 700 years, and it is considered a hybrid with moderate genetic diversity (Kahn *et al.*, 2001). It is a highly polyembronic cultivar and seed is used as a source of propagation in most countries, and therefore there is limited variability among the clones from different countries. It is the most extensively cultivated cultivar of lime across the world.

Tahiti lime/Persian lime/Bearss lime

Tahiti lime (*Citrus latifolia* Tanaka) is believed to be native to the old Persian region (modern-day Iran). It was transported to Brazil in the 1800s via Australia, where it is reported have been grown since 1824 (Bowman, 1955). It was introduced in the USA from Tahiti, and so given the name Tahiti lime. Now, the major producers include the USA, Brazil, Israel, Australia and Asian regions. A selection of Persian limes named 'Persian lime SPB-7' plants were distributed by a citrus clonal protection programme during 2007 to the farmers of California. This new selection is believed to be free from a genetic disorder known as wood pocket, which was common in large fruited limes. Currently it is grown in Brazil and the USA (Florida and California). Another cultivar named Page lime (*Citrus latifolia*) is also reported but that seems to be exactly like Bearss lime in tree and fruit characteristics. Similarly Pond's lime (*Citrus latifolia*), also exists, with certain differences from Bearss lime but principally it is considered to be similar to Bearss lime. It is believed to have originated in Hawaii, and is considered the best lime there. Roodan lime (*Citrus aurantifolia* Swingle) is the local cultivar of southern Iran, with unique characters of thornless stem and branches, which facilitates the harvesting process and it is preferred by the growers of Iran (Zandkarimi *et al.*, 2011). Its exact identification is obscure; however, its characteristics match the Tahiti lime, both being thornless.

Bearss seedless

The origin of this lime cultivar is not known. In the literature it has been reported to have originated in 1885 in the nursery of T. J Bearss at

Porterville, California (Webber, 1943). It came into origin by seed sowing of a lime fruit of Tahitian origin. It was described for the first time by Lelong (1902), and introduced and promoted by the Francher Creek Nursery Company of Frenso in 1905. The tree and fruit characteristics are somewhat similar to Tahiti lime. The fruit are seedless, and rarely seeded, but if seeded they are highly monoembryonic. The Bearss lime is reported to be triploid in its genetic constitution, so seedless (Bacchi, 1940). It is grown in Iran, Iraq, the USA and Asian countries.

Australian finger lime

Australian finger lime is formally known as *Citrus australasica or Microcitrus australasica.* It is believed to be native to rain forests of the border ranges of south-east Queensland and northern New South Wales. In 1965 it was imported to the USA from Sydney, Australia (Siebert *et al.*, 2010). Its mature fruit are about 15–30 cm long and round, and come in a wide range of colours including green, yellow, purple and pink to bright red. Several cultivars of finger lime are registered in Australia, which include (i) *Citrus australasica* var. *sanguine* 'Rainforest Pearl'; (ii) *Citrus australasica* 'Alstonville'; (iii) *Citrus australasica* 'Blunobia Pink Crystal'; (iv) *Citrus australasica* 'Durhams Emerald'; (v) *Citrus australasica* 'Judy's Everbearing'; and (vi) *Citrus australasica* 'Pink Ice'. Additionally some wild cultivars are also reported in Australia (Hardy *et al.*, 2010). Small-fruited lime cultivars are also reported but most of them have not gained commercial status and only present for germplasm conservation proposes. Some of them include Castelo lime, Chulo Key lime, Abhayapuri lime, Egyptian lime, Everglade lime, Giant Key lime, India lime and Soghi lime.

Another citrus cultivar, Rangpur lime, is sometimes confused for a lime cultivar; in fact it is not lime but a lemandrin (cross of mandarin and lemon). Some basic information about important lime cultivars and their hybrids is summarized in Table 3.1.

Lime hybrids

Several lime hybrids have been reported in the literature, these include Lakeland lime, Eustis limequat, Tavares limequat, Addanimma lime hybrid, limonime lime hybrid, Perrine lime hybrid and Warren limequat. The lime hybrids are small in size and are used as dooryard plants. The fruit are used as a substitute for lime. Keeping in view their importance, more hybrids with better fruit and plant characteristics are being developed. The scientists of the Citrus Research and Education Center (CREC), University of Florida, USA have recently released a seedless lime hybrid 'C4-5-27' (a cross of Key lime and tetraploid lemon) with superior fruit and plant characteristics (Grosser *et al.*, 2015). Some other

Table 3.1. Characteristics of some important lime cultivars and hybrids.

Name of cultivar	Scientific name	Origin	Important producers
Key lime, West Indian lime, Mexican lime	*Citrus aurantifolia* Swingle	India	México, India, Pakistan
Bearss lime/Tahiti lime/ Persian lime	*Citrus latifolia*	Old Persian region	Iran, Iraq, USA, Asian countries
Bearss seedless SPB-7	*Citrus latifolia*	USA	–
Australian finger lime	*Citrus australasica*		Australia
Roodan lime	*Citrus aurantifolia* Swingle	Southern Iran	Iran
Eustis limequat	*Citrus aurantifolia* × *Fortunella japonica*	USA	No commercial cultivations
Lakeland lime	*Citrus aurantifolia* × *Fortunella japonica*	USA	No commercial cultivations
Tavares limequat	*Citrus aurantifolia* × *Fortunella margarita*	USA	No commercial cultivations

Sources: Hodgson, 1967; Castle *et al.*, 1986; Zekri and Parsons, 1992; Chaudhary, 1994; Castle, 2010; Zandkarimi *et al.*, 2011; Castle, 2012; Crane, 2013 triploid seedless lime hybrids 'C4-5-20' and 'C4-5-33' were developed and released in 2016 (Grosser *et al.*, 2015). Similarly, the development of coloured Mexican lime (pigmented lime) has been reported recently through the introduction of transcription factor *Ruby* (Blood orange) and *VvmybA1* (Grapes) in Mexican lime (Grosser *et al.*, 2015; Dutt *et al.*, 2016). Keeping in view the importance of and demand for limes, it is expected that more cultivars with better fruit and plant characteristics will be developed and released in the lime-producing countries.

Cultivar Improvement through Breeding and Genetics

Rootstock and scion improvement

Several breeding strategies have been adopted for the improvement of lime fruit. Due to recent advancements in the science of genetics, plant breeding has seen new horizons followed by remarkable improvements in plants of economic importance. The improvement of citrus fruits has also been aided by these advancements with particular reference to limes, which themselves were the product of hybridization of basic citrus taxa (Grosser *et al.*, 1989). This hybridization resulted in more diversity in this group of citrus plants, which was exploited by the citrus breeders to select high yielding and good quality lime plants. Among the several methods for lime improvement, somatic hybridization and cybridization, somaclonal variation, induced mutation, manipulation of polyploidy, nucellar and zygotic embryony, use of cytogenetics, genetic engineering to develop transgenic plants and marker assisted selection (MAS) have largely been used (Khan and Kender, 2007). Similarly, zygotic embryony has also been tried in several diploid genotypes and seems promising in lime as compared to other citrus species (Cimen and Yesiloglu, 2016).

The improvement in different lime fruits is largely based on better synchronization of scions and their corresponding rootstocks. The successful rootstock and scion cultivars with desired characteristics are commercially grown all over the world. These rootstock and scion cultivars are screened out from several cultivars mainly on the basis of tree performance under that particular environment and quality of fruit produced by them. A combination of disease-free rootstock and scion of desired characteristics results in trees with high yield and good quality fruit.

Criteria for scion improvement

The main objectives of scion breeding and improvement are divided into three basic criteria, i.e. overall performance of the trees, fruit attributes and post-harvest characteristics of the fruit (Khan and Kender, 2007). The overall performance of the tree covers the fruit yield, cold hardiness and adaptability of the scion to withstand adverse biotic and abiotic stresses. Similarly, different fruit characteristics like the external appearance of the fruit, shape and size, juice content, flavour and colour, quality in the sense of soluble solids/acid ratio, seedlessness, ripening season and ease of peeling should be kept in mind when selecting the scion. Scions are also selected on the basis of their post-harvest attributes. These attributes include good handling and processing quality for fresh market as well as for storage purposes, longer storage life and better juice content and composition after storage.

Criteria for rootstock improvement

Many problems faced by the citrus industry could be solved by appropriate selection of locally adapted rootstocks (Wutscher and Hill, 1995; Bowman, 2000). There are several criteria for the selection of rootstocks for limes, e.g. early bearing, uniformity, ability to develop through apomixis (polyembryony), fruit with enhanced qualities like easy peeling, more juice, more seeds, and maximum resistance against biotic and severe climatic and post-harvest situations (Grosser and Gmitter, 2005). So in order to produce lime trees with the desired attributes, the above-mentioned points should be kept in mind before the selection of a particular rootstock to be used for grafting. Recently, Grosser and Gmitter (2011) have emphasized attributes like cold hardiness, resistance to various citrus diseases and potential for the production of new industrial oils as primary objectives in lime breeding and improvement programmes.

Development of improved cultivars by different techniques

Development of improved cultivars containing the desired attributes through conventional breeding is too laborious and time consuming: it requires 25–30 years for an improved cultivar to become available for commercial purposes. Nowadays, scientists work hard to develop techniques through which the time required to develop improved cultivars could be minimized along with enrichment of resistance against diseases, salt tolerance and fruit quality.

Somatic hybridization

The desired attributes of resistance against biotic and abiotic diseases are persisting in the citrus and related gene library. Development of improved cultivars via conventional breeding has a lower probability of producing a recombinant hybrid containing all the desired characteristics due to apomixis and heterozygosity of the parents (Grosser *et al.*, 2000). In contrast to conventional breeding, by adopting somatic hybridization, intergeneric allotetraploid somatic hybrid production could be accessible (Grosser *et al.*, 1996).

Somatic allotetraploid hybrids have been developed by fusing Key lime protoplasts from embryogenic suspension culture, with the leaf protoplasts of sweet orange (cv. Valencia). The resultant hybrids have shown very good vigour compared to their parents (Grosser *et al.*, 1989). The production of such allotetraploid hybrids has resulted in a revolution in lime breeding as it paves the way for using this technique to produce lots of interspecific and intergeneric hybrids. Based on these studies, Ollitrault *et al.* (1996) performed the first intergeneric somatic hybridization of *C. reticulate* protoplasts with those of *F. japonica*. Somatic hybrids of mandarin (*C. reticulata* Blanco) and Rangpur lime (*C. limonia* L. Osbeck) with sour orange have shown tolerance against bacterial blight and citrus tristeza virus (CTV) (Mendes *et al.*, 2001). Gloria *et al.* (2000) reported an effective drought-tolerant rootstock of Rangpur lime obtained by polyethylene glycol (PEG) fusion; this characteristic could be used for development of cultivars with enhanced drought tolerance in acid limes. Olivares-Fuster *et al.* (2005), introduced electrochemical protoplast fusion in which protoplasts of Mexican lime go through the combination of PEG and direct current (DC) impulses to enhance the rate of embryogenesis (somatic hybrids).

Cytogenetics

Cytogenetic research in lime is limited due to the very small size of the chromosomes. However, work on the development of polyploidization in citrus tree improvement has increased. In the past few decades, Mexican lime (*C. aurantifolia*) has been used as a model citrus plant for somatic hybridization, protoplast fusion, development of transgenics and to study plant–microbe interactions. Regardless of the enormous diversity in the citrus tribe, evolution of new species is largely restricted at the diploid level due to seed production through apomixis (Khan and Kender, 2007). This unique scenario signifies the use of somatic hybridization for citrus breeder to speed up citrus evolution with the help of polyploidization (Grosser *et al.*, 1996).

Very little research has been executed in acid limes using cytogenetics. Rangpur lime was first studied to investigate its structural differences, with Rough lemon and Chironja, on the basis of bivalent formation during meiosis. The studies revealed a lack of big structural differences in the genomes of these species. These studies also demonstrated that pollen sterility in Rangpur lime was conditioned by both genes and chromosomes, while in Rough lemon and Chironja it was only genic (Agarwal, 1987).

Cytogenetic studies have been used to study similarities and variations among citrus species at chromosome level. For instance, chromomycin A3 (CMA) banded chromosomes have been used to study genetic variation in different hybrids of mandarin and in the lime, lemon, citron and pummelo groups (Guerra, 1993; Befu *et al.*, 2001). Cytological studies based on the CMA+/DAPI− banding pattern of chromosomes and the distribution of the 5S and 45S rDNA sites of ten accessions of lime, lemon and citron have been investigated (Carvalho *et al.*, 2005). The results concluded that different limes species like *C. aurantifolia* and *Citrus limettioides* are related to each other to a lesser extent than thought previously because of the heterozygosity at most of the rDNA sites studied. Recently, Yamamoto *et al.* (2007) used fluorochrome staining with chromomycin A3 (CMA) to compare the CMA banded chromosomal patterns of 12 major diploid (2n = 18) citrus species. On the basis of seven CMA banding patterns, *C. aurantifolia* exhibited chromosome banding patterns with two chromosomes having one telomeric and one proximal band, nine chromosomes with one telomeric band only and seven chromosomes lacking any band.

Polyploidy

Somatic hybridization has resulted in allotetraploid breeding parents for the development of seedless triploids in citrus species including limes (Grosser and Gmitter, 2005). Polyploidy has been widely exploited for citrus improvement. Using electrofusion mediated somatic hybridization, several triploid and tetraploid hybrids were obtained by combining the diploid protoplasts of Mexican lime and eight other cultivars with haploid cell lines of clementine (Ollitrault *et al*., 2000). Recently Grosser and Gmitter (2011) have critically reviewed the implications of polyploidy for citrus improvement and development of polyploids generated from the somatic hybridization of various citrus genera and species including different varieties of limes. A schematic diagram is shown in Fig. 3.1 to demonstrate the development of different triploids and tetraploids in acid limes.

Several studies have demonstrated better rootstock/scion and fruit characteristics of polyploids as compared with their corresponding diploid species. In various polyploids, i.e. triploids and tetraploids, juice contents are more affected compared with diploid species, while the taste, acidity and other soluble contents are less affected. However, autotetraploids of Key lime with larger fruit and darker leaves have resulted in a flavour similar to those of the parental diploids (Barrett, 1992). Nonetheless, the polyploidy results in interesting germplasm for lime breeding in addition to giving rise to polyploids with good taste and fewer seeds. Similarly, Ni *et al*. (2009) demonstrated that allopolyploids of

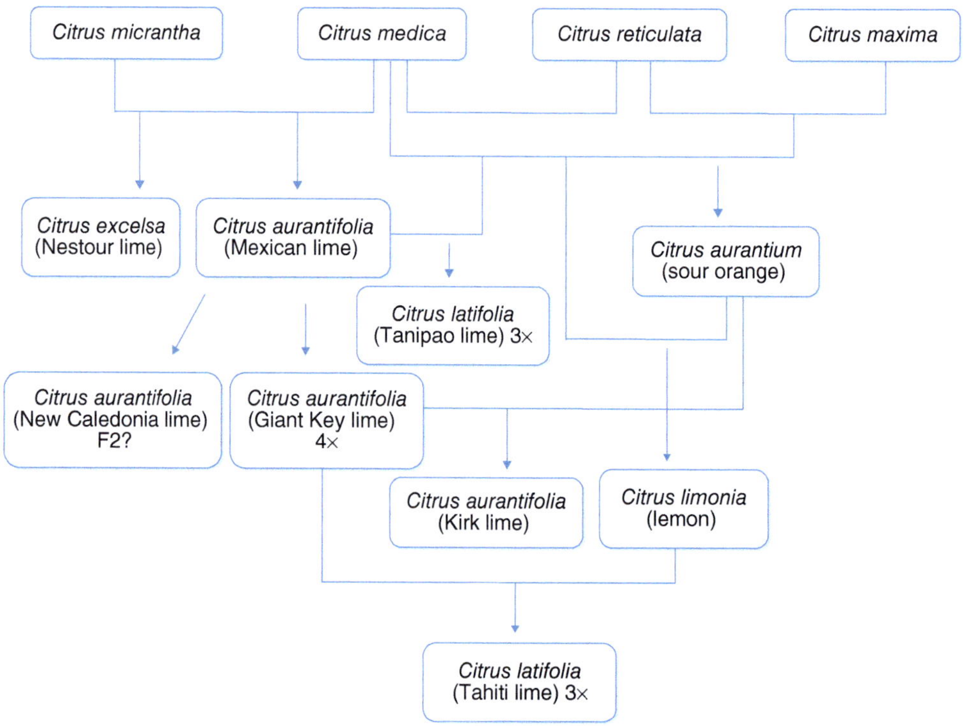

Fig. 3.1. Schematic diagram demonstrating the development of different diploid, triploid and tetraploid species in acid limes through hybridization of basic citrus taxa.

citrus lines have better control of circadian modulated metabolic and physiological pathways in addition to enhanced growth vigour and biomass production. Moreover, changes in the expression of genes in allotetraploids may be the function of genome hybridization instead of genome polyploidization (Auger *et al.*, 2005). The polyploid citrus genome expression studies suggest that changes in transcriptome response are not dependent on chromosome number but depend on the origin of the chromosome (Yu *et al.*, 2010). The diploid and autotetraploid clones of Rangpur lime rootstocks grafted with diploid sweet orange (cv. Valencia Delta) scions showed enhanced tolerance against drought stress (Allario *et al.*, 2013). This tolerance was largely based on differential expression of water stress responsive genes in the plant roots. Moreover, the tolerant plants exhibited lower stomatal conductance in addition to up-regulation of genes involved in biosynthesis of abscisic acid (ABA). These studies demonstrated that tetraploidy modulates the gene expression in Rangpur lime roots through ABA signalling (Allario *et al.*, 2013).

Additionally, polyploidization of lime and its hybrids caused better leaf morphology of tree leaves. For instance, tetraploids from lime have shown darker, thicker leaves and large leaf size as compared with their corresponding diploid citrus plants, while plant stature was reduced in the tetraploids (Allario *et al.*, 2011).Citrus breeders have limited approaches to develop sexual recombination that can result in progeny of seedless fruits. Many attempts have been made to produce hybrids by somatic hybridization that respond well to flowering and can be used as pollen parents to develop seedless lime fruits (Grosser *et al.*, 2000; Grosser and Gmitter, 2005). Polyploidization of citrus is also an important technique to develop seedless varieties of citrus. Several triploid citrus/limes cultivars have been successfully developed to accumulate the desirable traits of diploid and tetraploid parents.

The development of seedlessness in triploids lime plants is commonly linked to female and male sterility. The benefit of triploidy is the retention of larger fruit size that is not observed in other seedless variants of seedy cultivars. Tahiti lime is a natural triploid that yields large seedless fruit as compared with its diploid relative Mexican lime (Jackson and Sherman, 1975; Froelicher, 1999). Viloria and Grosser (2005) initiated a large breeding programme in 2000 using interploid hybridization to improve cold tolerance, disease resistance and seedlessness in limes and lemons. In this programme somatic hybridization of allotetraploid hybrids of different limes and lemons with autotetraploid lime and lemon parents was carried out to generate 35 cross-combinations. This resulted in 650 hybrids, out of which 500 were triploid. The authors suggested that these novel genetic combinations would lead towards genetic improvement of lemons and limes (Viloria and Grosser, 2005).

Use of biotechnology and genetic engineering for improvement of acid limes

Although conventional breeding has been practised for improvement of lime for decades, it has some unavoidable limitations like the production of large sized plants, apomixis, nucellar polyembryony, pollen or ovule sterility, high heterozygosity and the difficulty in accomplishing controlled crosses without contaminations (Vardi *et al.*, 1975; Vardi, 1981; Martin-Trillo and Martinez-Zapater, 2002; Singh and Rajam (2009). To overcome these limitations, use of genetic engineering aimed at the development of transgenic plants in acid limes seems promising. Transformation efficiency has remained a topic of concern for development of transgenic citrus fruits; nonetheless, transformation efficiency in lime has been studied by several scientists to introduce genes for biotic and abiotic stresses (Gutierrez *et al.*, 1997; Pena *et al.*, 1997; Pérez-Molphe and Ochoa-Alejo, 1998; Dominguez *et al.*, 2000; Ghorbel *et al.*, 2001). Various plant tissues have been transformed using heterologous genes in lime. For instance, epicotyl segments (Pena *et al.*, 1997; Dominguez *et al.*, 2000) and internodal stem segments (Ghorbel *et al.*, 2001) have been used to develop explants to be transformed. Moreover, lime plants have shown relatively higher transformation efficiencies in response to *Agrobacterium*-inoculated explants as compared with other citrus types except citranges (Pena *et al.*, 2004). Similarly, stable transformation frequency considerably increased in lime explants transformed with the extra copies of three genes, EHA105, pTiBo542 and virG, using *Agrobacterium*-mediated transformation (Pena *et al.*, 2004). The effect of four antioxidants (glycine betaine, lipoic acid, polyvinylpyrrolidone and glutathione) was studied on the *Agrobacterium*-mediated transformation efficiency of Mexican lime (Dutt *et al.*, 2011). Mexican lime cultivated with lipoic acid aids in callus development and also improves shoot growth of the epicotyl portion. The glycine betaine was found to be beneficial to a certain level while the other two antioxidants have no ability to cause changes in the transformation efficiency of Mexican lime in the manner that lipoic acid does (Dutt *et al.*, 2011). The genes for several characters have been transformed in lime trees with the main emphasis being on the fruit quality (i.e. seedlessness), disease resistance and abiotic stress tolerance. A detailed overview of fruit quality enhancement and development of stress tolerance has been given below using genetic transformations.

Fruit quality

The characteristic of seedlessness has been extensively addressed by genetic engineering for improvement in acid limes. Citrus transformation by exploiting *Agrobacterium* for 20 min to introduce genes responsible for low seed setting has been found to be successful for most of the cultivars including sweet orange and Nagpur mandarin (Fleming *et al.*, 2000; Koltunow *et al.*, 2000; Almeida *et al.*, 2003). Genes were introduced in West Indian lime, Mexican lime and Rangpur lime for decreased seed set, coat protein gene of CTV and bacterio-opsin, respectively (Dominguez *et al.*, 2000; Koltunow *et al.*, 2000; Azevedo *et al.*, 2006). Transgenic plants harbouring the genes responsible for low seed set were produced in West Indian lime by transforming the epicotyl and hypocotyl meristematic tissues through *Agrobacterium tumefaciens*-mediated transformation (Koltunow *et al.*, 2000). Recently, genetically engineered biofortification has been executed in Mexican lime. Genes like *VvmybA1* or *Ruby* from grapes and grapefruit have been introduced into Mexican lime to enable it to synthesize its own anthocyanins. Anthocyanins have good antioxidant characteristics that can help to overcome the risks of cancer, obesity, heart disease and diabetes (Dutt *et al.*, 2016). Moreover, it has opened up new avenues for the production of coloured limes as well as other citrus cultivars.

Recently some transcription factors like CitERF13 and CitVHA-c4 from citrus have been reported to be responsible for citric acid accumulation when transformed into citrus and Arabidopsis plants (Li *et al.*, 2016). These genes could be good candidates to create high citric acid content in acid limes.

Stress tolerance

Mexican lime (*C. aurantifolia*) has widely been used to study plant pathogen interactions in response to CTV (Gutierrez *et al.*, 1997; Dominguez *et al.*, 2000; Ghorbel *et al.*, 2000; Ghorbel *et al.*, 2001; Soler *et al.*, 2012; Flores *et al.*, 2013; Soler *et al.*, 2015). The Mexican lime plants were transformed with the coat protein coding gene of CTV, which resulted in 10–33% resistance initially against CTV (Gutierrez *et al.*, 1997; Dominguez *et al.*, 2000).

Transformation of the coat protein (CP) gene of CTV into key lime and sour orange was attempted by using internodal stem segments as explants (Gutierrez *et al.*, 1997). Mexican lime plant gave the best results to study coat protein resistance against CTV because of its sensitivity to the disease. Coat protein gene *p25* from CTV was used to induce resistance in Mexican lime plants. Moreover, after transformation only 33% of plants showed resistance to CTV, while the others remained prone to the development of the symptoms (Dominguez *et al.*, 2002). Another method to induce CTV resistance was introduced by Cervera *et al.* (2010) in which two single-chain variable antibody fragments (scFv) specific to CTV were inserted in Mexican lime either alone or in combination. Using this method only 40–60% of plants showed resistance, with reduced or lower development of symptoms. Plant-mediated RNAi-induced gene silencing of *p23* gene resulted in enhanced resistance in Mexican lime against CTV, which shows that *p23* silencing suppressor is very important for compatible interactions of CTV with lime plants (Fagoaga *et al.*, 2006). Similarly Mexican lime plants transformed with an intron-hairpin construct coding for untranslatable versions of three silencing suppressors *p25*, *p20* and *p23* from CTV strain T36 conferred complete resistance against CTV (Soler *et al.*, 2012). Azevedo *et al.* (2006) introduced a *bO* gene obtained from *Halobacteria halobium* in Rangpur lime to induce disease resistance against *Phytophthora nicotianae* through increased production of pathogenesis related (PR) proteins.

Dutt *et al.* (2012) transformed Mexican lime plant with promoter::*GUS* fusions by using phloem specific promoters from various genes like *Agrobacterium rhizogenes* rolC, *AtSUC2* gene from Arabidopsis, sucrose synthase 1 (*RSs1*) gene from rice and rice tungro bacilliform virus (RTBV) gene to drive expression of *GUS* in the phloem. They concluded that RTBV was the best one for phloem-specific expression of *GUS* protein and suggested that promoters like this are important candidates to control phloem-limited bacteria associated with huanglongbing/citrus greening disease of citrus (Dutt *et al.*, 2012). The acid lime group, along with pummelo and trifoliate orange, has an advantage over other citrus fruits in terms of resistance against citrus greening disease (Knapp *et al.*, 2004; Batool *et al.*, 2007). This is an important attribute of limes that can be used for the development of a huanglongbing resistant germplasm.

A recent study has shown that up-regulation of genes related to cell wall degradation and high metabolism in Mexican lime in response to witches' broom disease of lime (WBDL); however, down-regulated genes were mainly related to oxidative phosphorylation and ubiquitin proteolysis pathways, which are important for defence activation (Mardi *et al.*, 2015). These down-regulated genes are good candidates for overexpression to enhance disease resistance in lime plants.

The transcript levels of 12 genes have been studied in grapefruit using heat treatment and conditioning treatment (Sapitnitskaya *et al.*, 2006). Seven genes, *HSP19-I*, *HSP19-II*, *USP* (universal stress protein), dehydrin, *EIN2*, superoxide dismutase (*SOD*) and 1,3;4-b-D-glucanase, were particularly modulated by heat treatment. However, lipid transfer protein (*LTP*) and fatty acid desaturase2 (*FAD2*) were only regulated by the conditioning treatment, while a chaperonin, a translation initiation factor (*SUI1*) and alcohol dehydrogenase (*ADH*) were activated by both the treatments. This suggested that pre-storage treatment results in the activation of several molecular pathways to induce chilling tolerance in citrus plants; this phenomenon could be exploited for enhancement of chilling tolerance in acid limes.

Sanchez-Ballesta *et al.* (2003, 2004) reported different *WRKY* transcription factors and a dehydrin gene *CrCOR15* as being important for the regulation of heat-induced chilling tolerance in fortune mandarin (*Citrus clementina* Hort. Ex Tanaka × *Citrus reticulata*, Blanco). These genes could be used as candidates for enhancement of chilling tolerance in lime.

Genomes, transcriptomes and proteomes

Thanks to the recent advancements in molecular biology, sequencing technologies and bioinformatical applications have made it possible to study plant genomes in less time and with more accuracy. *Citrus sinensis* was the first species whose genome was published followed by several other species (Xu *et al.*, 2013; Wu *et al.*, 2014). Sequencing of genomes from various citrus species (i.e. sweet orange, clementine, pummelo, mandarin and sour orange) revealed that during the process of domestication, complex combinations of these citrus species occurred. These combinations resulted in enhanced fruit quality and resistance against diseases in the present-day citrus species like sweet orange (Wu *et al.*, 2014).

Similarly, the chloroplast genomes of various citrus species have been studied. Omani lime (*C. aurantifolia*) was first studied for its chloroplast genome, which was 159,893 bp in length (Su *et al.*, 2014). The gene contents and genome organization were found to be similar to most of the rosid lineages. There is a lot of scope in the field of genome sequencing in acid limes and to date none of the lime species has been sequenced for its genome.

The use of recent technologies like comparative transcriptome sequencing could be interesting to elucidate the genetic mechanisms underlying disease resistance in lime trees (Mardi *et al.*, 2015). Comparative high-throughput transcriptome sequencing of Mexican lime tree in response to WBDL revealed 2805 differentially expressed transcripts (Mardi *et al.*, 2015). The up-regulated transcripts were largely related to cell wall degradation and high metabolic rate, which aids the development of the disease on lime plants. However down-regulated transcripts were primarily related to oxidative phosphorylation and ubiquitin proteolysis pathways. This demonstrates that silencing of up-regulated genes essential for compatible plant–pathogen interaction and overexpression of down-regulated genes involved in plant defence pathways may result in enhanced disease resistance in plants as reported by Ali *et al.* (2013a, b; 2014) for plant–nematode interactions in Arabidopsis. Similarly, comparative expression profiles of microRNAs (miRNAS) in healthy Mexican lime trees and in plants infected with WBDL demonstrated that the miRNAs are differentially expressed in these comparisons. Moreover, miRNAs are involved in the regulation of nutritional, hormonal and stress signalling pathways in lime trees (Ehya *et al.*, 2013). Comparative genomic and transcriptome studies of various pathotypes of citrus canker causal pathogen *Xanthomonas citri* have demonstrated the up-regulation of proteases and pectate lyases to support pathogen development on the plants (Jalan *et al.*, 2013).

Mexican lime plants infected with WBDL were used to perform cDNA-amplified fragment length polymorphism (AFLP) analysis as compared with uninfected plants to study the gene expression, with the help of 43 primer combinations (Zamharir *et al.*, 2011). Zamharir and colleagues identified 36 differentially regulated genes out of which around 70% were down-regulated in the infected plants. Similar to other transcriptomes in response to plant pathogen infections, plant defence was shut down by the pathogen through suppressing the genes involved in the defence-related pathways of lime trees. Microarrays have been largely utilized to study comparative transcriptomes in response to pathogen infections. A cDNA microarray was also used in lime to study its transcriptional responses after infection with different isolates of CTV (Gandia *et al.*, 2007). Various transcriptomes from lime plant infected with different CTV isolates revealed differences in activation and suppression of plant metabolic and defence-related pathways.

Proteome studies using 2D gel electrophoresis from leaves of two allotetraploid hybrids (*C. deliciosa* + *C. aurantifolia* and *C. deliciosa*+ *F. margarita*) showed that these hybrids were very close to their mandarin parent as compared with other parents on the basis of absence or presence of protein spots and quantitative expression analysis (Gancel *et al.*, 2006). These studies revealed that 75% of the spots related to the non-mandarin parent were down-regulated in the somatic hybrids.

For the first time, the proteomics approach using 2D gel electrophoresis coupled with mass spectrophotometry was used to study the relative expression of various leaf proteins in lime trees inoculated with Candidatus *Phytoplasma aurantifolia*, the causal organism of WBDL (Taheri *et al.*, 2011). The authors reported that 55 out of 800 proteins showed a significant response to the disease. Mass spectrophotometry further identified 39 differentially regulated proteins, most of which were involved in metabolism, photosynthesis, oxidative stress defence and other stress responses (Taheri *et al.*, 2011).

Recently, the latest techniques have been used to study the proteome of lime trees infected with the causal agent of WBDL. Monavarfeshani *et al.* (2013) used label-free quantitative shotgun proteomics for this purpose in which they studied 990 proteins in total, out of which 448 proteins were differentially expressed. In the lime plants infected with WBDL, 274 proteins were down-regulated, while 174 had enhanced expression compared with the healthy plants. Contrary to other transcriptomic and proteomics studies, the gene-related defence mechanisms of the plants were induced in the infected plants as compared with uninfected lime plants.

Marker assisted selections

Marker assisted selection is a powerful tool that can enhance the speed and efficacy of cultivar breeding in limes. In this procedure, DNA markers are identified that are polymorphic in the target population and tightly linked to the segregating gene that is the target of selection (Fang and Roose, 1997). The seedlings that carry the marker are considered to have the gene coding for the character of interest. These seedlings are selected by the breeders more accurately on the basis of genetic markers associated closely with the target gene. Marker assisted selections are used for the trait of interest in plants and animals through breeding. This attribute makes marker assisted selection advantageous over the classical breeding strategy in citrus fruits and other crops. Genomic techniques have been widely used for the improvement of citrus plants (Talon and Gmitter, 2008; Singh and Rajam, 2009).

Polymerase chain reaction (PCR) based markers like AFLPs, random amplified polymorphic DNAs (RAPDs), inter-simple sequence repeats (ISSRs) along with co-dominant DNA markers like simple sequence repeats (SSRs), RFLPs, cleaved amplified polymorphisms (CAPS) and isozymes have been used in citrus (Fang *et al.*, 1997). Jannati *et al.* (2009) reported that microsatellites/SSR markers have shown satisfactory results because of their variable, easy-to-use and reproducible attributes. Very limited knowledge regarding the use of DNA markers and marker assisted selection is reported in acid limes. However, they have been used mainly for the evaluation of genetic diversity in acid limes (Robles-Gonzalez *et al.*, 2008; Al-Sadi *et al.*, 2012; Munankarmi *et al.*, 2014; Curk *et al.*, 2016).

Robles-Gonzalez *et al.* (2008) reported the diversity in five clonal selections of Mexican lime based on AFLP studies. All of them shared several bands after AFLP analysis. Moreover, AFLP studies recognized genetic differences among the five lime selections on the basis of AFLP and other plant characteristics. Genetic diversity or resistance against WBDL were studied among 143 acid lime samples from Oman by using four primer pair combinations followed by AFLP analysis (Al-Sadi *et al.*, 2012). These primer pairs resulted in the generation of 980 polymorphic loci (100%) and 146 AFLP genotypes. The authors suggested that low degree of genetic diversity coupled with regular movement of rootstocks and scion materials across various districts were the main reasons for the development of WBDL in Oman (Al-Sadi *et al.*, 2012). Recently, the use of a RAPD-PCR system was reported as an effective and rapid tool for determination of genetic diversity in acid lime landraces from Nepal (Munankarmi *et al.*, 2014). In this study, 60 acid lime landraces from various agro-ecological zones of Nepal showed 79 RAPD fragments amplified through nine arbitrarily selected primers out of which (94.94% (75)) were found to be polymorphic. This diversity was suggested to be useful to acid lime breeders for the development of elite lime cultivars for different ecological zones (Munankarmi *et al.*, 2014). Very recently, Curk *et al.* (2016) investigated 133 accessions of citrus for their genetic diversity by using 123 markers in SNP and indel markers. They reported that the group exhibiting various polyploids of limes and lemons was highly polymorphic, heterozygous with interspecific hybrids of two, three and even the four ancestral taxa genomes.

Germplasm conservation

Germplasm is biological information that has passed from generation to generation in the form of an unbroken chain (Wilkes, 1988). Once this chain is broken specific genetic resources are lost forever. This situation requires the conservation of germplasm for current and future use (Holden and Williams, 1984; Brown *et al.*, 1989; Holden *et al.*, 1993). Germplasm is used for the conservation of exciting plant types, and for the creation of new plant types (rootstocks and scions) through conventional as well as advanced biotechnological crop improvement approaches. Bearing in mind the importance of germplasm, several citrus germplasm protection programmes have been launched and operated in different citrus-producing countries. A list of currently ongoing citrus germplasm conservation programmes is given in Table 3.2.

Future Perspectives

The acid lime group of citrus fruits exhibits a considerable degree of genetic and phenotypic diversity that could be exploited as an important source of germplasm development and preservation. This germplasm could be further used to evolve rootstock and scion cultivars for biotic and abiotic stress tolerance. Additionally, the progress in biotechnologies like protoplast fusion followed by somatic hybridization could be used for the development of elite germplasm in acid limes for breeding scion and rootstock cultivars. It will not only increase the diversity of the present gene pool of acid limes but will also lead to the selection of different scions and rootstocks with various characteristics of interest ranging from adaptability to severe climatic conditions to high quality. Rangpur lime, which has a good degree of tolerance against drought stress, could be used as rootstock in arid regions of the world.

Similarly, the use of recent technologies like transcriptome sequencing could be interesting to allow the elucidation of the genetic mechanisms underlying disease resistance in lime trees (Mardi *et al.*, 2015). Tissue specific promoters have been used for the development of transgenic

Table 3.2. Germplasm conservation centres with lime accessions around the world.

Name of the centre	Country name	Year of establishment
Commonwealth Scientific and Industrial Research Organization (CSIRO)	Australia	1916
Australian Center for International Agricultural Research (ACIAR)	Australia	1982
National Research Centre for Cassava and Fruit Crops (CNPMF)	Brazil	1942
Fruit Crops Research Centre (FCRC)	Brazil	1972
Sylvio Moreira Citriculture Centre (CCSM)	Brazil	1928
National Citrus Germplasm Repository	China	1960
Ali Moubark Collection Citrus Germplasm Farm of Horticulture Research Institute	Egypt	–
National Research Centre for Citrus	India	1986
Global Citrus Germplasm Network (GCGN)	Italy	1997
Fruit Tree Research Station	Japan	1973
Citrus Germplasm Bank (IVIA)	Spain	1975
International Plant Genetic Resources Institute (IPGRI)	Thailand	1974
Citrus Research and Education Center (CREC)	USA	1917
National Center for Genetic Resources Preservation (NCGRP)	USA	1958
USDA-ARS National Clonal Germplasm Repository for Citrus and Dates Riverside, California	USA	1987
Citrus Variety Collection (CVC), California	USA	1909
Citrus Clonal Protection Program (CCPP), California	USA	1933

lime plants, which might be an important step forward to target the expression of anti-microbial proteins in the specific plant tissue to enhance disease resistance in lime (Dutt *et al.*, 2012). Similarly, the study of ultrastructure of wounded and non-wounded lime leaves infected with different diseases could be important to investigate the intra- and intercellular movements of pathogens in the leaves of lime plants (Lee *et al.*, 2009). A little work regarding marker assisted selection has been done in limes, so there is a lot of scope to identify markers associated with different attributes of interest. Moreover, the Citrus Genome Database has been established to provide a lot of genomic data related to citrus fruits including lime (Su *et al.*, 2014). A lot of information could be taken from this database for improving lime tree genetics. Although the chloroplast genome from Omani lime has been sequenced, there is still a lot of scope in the field of genome sequencing in acid limes as none of the lime species has been sequenced for its genome. The use of recent technologies like comparative transcriptome sequencing could be interesting to investigate the genetic mechanisms underlying stress tolerance in lime trees (Mardi *et al.*, 2015).

So altogether, by using cutting edge technologies and recent developments like protoplast fusion, polyploidization, genomics, transcriptomics, proteomics and genetic engineering, scion and rootstock cultivars in acid limes could be improved. This will ultimately lead to better fruit and plant characteristics, along with enhanced shelf life and tolerance to multiple biotic and abiotic stresses.

References

Agarwal, P.K. (1987) Cytogenetical investigations in Rutaceae. I. Meiotic studies in four Citrus species of hybrid origin. *Cytologia* 52, 753–756.

Ali, M.A., Plattner, S., Radakovic, Z., Wieczorek, K., Elashry, A., Grundler, F.M.W. and Bohlmann, H. (2013a) An Arabidopsis ATPase gene involved in nematode-induced syncytium development and abiotic stress responses. *The Plant Journal* 74, 852–866.

Ali, M.A., Abbas, A., Kreil, D.P. and Bohlmann, H. (2013b) Overexpression of the transcription factor RAP2. 6 leads to enhanced callose deposition in syncytia and enhanced resistance against the beet cyst nematode *Heterodera schachtii* in Arabidopsis roots. *BMC Plant Biology* 13, 47.

Ali, M.A., Wieczorek, K., Kreil, D.P. and Bohlmann, H. (2014) The beet cyst nematode *Heterodera schachtii* modulates the expression of WRKY transcription factors in syncytia to favour its development in Arabidopsis roots. *PloS One* 9, e102360.

Allario, T., Brumos, J., Colmenero-Flores, J.M., Tadeo, F., Froelicher, Y., Talon, M., Navarro, L., Ollitrault, P. and Morillon, R. (2011) Large changes in anatomy and physiology between diploid Rangpur lime (*Citrus limonia*) and its autotetraploid are not associated with large changes in leaf gene expression. *Journal of Experimental Botany* 62, 2507–2519.

Allario, T., Brumos, J., Colmenero-Flores, J.M., Iglesias, D.J., Pina, J.A., Navarro, L. and Morillon, R. (2013) Tetraploid rangpur lime rootstock increases drought tolerance via enhanced constitutive root abscisic acid production. *Plant, Cell and Environment* 36, 856–868.

Almeida, W.A.B., Filho, F.M., Mendes, B.M.J., Pavan, A. and Rodriguez, A.P.M. (2003) Agrobacterium-mediated transformation of *Citrus sinensis* and *C. limonia* epicotyl segments. *Scientia Agricola* 60, 23–29.

Al-Sadi, A.M., Al-Moqbali, H.S., Al-Yahyai, R.A. and Al-Said, F.A. (2012) AFLP data suggest a potential role for the low genetic diversity of acid lime (*Citrus aurantifolia* Swingle) in Oman in the outbreak of witches' broom disease of lime. *Euphytica* 188, 285–297.

Anderson, C. (2000) Citrus germplasm resources and their use in Argentina, Brazil, Chile, Cuba, and Uruguay. *Proceedings of International Society of Citriculture, IX Conference*, pp. 123–125.

Auger, D.L., Gray, A.D., Ream, T.S., Kato, A., Coe, E.H. and Birchler, J.A. (2005) Non additive gene expression in diploid and triploid hybrids of citrus. *Genetics* 169, 389–397.

Azevedo, F.A., Filho, F.M., Mendes, B.M.J., Almeida, W.A.B., Schinor, E.H., Pio, R., Barbosa, J.M., Gonzalez, S.G., Carrer, H. and Lam, E. (2006) Genetic transformation of Rangpur lime (*Citrus limonia* Osbeck) with the bO (bacterio-opsin) gene and its initial evaluation for *Phytophthora nicotianae* resistance. *Plant Molecular Biology Reporter* 24, 185–196.

Bacchi, O. (1940) Observacoes citologicasem citrus. I. Numero de cromosomos de algumasespecies e variedades. *Jornal de Agronomia Piracicaba* 3, 249–258.

Barrett, H. (1992) An autotetraploid of the Key lime, *Citrus aurantifolia*. *Fruit Varieties Journal* 46, 166–170.

Batool, A., Mughal, S.M., Khan, M.M., Jaskani, M.J., Abbas, M. and Khan, I.A. (2007) Citrus greening disease – a major cause of citrus decline in the world – a review. *Horticultural Science* 34, 159–166.

Befu, M., Kitajima, A. and Hasegawa, K. (2001) Chromosome composition of some citrus species and cultivars based on the chromomycin A3 (CMA) banding patterns. *Journal of the Japanese Society for Horticultural Science* 70, 83–88.

Bowman, F.T (1955) A history of citrus growing in Australia. 1788–1900. *Citrus News* 31, 102–103.

Bowman, K.D. (2000) New hybrid citrus developed by U.S. Department of Agriculture. *Proceedings of the International Society of Citriculture IX Congress*, 1, 51.

Brown, A.H.D., Frankel, O.H., Marshall, D.R. and Williams, J.T (1989) *The Use of Plant Genetic Resources*. Cambridge University Press, Cambridge, UK.

Carvalho, R., Filho, W.S., Brasileiro-Vidal, A.C. and Guerra, M. (2005) The relationships among lemons, limes and citron: a chromosomal comparison. *Cytogenetic Genome Research* 109, 1–3.

Castle, W.S. (2010) A career perspective on citrus rootstocks, their development, and commercialization. *HortScience* 45, 11–15.

Castle, W.S. (2012) *Choose the Right Rootstock*. Fact Sheet HS-932, Horticultural Sciences Department, Florida Cooperative Extension Service, Institute of Food and Agricultural Sciences, University of Florida, Florida, USA, pp. 1–3.

Castle, W.S., Youtest, C.O. and Hutchison, D.J. (1986) Rangpur lime × Troyer citrange, a hybrid citrus rootstock for closely spaced trees. *Proceedings of the Florida State Horticultural Society* 99, 33–35.

Cervera, M., Esteban, O., Gil, M., Gorris, M.T., Martinez, M.C., Pena, L. and Cambra, M (2010) Transgenic expression in citrus of single-chain antibody fragments specific to Citrus tristeza virus confers virus resistance. *Transgenic Research* 19, 1001–1015. Available at: http://doi.org/10.1007/s11248-010-9378-5 (accessed 18 April 2016).

Chaudhary, M.I. (1994) Fruit crops. In. Malik, M.N. (ed.) *Horticulture*. National Book Foundation, Islamabad, Pakistan.

Cimen, B. and Yesiloglu, T. (2016) Rootstock breeding for abiotic stress tolerance in citrus. In: Shanker, A.K. and Shanker, C. (eds) *Abiotic and Biotic Stress in Plants – Recent Advances and Future Perspectives*. InTech Publishers, Rijeka, Croatia.

Crane, J.H. (2013) *Key Lime Growing in the Florida Home Landscape*. Horticultural Sciences Department, UF/IFAS Extension, USA. FC47.

Curk, F., Ollitrault, F., Garcia-Lor, A., Luro, F., Navarro, L. and Ollitrault, P. (2016) Phylogenetic origin of limes and lemons revealed by cytoplasmic and nuclear markers. *Annals of Botany* 117, 565–583. Available at: http://doi.org/10.1093/aob/mcw005 (accessed 18 April 2016).

Dominguez, A., Guerri, J., Cambra, M., Navarro, L., Moreno, P. and Peña, L, (2000). Efficient production of citrus transgenic plants expressing the coat protein gene of citrus tristeza virus. *Plant Cell Reporter* 19, 427–433.

Dominguez, A., Hermoso de Mendoza, A., Guerri, J., Cambra, M., Navarro, L. and Peña, L. (2002) Pathogen-derived resistance to Citrus tristeza virus (CTV) in transgenic Mexican lime (*Citrus aurantifolia* (Christ.) Swing) plants expressing its p25 coat protein gene. *Molecular Breeding* 10, 1–10.

Dutt, M., Vasconcellos, M. and Grosser, J.W. (2011) Effects of antioxidants on Agrobacterium-mediated transformation and accelerated production of transgenic plants of Mexican lime (*Citrus aurantifolia* Swingle). *Plant Cell, Tissue and Organ Culture* 107, 79–89. Available at: http://doi.org/10.1007/s11240-011-9959-x (accessed 18 April 2016).

Dutt, M., Ananthakrishnan, G., Jaromin, M.K., Brlansky, R.H. and Grosser, J.W. (2012) Evaluation of four phloem-specific promoters in vegetative tissues of transgenic citrus plants. *Tree Physiology* 32, 83–93. Available at: http://doi.org/10.1093/treephys/tpr130 (accessed 3 May 2016).

Dutt, M., Stanton, D. and Grosser, J.W. (2016) Ornacitrus: development of genetically modified anthocyanin-expressing citrus with both ornamental and fresh fruit potential. *Journal of the American Society for Horticultural Science* 141, 54–61.

Ehya, F., Monavarfeshani, A., Fard, E.M., Farsad, L.K., Nekouei, M.K., Mardi, M. and Salekdeh, G.H. (2013) Phytoplasma-responsive microRNAs modulate hormonal, nutritional, and stress signalling pathways in Mexican lime trees. *PloS One* 8, e66372. Available at: http://doi.org/10.1371/journal.pone.0066372 (accessed 18 April 2016).

Fagoaga, C., Lopez, C., de Mendoza, A.H., Moreno, P., Navarro, L., Flores, R. and Pena, L. (2006) Post-transcriptional gene silencing of the p23 silencing suppressor of Citrus tristeza virus confers resistance to the virus in transgenic Mexican lime. *Plant Molecular Biology* 60, 153–165. Available at: http://doi.org/10.1007/s11103-005-3129-7 (accessed 18 April 2016).

Fang, D.Q. and Roose, M.L. (1997). Identification of closely related citrus cultivars with inter-simple sequence repeat markers. *Theoretical and Applied Genetics* 95, 408–417 (accessed 24 October 2016).

Fang, D.Q., Roose, M.L., Krueger, R.R. and Federici, C.T. (1997) Fingerprinting trifoliate orange germ plasm accessions with isozymes, RFLPs, and inter-simple sequence repeat markers. *Theoretical and Applied Genetics* 95, 211–219.

Fleming, G.H., Olivares-Fuster, O., Del-Bosco, S. and Grosser, J.W. (2000) An alternative method for the genetic transformation of sweet orange. *In Vitro Cellular and Developmental Biology – Plant* 36, 450–455.

Flores, R., Ruiz-Ruiz, S., Soler, N., Sanchez-Navarro, J., Fagoaga, C., Lopez, C. and Pena, L. (2013) Citrus tristeza virus p23: a unique protein mediating key virus-host interactions. *Frontiers in Microbiology* 4, 98. Available at: http://doi.org/10.3389/fmicb.2013.00098 (accessed 18 April 2016).

Froelicher, Y. (1999) *Organisation de la diversité dans le genre Clausena Burm. F. (Rutaceae)* Doctorat, Paris XI, Orsay, France.

Gancel, A.L., Grimplet, J, Sauvage, F.X., Ollitrault, P. and Brillouet, J.M. (2006) Predominant expression of diploid mandarin leaf proteome in two citrus mandarin-derived somatic allotetraploid hybrids. *Journal of Agricultural and Food Chemistry* 54, 6212–6218.

Gandia, M., Conesa, A., Ancillo, G., Gadea, J., Forment, J., Pallas, V. and Guerri, J. (2007) Transcriptional response of *Citrus aurantifolia* to infection by Citrus tristeza virus. *Virology* 367, 298–306. Available at: http://doi.org/10.1016/j.virol.2007.05.025 (accessed 14 April 2016).

Ghorbel, R., Dominguez, A., Navarro, L. and Peña, L. (2000) High efficiency genetic transformation of sour orange (*Citrus aurantium*) and production of transgenic trees containing the coat protein gene of citrus tristeza virus. *Tree Physiology* 20, 1183–1189.

Ghorbel, R., López, C., Fagoaga, C., Moreno, P., Navarro, L., Flores, R. and Peña, L. (2001) Transgenic citrus plants expressing the citrus tristeza virus p23 protein exhibit viral-like symptoms. *Molecular Plant Pathology* 2, 27–36.

Gloria, F.J.M., Mourao, F., Camargo, L.E.A. and Mendes, M.E.J. (2000) Caipira sweet orange + Rangpur lime: a potential somatic hybrid to be used as rootstock in the Brazilian citrus industry. *Genetics and Molecular Biology* 23, 661–669.

Grosser, J.W. and Gmitter, F.G. (2005) 2004 SIVB congress symposium proceedings 'Thinking Outside the Cell': applications of somatic hybridization and cybridization in crop improvement, with citrus as a model. *In Vitro Cellular and Developmental Biology – Plant* 41, 220–225. Available at: http://doi.org/10.1079/IVP2004634 (accessed 18 April 2016).

Grosser, J.W. and Gmitter, F.G. (2011) Protoplast fusion for production of tetraploids and triploids: applications for scion and rootstock breeding in citrus. *Plant Cell, Tissue and Organ Culture* 104, 343–357. Available at: http://doi.org/10.1007/s11240-010-9823-4 (accessed 14 April 2016).
Grosser, J.W., Moore, G.A. and Gmitter, F.G. (1989) Interspecific somatic hybrid plants from the fusion of 'Key' lime (*Citrus aurantifolia*) with 'Valencia' sweet orange (*Citrus sinensis*) protoplasts. *Scientia Horticulturae* 39, 23–29. Available at: http://doi.org/10.1016/0304-4238(89)90034-4 (accessed 24 March 2016).
Grosser, J.W., Mourao, F.A.A, Gmitter, F.G. Jr, Louzada, E.S., Jiang, J., Baergen, K., Quiros, A., Cabasson, C., Schell, J.L. and Chandler, J.L. (1996) Allotetraploid hybrids between Citrus and seven related genera produced by somatic hybridization. *Theoretical and Applied Genetics* 92, 577–582.
Grosser, J.W., Ollitrault, P. and Olivares-Fuster, O. (2000) Somatic hybridization in citrus: an effective tool to facilitate variety improvement. *In Vitro Cellular and Developmental Biology – Plant* 36, 434–449.
Grosser, J.W., Viloria, Z. and Dutt, M. (2015) New acid citrus selections for Florida. *Citrus Industry* July, 30–33.
Guerra, M. (1993) Chromosomal variability in Citrus species, 71(November), 234–241 (accessed 18 April 2016).
Gutierrez, M.A., Luth, D. and Moore, G.A. (1997) Factors affecting Agrobacterium-mediated transformation in Citrus and production of sour orange (*Citrus aurantium* L.) plants expressing the coat protein gene of citrus tristeza virus. *Plant Cell Reports* 16, 745–753.
Hardy, S., Wilk, P., Viola, J. and Rennie, S. (2010) Growing Australian native finger limes. *Primefact* 979, 1–11.
Hodgson, R.W. (1967) Horticultural varieties of citrus. In: Reuther, W. (ed.) *The Citrus Industry*, Vol. I. Division of Agricultural Sciences, University of California, Davis, California, pp. 431–592.
Holden, J.H.W. and Williams, J.T. (1984) *Crop Genetic Resources: Conservation and Evaluation*. George Allen and Unwin, London.
Holden, J., Peacock, J. and Williams, T. (1993) *Genes, Crops, and the Environment*. Cambridge University Press, Cambridge, UK.
Jackson, L.K. and Sherman, W.B. (1975) Chromosome counts in 'Tahiti' lime. *Florida State Horticultural Society* 38, 458–459.
Jalan, N., Kumar, D., Andrade, M.O., Yu, F., Jones, J.B., Graham, J.H. and Wang, N. (2013) Comparative genomic and transcriptome analyses of pathotypes *of Xanthomonas citri* subsp. *citri* provide insights into mechanisms of bacterial virulence and host range. *BMC Genomics* 14, 551. Available at: http://doi.org/10.1186/1471-2164-14-551 (accessed 14 April 2016).
Jannati, M., Fotouhi, R., Pourjan, A. and Salehi, A.Z. (2009) Genetic diversity analysis of Iranian citrus varieties using microsatellite (SSR) based markers. *Journal of Horticulture*, 1, 120–125.
Kahn, T.L., Krueger, R.R., Gumpf, D.J., Roose, M.L., Arpaia, M.L. *et al.* (2001) *Genetic Resources Conservation Program*. Division of Agriculture and Natural Resources, University of California, Davis, California pp. 1–58. Available at: http://www.grcp.ucdavis.edu (accessed 18 April 2016).
Khan, I.A. and Kender, W.J. (2007) Citrus breeding. In: Khan, I.A. (ed.) *Citrus Genetics, Breeding and Biotechnology*. CAB International, Wallingford, UK, pp. 1–8.
Knapp, J.L., Halbert, S., Lee, R., Hoy, M., Clark, R. and Kesinger, M. (2004) *The Asian Citrus Psyllid and Citrus Greening Disease*. Agricultural IPM: Fruit (citrus) Florida. IFAS, University of Florida, Gainesville, Florida. Available at: http:/www. ipm. ifas.ufl.edu (accessed 18 April 2016).
Koltunow, A.M., Brennan, P., Protopsaltis, S. and Nito, N. (2000) Regeneration of West Indian limes (*Citrus aurantifolia*) containing genes for decreased seed set. *Acta Horticulturae* 535, 81–91.
Lee, I.J., Kim, K.W., Hyun, J.W., Lee, Y.H. and Park, E.W. (2009) Comparative ultrastructure of nonwounded Mexican lime and Yuzu leaves infected with the citrus canker bacterium *Xanthomonas citri* pv. *citri*. *Microscopy Research and Technique* 72, 507–516. Available at: http://doi.org/10.1002/jemt.20707 (accessed 14 April 2016).
Lelong, B.M. (1902) *Culture of the Citrus in California*. Revised by the State Board of Horticulture. Supt. State Printing, Sacramento, California.
Li, S., Yin, X., Xie, X., Allan, A.C., Ge, H., Shen, S. and Chen, K. (2016) The Citrus transcription factor, CitERF13, regulates citric acid accumulation via a protein-protein interaction with the vacuolar proton pump, CitVHA-c4. *Scientific Reports* 6, 20151. Available at: http://doi.org/10.1038/srep20151 (accessed 18 April 2016).
Mardi, M., Farsad, L.K., Gharechahi, J. and Salekdeh, G.H. (2015) In-Depth transcriptome sequencing of Mexican lime trees infected with Candidatus *Phytoplasma aurantifolia*. *PloS One* 10, e0130425. Available at: http://doi.org/10.1371/journal.pone.0130425 (accessed 18 April 2016).
Martin-Trillo, M. and Martinez-Zapater, J.M. (2002) Growing up fast: manipulating the generation time of trees. *Current Opinion in Biotechnology* 13, 151–155.

Mendes, B.M.J., Mourao Filho, F.d.A.A., Farias, P.C.d.M. and Benedito, V.A. (2001) Citrus somatic hybridization with potential for improved blight and CTV resistance. *In Vitro Cellular and Developmental Biology – Plant* 37, 490–495.

Monavarfeshani, A., Mirzaei, M., Sarhadi, E., Amirkhani, A., Khayam Nekouei, M., Haynes, P.A. and Salekdeh, G.H. (2013) Shotgun proteomic analysis of the Mexican lime tree infected with 'Candidatus *Phytoplasma aurantifolia*'. *Journal of Proteome Research* 12, 785–795. Available at: http://doi.org/10.1021/pr300865t (accessed 14 April 2016).

Munankarmi, N.N., Shrestha, R.L., Rana, N., Shrestha, J.K.C, Shrestha, S., Koirala, R. and Shrestha, S. (2014) Genetic diversity assessment of acid lime (*Citrus aurantifolia* Swingle) landraces of eastern Nepal using RAPD markers. *American Journal of Plant Science* 3, 1674–1681.

Nawaz, M.A., Khan, M.M., Jaskani, M.J. and Iftikhar, Y. (2007a) Present status of citrus gene pool in Pakistan. *Proceedings of the International Symposium on Prospects of Horticulture Industry in Pakistan*, 28–30 March, 2007 at Institute of Horticultural Sciences, University of Agriculture, Faisalabad, Pakistan, pp. 70–74.

Nawaz, M.A., Khan, M.M., Jaskani, M.J., Iftikhar, Y. and Ahmed, W. (2007b) Citrus germplasm status in Pakistan and its role in citrus nurseries. *Proceedings of workshop on Production of Disease Free Citrus Nursery Plants (Production, Certification and Challenges)*, 12–13 September, 2007 at Institute of Horticultural Sciences, University of Agriculture, Faisalabad, Pakistan, pp. 91–98.

Ni, Z., Kim, E.-D., Ha, M., Lackey, E., Liu, J., Zhang, Y., Sun, Q. and Chen, Z.J. (2008) Altered circadian rhythms regulate growth vigour in hybrids and allopolyploids. *Nature* 457(7227), 327–331. doi:10.1038/nature07523.

Olivares-Fuster, O., Duran-Vila, N. and Navarro, L. (2005) Electrochemical protoplast fusion in citrus. *Plant Cell Reports* 24, 112–119.

Ollitrault, P., D. Dambier, and F. Luro, (1996) Somatic hybridization in Citrus: some new hybrids and alloplasmic plants. *Proceedings International Society Citriculture* 2, 907–912.

Ollitrault, P., Vanel, F., Froelicher, Y. and Dambier, D. (2000) Creation of triploid citrus hybrids by electrofusion of haploid and diploid protoplasts. *Proceedings of the First International Symposium on Citrus Biotechnology*. Eilat, Israel, pp. 191–197.

Pena, L., Cervera, M., Juárez, J., Navarro, A., Pina, J.A. and Navarro, L. (1997) Genetic transformation of lime (*Citrus aurantifolia* Swing.): factors affecting transformation and regeneration. *Plant Cell Reports* 16, 731–737.

Pena, L., Perez, R.M., Cervera, M., Juarez, J. and Navarro, L. (2004) Early events in Agro-bacterium-mediated genetic transformation of Citrus explants. *Annals of Botany* 94, 67–74.

Pérez-Molphe, E. and Ochoa-Alejo, N. (1998) Regeneration of transgenic plants of Mexican lime from *Agrobacterium rhizogenes*-transformed tissues. *Plant Cell Reports* 17, 591–596.

Reuther, W., Webber, H.J. and Batchelor, L.D. (1967) *The Citrus Industry. Vol. 1. History, World distribution, Botany and Varieties*. Revised Edition, University of California, Davis, California, Division of Agricultural Sciences, pp. 431–588.

Robles-Gonzalez, M.M., Medina-Urrutia, V.M., Velazquez-Monreal, J.J. and Simpson, J. (2008) Field performance and molecular profiles of Mexican lime selection. *Euphytica* 161, 401–411.

Sanchez-Ballesta, M.T., Lunch, Y., Gosalbes, M.J., Zacarias, L., Granell, A. and Lafuente, M.T. (2003) A survey of genes differentially expressed during long-term heat-induced chilling tolerance in citrus fruit. *Planta* 218, 65–70.

Sanchez-Ballesta, M.T., Rodrigo, M.J., Lafuente, M.T., Granell, A. and Zacarias, L. (2004) Dehydrin from citrus, which confers in vitro dehydration and freezing protection activity, is constitutive and highly expressed in the flavedo of fruit but responsive to cold and water stress in leaves. *Journal of Agricultural and Food Chemistry* 52, 1950–1957.

Sapitnitskaya, M., Maul, P., McCollum, G.T., Guy, C.L., Weiss, B., Samach, A. and Porat, R. (2006) Postharvest heat and conditioning treatments activate different molecular responses and reduce chilling injuries in grapefruit. *Journal of Experimental Botany* 57, 2943–2953.

Siebert, T., Krueger, R., Kahn, T., Bash, J. and Vidalakis, G. (2010) Descriptions of new varieties recently distributed from the citrus clonal protection program. *Citrograph* March/April 2010, 20–26.

Singh, S. and Rajam, M.V. (2009) Citrus biotechnology: achievements, limitations and future directions. *Physiology and Molecular Biology of Plants* 15, 3–22. Available at: http://doi.org/10.1007/s12298-009-0001-2 (accessed 14 April 2016).

Soler, N., Plomer, M., Fagoaga, C., Moreno, P., Navarro, L., Flores, R. and Pena, L. (2012) Transformation of Mexican lime with an intron-hairpin construct expressing untranslatable versions of the genes

coding for the three silencing suppressors of Citrus tristeza virus confers complete resistance to the virus. *Plant Biotechnology Journal* 10, 597–608. Available at: http://doi.org/10.1111/j.1467-7652.2012.00691.x (accessed 18 April 2016).
Soler, N., Fagoaga, C., Lopez, C., Moreno, P., Navarro, L., Flores, R. and Pena, L. (2015) Symptoms induced by transgenic expression of p23 from Citrus tristeza virus in phloem-associated cells of Mexican lime mimic virus infection without the aberrations accompanying constitutive expression. *Molecular Plant Pathology* 16, 388–399. Available at: http://doi.org/10.1111/mpp.12188 (accessed 3 May 2016).
Su, H.J., Hogenhout, S.A., Al-Sadi, A.M. and Kuo, C.H. (2014) Complete chloroplast genome sequence of omani lime (*Citrus aurantiifolia*) and comparative analysis within the rosids. *PLoS ONE* 9. Available at: http://doi.org/10.1371/journal.pone.0113049/ (accessed 3 May 2016).
Taheri, F., Nematzadeh, G., Zamharir, M.G., Nekouei, M.K., Naghavi, M., Mardi, M. and Salekdeh, G.H. (2011) Proteomic analysis of the Mexican lime tree response to 'Candidatus Phytoplasma aurantifolia' infection. *Molecular Bio Systems* 7, 3028–3035. Available at: http://doi.org/10.1039/c1mb05268c (accessed 3 May 2016).
Talon, M. and Gmitter Jr, G.F. (2008) Citrus genomics. *International Journal of Plant Genomics*, 1–17.
Vardi, A. (1981) Protoplast derived from different citrus species and cultivars. *Proceedings of International Society of Citriculture* 1, 149–152.
Vardi, A., Spiegel-Roy, P. and Galun, E. (1975) Citrus cell culture: isolation of protoplasts, plating densities, effect of mutagens and regeneration of embryos. *Plant Science Letters* 4, 231–236.
Viloria, Z. and Grosser, J.W. (2005) Acid citrus fruit improvement via interploid hybridization using allotetraploid somatic hybrid and autotetraploid breeding parents. *Journal of the American Society for Horticultural Science* 130, 392–402.
Webber, H.J. (1943) Cultivated varieties of citrus. In: Webber, H.J. and Batchelor, L.D. (eds) *The Citrus Industry*, 1. University of California Press, Berkeley and Los Angeles, California, pp. 475–466.
Wilkes, G. (1988) Germplasm preservation: objectives and needs. In: Knutson, L. and Stoner, A.K. (eds) *Biotic Diversity and Germplasm Preservation. XIII Beltsville Symposium*. USDA-ARS, Beltsville, Maryland, pp. 13–42.
Wu, G.A., Prochnik, S., Jenkins, J., Salse, J., Hellsten, U. and Murat, F. (2014) Sequencing of diverse mandarin, pummelo and orange genomes reveals complex history of admixture during citrus domestication. *Nature Biotechnology* 32, 656–662. doi:10.1038/nbt.2906.
Wutscher, H.K. and Hill, L.L. (1995) Performance of 'Hamlin' orange on 16 rootstocks in east-central Florida. *HortScience* 30, 41–43.
Xu, Q., Chen, L.L., Ruan, X., Chen, D., Zhu, A., Chen, C., Bertrand, D., Jiao, W., *et al.* (2013) The draft genome of sweet orange (*Citrus sinensis*). *Nature Genetics* 45, 59–66.
Yamamoto, M. (2007) Application of fluorescent staining of chromosomes to genetic studies in citrus. *Japenese Journal of Plant Science* 1(1), 12–19.
Yu, Z., Haberer, G., Matthes, M.T., Rattei, K.F., Gierl, A.M. and Torres-Ruiz, R.A. (2010) Impact of natural genetic variation on the transcriptome of autotetraploid limes. *Proceedings of the National Academy of Sciences* 107, 17809–17814.
Zamharir, M.G., Mardi, M., Alavi, S.M., Hasanzadeh, N., Nekouei, M.K., Zamanizadeh, H.R. and Salekdeh, G.H. (2011) Identification of genes differentially expressed during interaction of Mexican lime tree infected with 'Candidatus Phytoplasma aurantifolia'. *BMC Microbiology* 11, 1. Available at: http://doi.org/10.1186/1471-2180-11-1 (accessed 3 May 2016).
Zandkarimi, H., Talaie, A. and Fatahi, R. (2011) Evaluation of cultivated lime and lemon cultivars in Southern Iran for some biological compounds. *Food* 5(Special Issue 1), Global Science Books, 84–88.
Zekri, M. and Parsons, L.R. (1992) Salinity tolerance of citrus rootstocks: effects of salt on root and leaf mineral concentrations. *Plant and Soil* 147, 171–181.

4 Growth, Development and Reproductive Physiology of the Mexican Lime (*Citrus aurantifolia* Christm (Swingle))

G. Almaguer-Vargas[1]*, S.R. Márquez-Berber[1], V. González-Hernández[2], N. Cruz-Huerta[2] and I. Ramírez-Ramírez[2]

[1]*Departamento de Fitotecnia, Universidad Autónoma Chapingo, Chapingo, México;* [2]*Colegio de Posgraduados, Campus Montecillo, Montecillo, México*

Introduction

Lime trees have average vigour. They develop as shrubs, with a dense canopy and small, abundant sharp spines on all their branches (Hodgson, 1967; Flores *et al.*, 2005). If lime trees are allowed to grow freely in deep, sandy, loam soil, with ample nutrient availability, they can reach up to 8 m tall in 12 years. Ideally, orchard management keeps trees pruned at high planting density, in a rectangular arrangement with 7 m between rows and 5 m between trees. Under this system, a 5-year-old tree on a *Macrophila* rootstock has an average height of 2.0 m, a treetop diameter of 3.6 m and 0.09 m of stem diameter at 0.2 m above ground. Aside from manual pruning, tree growth might be reduced by application of paclobutrazol, a gibberellin inhibitor. Paclobutrazol reduces tree height and promotes canopy spread horizontally (Baskaran *et al.*, 2011).

Lime trees are evergreen and have functional leaves all year round. New leaves are substituted for old leaves year-round. Each leaf has a bud at its base. Its insertion point is called a node. The distance between two leaves is the internode. When there is strong branch growth, internodes tend to be longer. When a terminal bud appears on a branch, thus halting new leaf production and growth of any existing ones, internodes become shorter. These are fundamental concepts that aid differentiation of the kinds of lime branches.

Branches

Branches originate from apical or axillary buds, and they are either structural or functional branches (see Tables 4.1 and 4.2).

Types of buds

Mature and differentiated branches have either:

1. Mixed buds. One or more flowers present, with a vegetative bud.
2. Flowering buds. One or more flowers present, without a vegetative bud.
3. Vegetative buds. No flowers present.

*E-mail: almaguervargas@hotmail.com

Table 4.1. Lime structural branches.

Type	Characteristics
Main stem	It is the first branch produced by the tree after it has successfully established and development has started
Primary, mother or support branches	The main stem ramifies for the first time. This event produces primary branches, also called mother branches. These branches are the scaffolding and support for the canopy and carry the load of other branches, leaves, flowers and fruit
Secondary branches	These branches derive from primary ones; both types of branches shape the tree frame. Secondary branches are distributed among the main branches and consolidate the tree structure
Tertiary branches	These branches form from buds located on the secondary branches
Vegetative and reproductive branches	These branches grow from tertiary branches. Owing to reduced vigour and maturity, these branches flower and produce fruit

Table 4.2. Lime functional branches. The branches classified in this category usually allow tree development and fruit production.

Type	Characteristics
New or recently emerged branch	This type of branch is up to a month old and round in shape. They are small branches, usually less than 20 cm long, and with medium internode length. Buds on a new branch have developing vegetative buds
Vegetative branch up to 4 months old	This branch is growing, and it only has vegetative buds of various sizes. Its length varies from 15 cm to 1 m. It is round in shape and with medium internode length
Mature branch up to 6–7 months old	Mature branches are 15–40 cm long. Vegetative growth in these branches has stopped, and a terminal bud is already on its end. If climate conditions are adequate, floral induction is promoted on most occasions. Branches are round in shape, with medium internode length. Flowers or fruit are not present yet
Productive branch	Productive branches have flowers or fruit. Usually they are more than 6 months old and evolved from mature branches. Their length is 15–40 cm
Sucker branch	Sucker branches have high vigour and are more than 1 m long when they finish their growth. They grow from the neck of the tree, on main branches; even in secondary ones; and are flat in shape, with long internodes, and with a soft consistency. Usually they have vegetative buds, without branching. Pruning of sucker branches is recommended
Flag branch	A flag branch is a sucker branch that extends above the canopy and grows above other branches. Under extreme occasions, they can produce fruit. Pruning of flag branches is recommended
Non-productive branch	Many types and ages of branches can be classified as non-productive. Usually these branches grow in shadows made by the surrounding branches, inside the tree canopy. Only vegetative buds are on these branches. Sucker and flag branches can be included in this category
Crowfoot branch	This branch derives from two or more branches growing from the same node. The insertion angle is narrow, and there is high possibility of tearing

Terminal and lateral buds on a branch fall within any of the three types cited above; however, the terminal bud is usually a mixed or floral bud. When buds open, different shoot types can grow from them (Table 4.3).

These shoot types have been reported in other citrus trees (Sauer, 1951; Lenz, 1966; Moss, 1969; Agustí *et al.*, 1982; Spiegel and Goldschmidt, 1996).

Table 4.3. Lime shoot classification.

Type	Characteristics
Multi-flower with leaves or mixed	It has more than two flowers, and one or more leaves. If more leaves are present, fruit set is better
Multi-flower without leaves or flower bundle	It has more than one flower, but no leaves
Mono-flower with leaves or ringer	Only one flower with one or more leaves
Uni-flower without leaves or lonely flower	Each bud produces only one flower without leaves
Vegetative	It only has leaves

Table 4.4. Lime leaf characteristics.

Characteristics[a]	Leaves on the outer canopy	Leaves in the middle of the tree
Leaf density	0.68 g/ml[3]	0.81 g/ml[3]
Average single leaf area	18.73 cm^2	17.23 cm^2
Number of stomata per mm^2 of leaf	97.8	105.4

[a]Fifty leaves average for each of ten 5-year-old trees growing in dry tropical conditions in central-east México in a sandy clay soil. Stomata were found only on the underside. They were counted with an Olympus® microscope and a 40× lens. A LICOR® model LI3000C was used to measure leaf area.

Leaves

Lime leaves are small, 5–15 cm long, and 2–7 cm wide. They have an oval-elliptic, lanceolate shape. The tip is usually obtuse, and its base is round, with a slightly winged petiole. The edges are usually smooth. The central vein and the lateral veins are not prominent. When the leaves are mature, their face is a shiny, pale green colour. The underside is greenish (Flores *et al.*, 2005). Average leaf age is 14 months, but this depends on various factors. Leaf growth is determined by a wide variety of factors, such as soil conditions (moisture, fertility), weather and orchard management, among others.

Leaf number per tree varies according to age, cultivar, weather, climate, soil conditions and orchard management. In México, a sample of ten 5-year-old trees growing under dry tropical conditions averaged 24,430 leaves and a total leaf area per tree of 43.88 m^2. Erickson (1968) reported higher values in Valencia orange trees; the total average leaf surface was 59.98 m^2.

Leaves on the lime tree differ depending on their location. The outer canopy has lower leaf density; leaves here tend to be bigger and have fewer stomata than those in the middle of the tree (Table 4.4). Curiously, Mexican lime leaves are smaller, with lower density and fewer stomata than orange (Erickson, 1968).

As in any other tree, the leaves carry out photosynthesis, transpiration, respiration, etc., and also participate in synthesis and storage of photosynthesis-derived compounds. The canopy protects flowers, fruit and the whole tree from the rays of the sun. It is therefore important to control attacks from pests and diseases, since any reduction in leaf area could diminish fruit production.

Lime Tree Phenology

Vegetative sprouting

A study was done in a Mexican lime orchard in a mono-crop system in the coastal region of central-east México. This area has a warm dry climate (BS1 (h1) w (w) i'). When the number of vegetative and reproductive sprouts per square metre was counted around the year, four vegetative sprouting periods were found. The first vegetative sprouting period began in January and finished in March. The second one spanned from mid-April to mid-June. The third one was the most intense, because it coincided with the rainy season, and it started in mid-June and lasted until July. The last period started in September (Arias, 1988). Other researchers, such as Martínez (2002) and Flores *et al.* (2005), found vegetative sprouting starting in November, when irrigation started, and ended in March. During the length of the reported period, two to three vegetative sprouting fluxes occurred. A second sprouting season spanned from June to September. The rainy season promoted one or more sprouting fluxes. The number of new sprouts in summer tends to be higher than during winter or spring. Irrigation stimulates vegetative sprouting.

Flower sprouting

Two periods of flower sprouting in Mexican lime trees were reported by Arias (1988). The first one began in January and continued until June. The second period ran from September to mid-October. It was most intense in April, with 26.1 sprouts/m^2, followed by 16.4 sprouts/m^2 in February. Practically, there was no flowering in July, August or September, since many fruit are developing at this time, and the rainy climate and high temperature conditions during these months do not allow flower differentiation.

In the coastal region of central-east México, sprouting flowers on lime trees are more abundant, with several fluxes from November to April (Martínez, 2002). The flux intensity period increases as the drought season moves forward; before the first irrigation, a flux lasts. The number of flowerings depends on the number of developed, 5-month-old branches in April or May. From the end of summer until the beginning of winter, one or two flowerings occur. The fruit from this late flowering will be the winter harvest, if they are not attacked by anthracnose, which is common during the winter rainy season, or as it is called 'northerlies' (Martínez, 2002; Flores *et al*., 2005).

In humid regions, either hot or warm, the highest vegetative and reproductive sprouting is during spring, after the low temperature, winter period. It has been suggested that citrus flowering in the tropics is mainly affected by water stress, while subtropical areas have low temperature winters (Guardiola *et al*., 1977; Monselise, 1985; Davenport, 1990).

Arias (1988) observed up to 18 sprouting fluxes during 1 year at the coastal region of central-east México. This observation indicates that citrus in tropical climates can sprout all year round (Erickson, 1968).

Fruit development

Arias (1988) tagged flower buds of Mexican lime trees in February, April, June and September, and then measured fruit diameter weekly until harvest size (33 mm). He found that flowering sprouts in February and April took 129 and 128 days to develop from flower to market size, respectively, while sprouts from June and September developed in 102 and 100 days, respectively. Average fruit growth rate was 0.29 mm/day.

Martínez (2002) found high harvest from April to October. The first fruit came from the November–December flowering and were harvested in April–May. Therefore, they needed 140 days to reach market size. Fruit from the spring flowering, between March and April, needed 110–130 days to be harvested. Summer fruit develop faster, and they reach their final size in only 100 days.

Fruit maturation

When the lime fruit mature, they are round, or slightly elliptic, with a round base and a little residual protuberance from the style. Lime fruit are highly polyembryonic and have regular seed quantity. When ripe, they have a green yellowish colour; the fruit is juicy, divided into 10–12 segments, highly acidic with a distinctive scent (Table 4.5). Ripe fruit tends to fall (Hodgson, 1967). Mexican lime and Tahiti or Persian lime fruit have similar form. However, lime fruit are smaller in polar and equatorial diameters. They have higher titratable acidity and slightly lower albedo thickness and juice content (Table 4.5).

Fertilization and growth regulators can also affect fruit development and retention. Micronutrients, auxins and gibberellins reduce

Table 4.5. Comparison of different characteristics between Mexican lime and Tahiti lime fruits.

	Polar diameter (cm)	Equatorial diameter (cm)	Number of segments	Albedo thickness (mm)	Flavedo thickness (mm)	Brix degrees	pH	Titratable acidity	Humidity %	Juice %
Mexican lime	4.5	4.21	10.5	0.9	1	7.58	2.81	8.25	86.09	46.06
Tahiti lime	7	5.93	10	1	1	7.46	2.83	5.63	84.41	51.5

the time required for fruit set (Haribabu, 1980; Kachave and Bhosale, 2007). A mixture of micronutrients and auxins (NAA 200 ppm) can reduce fruit drop and increase fruit and yield retention. This treatment has been shown to increase the number of fruit per tree, fruit volume, fruit weight and juice percentage (Kachave and Bhosale, 2007). Potassium nitrate (2%), auxins (NAA 300 ppm) and gibberellic acid (100 ppm) can also increase fruit weight and volume, juice percentage and total soluble solids (Debaje *et al.*, 2011). Paclobutrazol, a gibberellin synthesis inhibitor and a broad spectrum growth retardant cited above, can increase the number of flowers per shoot irrespective of the season (Baskaran *et al.*, 2010).

Lime Flowering Management

It is possible to increase lime flowering in some months of the year, especially with drought management. It has been suggested that water stress is the main environmental factor affecting citrus flowering in the tropics (Guardiola *et al.*, 1977; Monselise, 1985; Davenport, 1990; Almaguer-Vargas *et al.*, 2011).

Flowering management is important for México. In 2010, from February to April, lime fruit reached up to US$3/kg, during the same season in 2014, the price climbed to $7/kg. However, by June, the producer only got $0.30/kg. An initial foray into flowering management was started in 2012, though data have not been published. The objective was to test the viability of manipulation of lime flowering and fructification and to increase winter harvest. The experiment was set in the central-east region of México (Gabriel Zamora, Michoacán; 19°01′ longitude N; 102°06′ latitude W). The average annual rainfall in this zone is 698.2 mm, and its annual average temperature is 27.3°C. The climate in the area is classified as BS1 (h') w (w) ig (García, 1988; modified from Köppen). For this experiment, the treatments applied were: (i) irrigation withheld from 15 June to 15 September and foliar application of 6% urea on 20 September; (ii) irrigation withheld from 15 June to 15 September; (iii) foliar application of 6% common urea on 20 September; and (iv) control. Orchard management was the same for all treatments, with irrigation every 20–25 days, except for treatments 1 and 2. Fertilization to the soil included 1 kg of ammonium sulfate per tree. A total of four weeding procedures were done per year. Pest and diseases were controlled. The results are shown in Table 4.6.

Fruit yield at the end of January increased greatly with the water stress treatments: 5.6 t/ha (treatment 1), and 4.9 t/ha (treatment 2). Higher flowering and fruit set as a result of those treatments explain the increased yield (Table 4.6).

Table 4.6. Effects of several treatments to manipulate lime flowering and fructification with the aim of increasing winter harvest in the central-east region of México.

		15/10/2012		02/11/2012		15/11/2012		02/12/2013	
Treatment	Description	FLO	FRUS	FLO	FRUS	FLO	FRUS	FLO	FRUS
1	Irrigation withheld from 15 June to 15 September, and foliar application of 6% urea with biuret	29.1 a[a]	0.0 a	51.5 a	16.5 a	55.8 a	56.8 a	17.2 b	35.9 a
2	Irrigation withheld from 15 June to 15 September	30.0 a	0.0 a	48.9 a	16.6 a	50.2 a	47.3 b	17.4 b	32.3 a
3	Foliar application of 6% urea with biuret	2.0 b	0.0 a	8.1 b	5.5 b	26.6 b	20.2 c	15.3 b	30.0 a
4	Control	0.0 b	0.0 a	7.6 b	2.6 b	21.0 b	5.0 d	88.2 a	19.8 a
Minimum Significant Difference		20.3	0.0	22.8	7.4	8.3	7.5	10.5	27.2

FLO: Flower number/m^2; FRUS: Number of fruit set/m^2; [a]Means with the same letter within the same column are not significantly different (Tukey $\alpha \leq 0.05$).

These results confirm the important role of water stress in promoting flowering in the tropics. Foliar fertilization doubled yield (2.8 t/ha in treatment 3 vs. 1.3 t/ ha for the control treatment).

In another experiment, Almaguer-Vargas *et al.* (2011) obtained 20.2 fruits/m^2 in Tahiti lime, when 10% urea was applied to the foliage, and fruit-producing branches were pruned by 30%. The control had 2.33 fruits/m^2.

References

Agustí, M., Garcia-Mari, F. and Guardiola, J.L. (1982) The influence of flowering intensity on the shedding of reproductive structures in sweet orange. *Scientia Horticulturae* 17, 343–352.

Almaguer-Vargas, G., Espinoza-Espinoza, J.R. and Quirós-García, J.L. (2011) Desfasamiento de cosecha de limón persa. *Revista Chapingo Serie Horticultura* 17(3), 197–205.

Arias T.A. (1988) Fenología de limón Mexicano (*Citrus aurantifolia* Christm (Swingle)), en Colima. Professional thesis. Universidad Autónoma Chapingo, Chapingo, México.

Baskaran, A., Renuga, R. and Saraswathy, S. (2010) Advancement of flowering in acid lime (*Citrus aurantifolia* Swingle) by soil applied Paclobutrazol. *Madras Agriculture Journal* 97 (10–12), 388–389.

Baskaran, A., Parthiban, S. and Sundharaiya, K. (2011) Influence of soil applied Paclobutrazol on tree growth and canopy spread in acid lime (*Citrus aurantifolia* Swingle). *Madras Agriculture Journal* 98(1–3), 79–81.

Davenport, T.L. (1990) Citrus flowering. *Horticultural Review* 12, 349–408.

Debaje, P.P., Shinde, E.D. and Ingale H.V. (2011) Effect of plant growth regulators and nutrients on quality of acid lime (*Citrus aurantifolia* Swingle). *The Asian Journal of Horticulture* 6(1), 253–255.

Erickson, L.C. (1968) The general physiology of Citrus, Chapter 2. In: Reuther, W., Webber, H.J. and Batchelor, L.D. (eds) *Citrus Industry. Vol. 2. Anatomy, Physiology, Genetics, and Reproduction*. University of California, Riverside, California, E. U. A., pp. 86–126.

Flores V.R., Romero, C.A., Figueroa, V.J. and Munro Olmos, D. (2005) Paquete tecnológico para el cultivo del limón en el Estado de Colima. Secretaria de Desarrollo Rural (ed.) Gobierno del Estado de Colima.

García, E. (1988) *Modificaciones al sistema de clasificación climática de Köppen (para adaptarlo a las condiciones de la República Mexicana)*, Ed. Universidad Nacional de México, Offset Larios S.A., México D.F., pp. 46–52.

Guardiola, J.L., Agustí, M. and García-Marí, F. (1977) Gibberellic acid and flower bud development in sweet orange. *Proceedings of the International Society of Citriculture* 2, 696–699.

Haribabu, R.S. (1980) Effect of zinc, 2-4-D and GA3 on the flowering yield and quality of *Kagzi* lime (*Citrus aurantifolia* Swingle). PhD Thesis. B.H.U.,Varanasi, India.

Hodgson, R.W. (1967) Horticultural varieties of citrus, Chapter 4. In: Reuther, W., Webber, H.J. and Batchelor, L.D. (eds) *Citrus Industry. Vol. 1. History, World Distribution, Botany, and Varieties*. University of California, Riverside, California, E. U. A., pp. 431–591.

Kachave, D.B. and Bhosale, A.M. (2007) Effect of plant growth regulators and micronutrients on fruiting and yield parameters of Kagzi lime (*Citrus aurantifolia* Swingle) fruits. *The Asian Journal of Horticulture* 2(2), 75–79.

Lenz, F. (1966) Flower and fruit development in 'Valencia Late' orange as affected by type of inflorescence and nutritional status. *Horticultural Review* 6, 65–78.

Martínez, G.S. (2002) Diagnóstico del sistema producto limón mexicano en Colima, Ed. *Consejo Estatal del Limón Mexicano de Colima (Coelim-Col)*. Colima, Colima, México, p. 145.

Monselise, S.P. (1985) Citrus and related genera. In: Halevy, A.H. (ed.) *CRC Handbook of flowering.* CRC Press Boca Raton, Florida, USA, 2, pp. 257–294.

Moss, G.I. (1969) Influence of temperature and photoperiod on flower induction and inflorescence development in sweet orange (*Citrus sinensis* L. (Osbeck)). *Journal of Horticultural Science* 44(4), 1–20.

Sauer, M.R. (1951) Growth of orange shoots. *Australian Journal of Agricultural Research* 2, 105–117.

Spiegel R. P. and Goldschmidt, E. (1996) *Biology of Citrus*. Cambridge University Press, Cambridge, UK, p. 230.

5 Propagation and Nursery Certification

Muhammad Azher Nawaz[1,2]*, Yong-Zhong Liu[1], M. Mumtaz Khan[3] and Waqar Ahmed[4]
[1]*College of Horticulture and Forestry Sciences, Huazhong Agricultural University/ Key Laboratory of Horticultural Plant Biology, Ministry of Education, Wuhan, China;* [2]*Department of Horticulture, University College of Agriculture, University of Sargodha, Sargodha, Pakistan;* [3]*Department of Crop Sciences, College of Agriculture and Marine Science, Sultan Qaboos University, Al-khod, Sultanate of Oman;* [4]*Sector Advisor-Horticulture, USAID-CNFA, Lahore, Pakistan*

Introduction

Plant propagation is the reproduction, duplication or multiplication of plants from the parent plant. The prime objective of the propagation is to produce more plants exactly like the parent plant. It is performed to save the germplasm, as well as for crop production on a commercial scale. Orchard establishment is a long-term investment and it has lifelong financial impacts. The availability of true to type certified, healthy and disease-free nursery plants leads to the success of an orchard. Under these circumstances, the propagation of perennial fruit plants such as lime becomes more important. In limes, propagation is more critical due to the presence of cross-pollination, cross-compatibility and polyembryony. Similarly the use of appropriate rootstocks is another unique aspect that must be considered during the process. In this chapter we have summarized the available information about propagation, nursery management and certification, and registration systems for lime nurseries.

Propagation Methods

The propagation of lime, as well as other citrus species, occurs mainly through sexual and asexual methods. Asexual methods are particularly important in the propagation of lime, as with other perennial fruit crops.

Sexual propagation

Propagation through seeds is called sexual propagation. It is used as a means of propagation and cultivation for a large number of agronomic and horticultural crops (vegetables and ornamentals), and forest trees. The merits of sexual propagation include evolution of new varieties (new commercial scion and root stock cultivars), simplicity of multiplication, production of a large number of plants from a single parent, production of plants tolerant to environmental stresses, long-lived and heavy bearing plants, and restricted transmission of some diseases to the next generation; while the demerits include

*E-mail: azher490@hotmail.com

the lack of true to type plants (cross-pollinated species), long juvenile periods and tall plants where canopy management and plant protection measures become a problem.

Although some citrus cultivars ('Temple' (tangelo), 'Clementine' (tangerine), grapefruit, Bearss lime and Tahiti lime) are monoembryonic and seed cannot be used for the production of true to type plants, luckily some lime cultivars (Mexican lime/West Indian lime/Kaghzi lime) are highly polyembryonic, so they can be propagated through seeds on a commercial scale. Additionally, to grow rootstock for asexual propagation (budding/grafting) seeds are also required.

Seed extraction

Seeds are extensively used in the propagation of lime cultivars (sexual and asexual propagation). Fruit for seed extraction should be collected from healthy, vigorous and disease-free plants of the representative cultivar (scion/rootstock). Seeds are extracted manually by circumcising the fruit into two halves with the help of a knife and then squeezing it. Care must be taken that seeds are not damaged while circumcising. Now mechanical seed extractors are also available, which can be utilized to facilitate the seed extraction process on a commercial scale. After extraction, seeds should be washed with water to remove mucilage (gelatinous material) from the seed surface.

Seed storage

There is considerable variation in seed storage behaviour among various citrus cultivars (Khan *et al.*, 2002). Citrus seeds have recalcitrant storage behaviour, and the loss of moisture over 7% can damage seed vigour (Khan, 2002). So, freshly harvested seeds have a better germination rate compared with stored seeds. However, some studies suggest that after extraction, washing and fungicide treatment, if seeds are stored at 5°C they can be stored up to a period of 8 months. Some other scientists have suggested that if seed storage is necessary, they should be stored in ground charcoal or in vacuum tin packs at 3–10°C (Chaudhary, 1994). If seeds are used for sowing within 2 to 3 days after extraction, they can be placed (temporary storage) in the juice and other fruit material squeezed during seed extraction. It is considered better than storage after washing at room temperature. However, freshly harvested seed is always considered better than stored seed. In fact there is limited scientific information available about lime seed storage behaviour. This subject needs to be taken in hand by researchers.

Seed sowing and pre-sowing seed treatment

In addition to other factors such as proper seed bed/media preparation, type of media and moisture content, temperature plays a vital role in regulating seed germination. Citrus seeds can germinate over a wide range of temperatures. The citrus cultivars can germinate at minimum temperatures of 6–15°C, whereas the maximum temperature goes up to 39°C (Mobayen, 1980). The optimum temperature for seed germination varies from cultivar to cultivar: *Poncirus trifoliata* seed shows the best germination levels at 25°C, while a hybrid of Rangpur lime × Troyer citrange germinates best at 32.5°C. However, based upon germination performance of the majority of citrus cultivars, 29.7°C seems optimum (Rouse and Sherrod, 1996).

In addition to temperature, lime seed requires a long time for germination, and there is a great variation among the cultivars. It ranges from 5 to 28 days; however, it can be reduced by providing optimum conditions. Additionally, some other approaches have been devised to reduce the germination time: for instance, removal of the seed coat before sowing helps to reduce germination time. Similarly, soaking seeds in aerated water for 8 h just before sowing reduces germination and seedling emergence time (Chilembwe *et al.*, 1992).

Before sowing, seed is treated with fungicide to protect it from fungus attacks, as it needs a long time for germination. Kaghzi lime is the leading lime cultivar cultivated in India, and is propagated through seed. It has been reported that a 0.2% application of Mancozeb + Carbendazime significantly increases germination rate and early seedling growth (Shukla *et al.*, 2012). Moreover, the combined application of 0.2% application of Mancozeb + Carbendazime, 50 ppm naphthalene acid (NAA) and 10 ppm gibberellic acid (GA) has been shown to further increase germination rate and early seedling growth, and reduced time for germination (Shukla *et al.*, 2012). Previously, many fungicides such as benlate, vitavax and thiram were being used but environmental safety concerns related to human, mammals, and birds led to the development of a new class of fungicides and organic fungicides. US scientists recently recommended Mefenoxam for citrus seed treatment (Fishel, 2014). Similarly, application of Iturin at 50–100 ppm has also proved its ability to reduce the total seed mycoflora of many stored grains (Klich *et al.*, 1994). For seed treatment it may be a potential candidate; however, the dose of Iturin needs to be standardized for lime and other citrus species.

In field nurseries, seed sowing is done in well prepared seed beds. Drainage is very important for growing lime seedlings, so the sowing is always done on raised seed beds. The seeds are sown in lines at a distance of at least 15 cm apart. The width of seed beds should be kept narrow (about 1 m) to facilitate cultural practices.

Care of nursery plants

IRRIGATION. The young growing lime plants are sensitive to environmental stresses, and require utmost care. The plants should have a regular supply of moisture from the root zone. The maintenance of soil moisture determines the growth of nursery plants. Drought conditions lead to the restriction of growth. On the other hand, over-irrigation leads to poorly growing seedlings with yellow leaves. Nursery beds should be irrigated with 2.5–5 cm of water each week under field conditions. However, the amount and frequency of irrigation may be modified according to the prevailing soil and environmental conditions of the area. The quality of water should also be considered, although limes are comparatively salt tolerant, but even then high amounts of salts in the soil or in irrigation water will adversely affect the seedlings/plants. Moreover, there is a considerable difference in salt tolerance between the seedlings and field grown mature plants–due to limited root zones and pot bound root systems in the case of container-grown citrus nurseries (Zekri and Parsons, 1992).

The water may contain fungal spores and lead to infection among the nursery plants. This has been particularly observed in nurseries following the recirculation of the irrigation system. Thomson and Allen (1974) separated *Phytophthora parasitica*, *Phytophthora citrophthora* and unidentified *Phytophthora*. spp. along with fungi of 47 other genera from the water of a recycling irrigation system of a citrus nursery. Similarly, in another study, 17 species of *Phytophthora*, 26 of *Pythium*, 27 genera of fungi, 8 species of bacteria, 10 viruses, and 13 species of plant parasitic nematodes were detected in water resources (Hong and Moorman, 2007). These conditions indicate how important it is to monitor the quality of irrigation water to keep nursery plants healthy and free from diseases. Many procedures including the use of heat, ultraviolet radiation, chlorine, ozone, and hydrogen peroxide are used to disinfect irrigation water, especially in recycling irrigation water systems of nurseries (Newman, 2004) to ensure the quality of the irrigation water.

FERTILIZER APPLICATION. Fertilizers play a major role in the production of lime nursery plants. Nurseries require high amounts of fertilizers due to plant densities and rapid vegetative growth. However, the application of fertilizer depends on the type of nursery (field nursery/container-grown), soil type, cultivar, rootstock and stage of plant growth. Container-grown nurseries use slow release fertilizer in the substrate. Water-soluble fertilizers are used with irrigation water, and a regular application of 75–100 ppm nitrogen or 200–400 ppm weekly nitrogen application is considered sufficient for the young growing nursery plants (Sauls, 2008). Similarly, Mattos *et al.* (2010) recommended a fertigation solution having 240 ppm nitrogen with 5–10 ppm copper for proper growth of nursery plants. However, Maust and Williamson (1994) advised a very low level of 15–19 ppm nitrogen application in irrigation water on a regular basis, which leads to satisfactory plant growth. The maintenance of a citrus nursery under this low level of nitrogen application is expected to increase plant survival under field conditions after transplanting (Maust and Williamson, 1994), reduce insect attack and lead to economic benefits. Another fertigation medium containing nitrogen (200 mg/l), phosphorus (30 mg/l), potassium (180 mg/l), calcium (150 mg/l), magnesium (30 mg/l), sulfur (40 mg/l), boron (0.3 mg/l), copper (0.5 mg/l), iron (2 mg/l), manganese (0.5 mg/l), zinc (0.3 mg/l), and molybdenum (0.1 mg/l) with an electrical conductivity of 2.0–2.5 dS/m seems more appropriate under soilless cultivation or cultivation in nutrient-free media (Furlani *et al.*, 2009). However, some modifications can be done according to the season, and scion and rootstock requirements.

In case of field nurseries, an application of 170–227 kg of nitrogen per hectare per annum is advisable, in split doses with an interval of 30–45 days starting from late winter to continue throughout the summer season (Sauls, 2008).

WEED CONTROL. Weeds compete with nursery plants for water, nutrients, space and light. Additionally, they serve as a host for some insects and diseases, and cause significant losses. If seed sowing is done under field conditions on raised seed beds, citrus seed takes a long time to germinate, while weeds germinate quickly and become a major problem. In most countries, the weeds in lime nurseries are controlled manually by using hand tools. Mechanical control of weeds is possible to some extent. However, some studies have suggested promising results using herbicides. Pre-emergence application of oxyfluorfen at concentrations of 1.1 or 2.2 kg/ha provided 96% and nearly 100% weed control in three rootstocks (Carrizo citrange, trifoliate orange, and Swingle citrumelo) with only minor toxicity to seedlings. Similarly, trifluralin and oxadiazon (1.1 or 2.2 kg/ha), and oryzalin and norflurazon (2.2 kg/ha) provided over 70% weed control without any toxicity to seedlings (Singh and Achhireddy, 1984). Later on, post-emergence narrow leaved selective herbicides can prove a good option, but due to lack of information about the use of herbicides in lime nurseries, it is suggested to be very careful while selecting herbicides to avoid any possible injury. Various citrus cultivars respond differentially to the same herbicide at the same concentrations, and herbicides can show different levels of toxicity to seedlings (Singh and Achhireddy, 1984). Moreover, the selection of proper herbicide depends on the types of weeds (annual or perennial weeds, and narrow or broad leaved weeds). Similarly, it is very important to know the mechanism of action of herbicides. Herbicides are categorized as systemic (e.g. glyphosate) and non-systemic (e.g. paraquat). Different herbicides are sometime used in the tank mix but at the same time it is very important to consider that a systemic and non-systemic herbicide cannot be applied in the tank mix. If this is practised, only the result of the non-systemic herbicide will be observed, while the systemic herbicide will be merely a waste of resources. Systemic herbicide needs sufficient time for absorption. With the continuous evolvement and increasing adaptation of container-grown nurseries, and the use of clean and sterilized weed seed-free media, weeds are not a problem anymore. While in lime nurseries grown under field conditions, mechanical, manual (hand tools) and chemical methods of weed control can be applied justifiably to obtain the desired results.

INSECTS AND DISEASE CONTROL. Typical citrus orchard pests of each area can affect lime nurseries growing in that area. Nurseries can be damaged by caterpillars, cutworms, slugs, snails and rodents. Insects not only directly damage the plants by chewing or sucking cell sap, but can also act as vectors and transfer some diseases; for example, aphid is the efficient vector of citrus tristeza virus (CTV). In container-grown nurseries, screen houses and other insect protecting structures are used to control insect attacks. Additionally, to effectively monitor intrusion of insects and to control common insects, yellow sticky tapes are used. In the case of more infestation, appropriate insecticides are applied. However, field nurseries are easily prone to insects such as aphid, citrus psylla, leaf miner, whitefly, thrips, black fly, lemon butterfly, mealy bug, mites, scale insects and certain other insects, which may be controlled by the application of proper insecticides. The most common insecticides being used in most countries include Imidacloprid (citrus psylla, leaf miner), Abamectin (citrus psylla, leaf miner and mites), Bifenthrin (citrus psylla, leaf miner, whitefly, black fly, mealybug and lemon butterfly), Spinosad (lemon butterfly, mites and leaf miner), Fenpropathrin (citrus psylla and root weevil), Cyfluthrin (citrus psylla, root weevil and lemon butterfly), Zeta cypermethrin (citrus psylla, root weevil, leaf miner and lemon butterfly), Dimethoate (citrus psylla and scale insects) and Chlorpyrifos (citrus psylla, scale insects, mealybug and leaf miner). The use of horticulture mineral oil along with some insecticides shows a better result against insects such as scales, mealybugs and root weevils (Rogers *et al.*, 2014).

In addition to insects, lime nurseries should be free of diseases, as this provides the basis for the establishment of a healthy orchard. Lime plants are affected by a number of serious diseases such as root rot, foot rot, wither tip, citrus canker, and CTV and other virus and viroid diseases. In container-grown nurseries, diseases are no longer an issue and are controlled through proper hygiene and sanitation practices, use of screen houses and use of certified bud wood. However, in field nurseries, root rot, damping off and citrus canker require special consideration and cause considerable losses. Damping off and root rot are the fungal diseases caused by *Phytophthora* spp., which can be controlled by using sterilized media, soil fumigation, maintenance of proper moisture conditions in the seed beds (too much soil moisture promotes the diseases), fungicide seed treatment and foliar application of fungicides such as Metalaxyl + Mancozeb, copper oxychloride, or copper hydroxide. Citrus canker and viral diseases can be controlled by using healthy lime bud wood, control of insects (vectors), separation distance from citrus orchards and other citrus nurseries. It is important to mention here that lime plants are susceptible to canker compared with other citrus species, so it should be given proper attention otherwise considerable losses of young plants may occur.

Asexual propagation

Multiplication of the plants from any part of the plant other than the seed is referred to as asexual propagation. Similar to other perennial fruit crops and citrus cultivars, asexual propagation is used for the multiplication of lime cultivars on a commercial scale. The merits of asexual propagation include production of true to type plants, short juvenile periods, ease of all cultural operations due to reduced plant height and resistance to adverse climatic and soil conditions through the use of suitable rootstock. In the monoembryonic cultivars of lime such as Bearss lime/Tahiti lime/Persian lime, and the all seedless lime cultivars, asexual propagation is the widely adopted method to get true to type plants.

Grafting

Grafting is a special type of plant propagation in which one part of plant (the scion) is inserted onto another plant (rootstock or stock) in such a way that both grow and form a new plant. Grafting is used to modify plant architecture, plant vigour and to increase disease resistance across the world (Stegemann and Bock, 2009). Many important cultivated fruit plants such as apples, pears, plums, mango and citrus cannot be propagated through seeds, as they do not produce true to type seedlings, therefore, grafting is used for propagation (Hartmann *et al.*, 1989; Ahmed, 1994; Melnyk and Meyerowitz, 2015). Budding is a type of grafting in which scion wood consists of a single bud with a piece of bark, and sometimes a slice of wood beneath the bark. It is chosen for most of the commercial citrus cultivars across the world except those cultivars with thorns or stout spines. Most of the lime cultivars have long and stout spines and it is difficult to remove buds from the scion wood, so budding is not commonly practised for lime cultivars, and grafting is preferred for the propagation of lime and lemon cultivars (Ahmed, 1994; Chaudhary, 1994).

A typical grafted citrus plant consists of rootstock and scion. The underground portion of the plant that provides roots, anchorage and support to the plant is called rootstock, while the above ground portion of the plant, which provides the fruit bearing portion, is referred to as the scion. Both portions (scion and stock) provide the basis for many fruit industries across the world. Both are important, but rootstock can be considered more critical. No failure of citrus industry in any country has been reported due to failure of scion variety, but a number of examples exist of failure due to lack of appropriate rootstock (Castle, 2010). In the production system of grafted plants on a commercial scale, continuous supply of suitable rootstock seedlings is always required.

SELECTION OF ROOTSTOCK. The selection of appropriate rootstock is of extreme importance, primarily as it reduces the juvenile period and leads to early fruiting of commercial scion cultivars. A successful rootstock should have compatibility between scion and rootstock as well as tolerance against prevalent edaphic and environmental conditions (Ahmed *et al.*, 2006). Rootstock affects various horticultural traits and provides tolerance to pests and diseases (Castle, 2010). It provides growers with a useful tool to manipulate the vigour and production of orchard trees. Its effects on tree size, fruit quality and taste, precocity (Albrigo, 1977; Castle, 2010), nutrient and water uptake (Ahmed *et al.*, 2007), fruit production and maturity are achieved through the complex interrelationship between the roots and canopy (fruit bearing portion) of the plant (Ahmed *et al.*, 2006). There is no rootstock available with all the desired characteristics, so selection of the appropriate rootstock is an important decision, and it has long-term significance.

Rootstock should be compatible with the scion cultivars. Overgrowth of scion over rootstock and overgrowth of rootstock over scion is sometime observed due to incompatibility. Similarly, bud union crease formation is another issue leading to disturbance in translocation of water, nutrients and prepared food at the graft junction.

Prevailing soil chemical and physical properties should also be considered when deciding on which rootstock to use. Soil type, texture, organic matter contents, depth to hard pan, pH, water holding capacity, drainage and nutrient status of soil should be taken into account when choosing the rootstock. Lemon and lime rootstocks perform well in alkaline soils with higher pH compared with sour orange, trifoliate and citranges. However, in acidic soils, the performance of sour orange is better. There is another important consideration if the soil condition is very good, and organic matter and nutrient levels are sufficiently high: the use of lemon or lime rootstocks may lead to overgrowth of lime scion cultivars because of their strong vigour. In such conditions, comparatively less vigorous rootstocks like citranges can be used to control the plant height and to facilitate canopy management. In calcareous soils, for the production of Key lime Alemov (*Citrus macrophylla*), Rough lemon (*Citrus jambhiri*), Volkamer lemon (*Citrus volkamariana*) and Rangpur lime (*Citrus limonia*) can prove appropriate rootstocks (Crane, 2013). In a study, Tahiti lime (*Citrus latifolia*) exhibited a varying response to 35 days of flooding conditions; tree mortality was greatest in plants grafted onto sour orange followed by *C. macrophylla*, whereas no mortality was observed in plants grafted onto grapefruit cv. 'Pine Island' rootstocks or through air layering (Schaffer and Moon, 1991). This situation again suggests care should be taken with the selection of appropriate propagation methods as well as suitable rootstock.

The environmental condition of the area is another important consideration: trifoliate and its hybrids are relatively tolerant to low temperatures, while lemon and lime rootstocks are susceptible to low temperature but perform better under high temperatures and salinity. Additionally, rootstock resistance to pests and diseases should be potentially considered. Sweet orange rootstocks are susceptible to foot rot or root rot; sour orange rootstocks are susceptible to CTV, while Carrizo citrange is not suitable for use when irrigation water has a higher concentration of salts (Castle, 2012). A hybrid of Rangpur lime (*C. limonia* Osb.) and Troyer citrange (*Citrus sinensis* (L.) Osb. × *Poncirus trifolaiata* (L.) Raf.) is resistant to CTV and *phytophthora* foot rot (Castle *et al.*, 1986).

Plant spacing is also another aspect to be considered: lime and lemons are vigorous rootstocks and impart vigour to the scions, while trifoliate species are slower growing and less vigorous. Castle *et al.* (1986) reported that a hybrid of Rangpur lime (*C. limonia* Osb.) and Troyer citrange (*C. sinensis* (L.) Osb. × *P. trifolaiata* (L.) Raf.] is suitable with most of the commercial scions under dense planting (625 plants/ha).

Rootstocks should have sufficient seed availability and viability; polyembryonic rootstocks have the edge over monoembryonic rootstocks due to the simplicity of propagation. So it can be concluded that while selecting rootstock for a specific area the whole set of circumstances should be considered. There is less specific information about the suitability of lime rootstock compared with other citrus cultivars; however, the characteristics of important rootstocks are summarized in Table 5.1, which could be helpful for the selection of appropriate rootstock for lime cultivars.

SELECTION OF SCION WOOD. The selection of scion wood should be done very carefully from healthy, vigorous and true to type plants with desirable yield and fruit characteristics. Plants should be apparently free from diseases and insect attack. The scion should be selected from relatively mature wood, round twigs from the last flush. Round wood is easier to handle, moreover it has already spent some of its life time and helps lead the grafted plants to early fruiting. The buds on the scion should be well developed, plump and dormant (Ahmed, 1994). The selection of scion cultivars depends on market demand (production for fresh use or processing), and climatic and soil conditions of the area. Some important cultivars of lime along with their producing

Table 5.1. Characteristics of some important rootstocks.

Rootstock	Scientific name	Important characteristics	Diseases	
			Tolerance	Susceptibility
Sour orange	*Citrus sinensis*	Polyembryonic, moderately vigorous, most widely planted stock in the world, low temperature tolerant	*Phytophthora*, citrus xyloporosis viroid, citrus exocortis viroid	Citrus tristeza virus, nematodes
Rough lemon	*Citrus jambhiri*	Polyembryonic, vigorous, salt tolerant, lacks freeze tolerance	Citrus tristeza virus, citrus exocortis viroid, citrus xyloporosis viroid	*Phytophthora*, blight, nematodes
Volkamer lemon	*Citrus volkamariana*	Polyembryonic, vigorous, salt tolerant, used in Australia	Citrus tristeza virus	*Phytophthora*
Rangpur lime, Sylhet lime	*Citrus limonia* Osbeck	Vigorous, salt tolerant, lacks freeze tolerance	Citrus tristeza virus	*Phytophthora*, citrus xyloporosis viroid, citrus exocortis viroid
Cleopatra mandarin	*Citrus reshni* Hort. ex Tan.	Polyembryonic, salt tolerant, low temperature tolerant	Citrus tristeza virus, citrus exocortis viroid, citrus xyloporosis viroid	*Phytophthora*, nematodes
Citranges	*Citrus sinensis* × *Poncirus trifoliata*	Cold tolerant, citrus nematode resistant, sensitive to salinity, scion overgrows on it (compatibility issues), less vigorous	Citrus tristeza virus, *phytophthora*, citrus exocortis viroid	Nematodes
Citrumelos	*Citrus paradisi* × *Poncirus trifoliata*	Polyembryonic, low temperature tolerant, widely used in Florida, less vigorous	Citrus tristeza virus, citrus exocortis viroid, citrus xyloporosis viroid, nematode	Burrowing nematodes
Alemow	*Citrus macrophylla* Wester	Polyembryonic, suitable for lemons and limes, freeze sensitive, salt tolerant, similar to lemons and limes	*Phytophthora*, citrus exocortis viroid	Citrus tristeza virus, nematodes

Source: Hodgson, 1967; Crane, 2013; Crane and Osborne, 2013

countries, methods of propagation and uses are described in Table 5.2.

There are different types of grafting like approach, cleft, side, veneer and bridge grafting, but for lime propagation side T-grafting is practised on a commercial scale. With this method a T-shaped cut is made in the bark of rootstock with the help of a sharp knife and a scion wood having at least three to five buds inserted into it. While preparing the scion, a slanting cut is given on the lower end of the scion wood with a sharp knife, preferably in a single stroke. The lengths of

Table 5.2. Characteristics of some important lime cultivars.

Name of cultivar	Scientific name	Important producers	Rootstock/ scion	Polyembryony (present, absent)	Seeds (high, moderate, few, seedless)	Propagation	Uses
Key lime/West Indian Lime/Mexican lime/ Kaghzi lime	*Citrus aurantifolia* Swingle	México, India, Pakistan	Scion	Present	Moderate	Seed, cuttings, grafting, layering	Fresh use, processing
Bearss lime/Tahiti lime/ Persian lime	*Citrus latifolia*	Iran, Iraq, USA, Brazil, Asian countries	Scion	Absent (rarely seeded)	Seedless	Cuttings, grafting, budding, layering	Fresh use, processing
Eustis limequat	*Citrus aurantifolia* × *Fortunella japonica*	India, Pakistan, USA, Japan, Israel, Malaysia	Scion	–	High	Cuttings, grafting, budding	Cold resistant, not commercially important, ornamental, home use
Lakeland lime	*Citrus aurantifolia* × *Fortunella japonica*	India, Pakistan, USA, Japan, Israel, Malaysia	Scion	–	High	Cuttings, grafting, budding	Cold resistant, not commercially important, ornamental, home use
Australian finger lime	*Citrus australasica*	Australia	Scion	–	Moderate	Budding, grafting	Garnish for seafood, pulp used in sauces, jams and jellies

Source: Hodgson, 1967; Castle, 1986; Zekri and Parsons, 1992; Chaudhary, 1994; Castle, 2010; Castle, 2012; Crane, 2013 cut on stock and scion are kept similar to facilitate union. Then to firmly hold the scion against the stock, it is tightly covered with waxed tape, and the scion and corresponding area of rootstock are wrapped with plastic film to increase humidity. Grafting height is important to save the plants from soil-borne diseases under field conditions; a grafting height of 15–20 cm is appropriate (Sauls, 2008).

Cuttings

Lime cultivars can be propagated through cuttings. Additionally, this method provides the possibility of propagation for rootstocks that are monoembryonic or have limited seed availability. Moreover, cuttings can be used to regulate the nursery propagation cycle, as seeds of rootstock are available only at a specific time of the year. With this method, the cuttings take root, and the process is technically called adventitious root formation, then the roots absorb water and nutrients from the soil, which are transported to the upper parts and leading to the transformation of a cutting to a young plant (Hartmann *et al.*, 1989).

Keeping in view the results of various studies, soft and semi-hard wood cuttings, including the current hardened flush or previous flush should be used. The young succulent stems should be avoided because they have poor rooting capacity due to excess of nitrogen and lack of sufficient quantity of carbohydrates (Ferguson and Young, 1985). Moreover, plants with defoliated branches/twigs are not suitable for cuttings. The plant selected for making cuttings should be healthy, vigorous, well fertilized, and free from insects, frost damage and disease attack (especially virus and citrus canker). Phosphorus nutrition of the parent plant can also affect the rooting ability of cuttings, so proper management of the parent plants is necessary. Cuttings should be made during the early day time when the leaves are fully turgid. In various citrus species and genera, it has been observed that cuttings made during the flowering season have less root generation capacity. So cuttings should be made before or after the flowering season to get better results (Dore, 1953). The length of cutting should be kept about to 15–27 cm with 8–14 intact leaves. Little is known about the genetic basis of rooting of cuttings, although Siviero *et al.* (2003) identified two quantitative traits loci (QTLs) for rooting traits in plant cuttings of *Poncirus trifoliata* hybrids. However detailed studies are required in different cultivars of limes, lemons and other citrus species.

The rooting of the cuttings depends on leaf area, maturity of the parent plant, media, propagation environment and the rooting hormones. Juvenile cuttings with leaves produce more roots than mature cuttings with leaves, perhaps due to more availability of auxin. Rooting of cuttings takes time. Interestingly, lemons and limes root easily (4–6 weeks) compared to other citrus groups including oranges, grapefruits, citranges and mandarins (Ferguson and Young, 1985). Another study found that 100% of the Rangpur lime (*Citrus limonia* Osb.) cuttings used took root after 36 days, while only 12% of the Oroblanco, a triploid pummelo–grapefruit hybrid (Citrus *grandis* Osb. × *Citrus paradisi* Macf.) cuttings took root (Sagee *et al.*, 1992). Many studies suggested that auxin plays a critical role in root initiation and final architecture of lateral roots in plants (Jarvis, 1986; Grieneisen *et al.*, 2007). The level of indole-3-acetic acid (IAA) in the bark of Rangpur lime was 3.6 times higher than in Oroblanco cuttings 19 days after planting (Sagee *et al.*, 1992).

Application of IAA at 2500 ppm, indole butyric acid (IBA) and NAA at 3000 ppm each induced maximum roots in juvenile as well as mature cuttings in citrus species (Ferguson and Young, 1985; Seran and Umadevi, 2011). While making cuttings, leaves should not be removed as it has been observed that cuttings with leaves show better success compared to those without leaves (Ferguson and Young, 1985). Although endogenous IAA is considered to play a central role in adventitious root formation, exogenous application IBA or NAA is preferred as IAA is less stable. In some studies, IBA was found to be more effective to enhance root formation compared with NAA at the same concentrations (Ferguson and Young, 1985). Limes are categorized as easy to root. A total of 500 ppm IBA could give better results compared with higher concentrations (1000 and 1500 ppm) in Kaghzi lime (*C. auriantifolia* Swingle) (Bhatt and Tomar, 2010). The availability of plant rooting hormones (IBA) on a commercial scale has made it easier to get the desired results. The application of root promoting hormones increases the percentage of the cuttings that take root, reduces time for root formation, increases the number and quality of roots per cutting, and aids with uniformity of rooting.

In addition to hormone applications, temperature is another critical factor for the success of cuttings. A temperature of 24–32°C is considered suitable for rooting. A temperature above 35°C may lead to leaf wilting and burning, which reduces the survival rate of the cuttings. To maintain temperature and induce rooting in greenhouses, copper pipes containing warm water are used for heating beds. During the rooting of stem cuttings, leaves continuously transpire and lose water. Water loss has a bad impact on cutting survival rates, so water misting is also essential to raise humidity and to get better results. In glasshouses or polyhouses, humidity can be maintained; however, temperature may rise at the same time to exceed the limits, which should be managed by the installation of cooling pads, by covering with plastic sheeting or green or black sheeting, by installing exhaust fans or simply by opening the doors for a specific time. It has been observed that sub-irrigation of nitrogen improves the rooting percentage and root biomass of cuttings, and proves better than misting (Zhang and Graves, 1995). However, misting is important to maintain high humidity. Therefore, sub-irrigation with nitrogen application and misting in combination can produce better results. Fungicides should be applied to improve root quality and survival because the cuttings are susceptible to a number of pathogens during and after the process of taking root.

Despite all the merits and simplicity of propagation, cuttings use a lot of scion material. For instance, if commercial lime varieties are directly propagated through cuttings, rootstock benefits such as size regulation, resistance to soil-borne diseases and adverse climatic conditions cannot be obtained. The selection of parent material needs more care. If diseases infect the parent material, it is easily transmitted to the plants produced from cuttings. On the other hand, rootstocks produced through seeds have low chances of any disease infection.

Layering

Rooting of shoots while they are still attached to the parent plants is called layering. It is an old method of propagation but still used for propagation of lime cultivars in the USA, India, Pakistan, Egypt, the West Indies and some other countries, on a small scale. It can be used for the propagation of self-rooted scion cultivars or those plants can be used as rootstock. The advantage of layering is that the young plants continue to receive food from the parent plant. It has been observed that plants propagated through air layering have extraordinary vigour, fruit quality and bear fruit earlier compared with budded plants (Sutton, 1954). Layering can be done in spring or late summer months to facilitate rooting. Both ground layering and air layering can be performed for the propagation of lime cultivars (Crane, 2013). Similar care should be taken while selecting shoots for air or ground layering, as in selecting scion material for grafting or budding. In the case of air layering, the shoot should remain on the parent plant for at least about 6–8 weeks to establish sufficient roots; shoots are separated from the parent plant and transferred to pots for 3–4 months before transplanting to groves. The rooting ability of different parts of the same plants is different for vegetative propagation; for example, the plant material from the upper part of the trees has higher rooting ability compared with older parts (Garner, 1976). The air layering of Spanish lime (*Malicoccus bijugatus* Jacq.) from more vigorous upright shoots gave higher success rates than less vigorous lateral shoots (Larson *et al.*, 1991). However, propagation through water shoots/suckers (vigorously growing upright branches emerging from the main stem or scaffolds) should be avoided, as they bear fruit very late and fruit quality is also inferior. Therefore, they are not a good option for air layering. It has been observed that girdling of shoots before air layering stimulates root formation. Herrin and Carter (1995) investigated the effect of different methods of girdling including slit, copper wire and debarking on root formation, and found that debarking proved superior compared with the control and other methods in the propagation of rubber plant (*Ficus elastic* L.) under greenhouse conditions.

Ground layering is usually performed in spring, and the partially buried branch is allowed to develop roots until next spring (Platt and Optiz, 1973). Moreover, the portion of the branch that is buried in the soil is cut from the lower side to keep the expected new plant in an upright position, and is directly exposed to the soil, which could lead to soil-borne fungal infections. However, this can be controlled in air layering or marcotting by using sterilized covering materials.

Marcotting is preferred in humid areas of the world. When marcotting, an easily accessed twig or branch is always selected and it is usually girdled at a point 30–45 cm from the top end. In the process, a centimetre-wide strip of bark is removed or tonguing is done using an upward cut, and a small piece of wood is inserted to keep it apart, to avoid reunion. The girdled area is immediately covered with sphagnum moss, coconut fibre or clay, forming a ball of 10–12.5 cm diameter. Then this medium is encased with a cloth, burlap, aluminium foil, plastic sheet or rubber tube. The rooting medium is kept moist, as marcotting will fail if it dries out. About 6–10 weeks are required for the marcott to develop roots. After separation from the parent plant, the marcotts should be planted in pots for further growth and then transplanted to fields (Platt and Optiz, 1973). Here, we suggest using proper hygiene practices, and the use of sterilized covering media to avoid infection to the parent as well as the new plant.

Plants propagated through air layering should be planted deeper compared with grafted plants, and a portion of plants from the top should be pruned back to maintain a root–shoot balance. The cost of lime propagation (on a per plant basis) is more if plants are propagated through layering because, compared with other methods such as grafting, more time is required for the propagation of the plants. Hence, it is less employed on a commercial scale. However, it is used by some nurserymen and gardeners for small-scale production of plants.

Propagation through tissue culture

In vitro multiplication of higher plants, under aseptic conditions on a nutrient medium is called tissue culture. With this method, different plant parts including seeds, embryos, anthers, leaves, nodes, roots and shoot tips are used as explants to grow plants. Owing to the potential applications for this technique, the number of commercial tissue culture laboratories is increasing across the world. For example, in the Netherlands, the number of commercial tissue culture laboratories increased from 28 in 1983 to 50 in 1986, and the number of cloned plants increased from 21 to 43 million in the same period (Pierik, 1988). The adoption of this technique is continuously increasing due to its benefits, the most important being the rapid clonal propagation and production of virus- and mycoplasma-free plants (Murashige, 1974).

Like other plants, lime plants are affected by many diseases in field nurseries. Tissue culture provides a suitable solution to produce disease-free and good quality lime plants. This technique can be used for the production of self-rooted as well as grafted lime plants. Here, we will focus on how lime plants can be propagated by using this technique. There are three major steps: (i) establishment of aseptic cultures; (ii) multiplication of propagula (plant part used for multiplication); and (iii) preparation for re-establishment of plants in the soil (Murashige, 1974). Each step has its own requirements. To fulfil these requirements and to provide appropriate growing conditions, a tissue culture facility must consist of an inoculation room, growth room and greenhouse (for acclimatization of plants).

Establishment of aseptic cultures

The explants taken from the mother plant are sterilized in disinfectant solution (Table 5.3) and placed in a nutrient medium to confirm and maintain sterility, and to start the growth of explants. It usually takes 4–8 weeks. The explants for the propagation of limes include seeds,

Table 5.3. Chemicals used for disinfection of explants.

Disinfectant	Concentration	Treatment time
Sodium hypochlorite	0.5–5%	5–20 min
Ethyl alcohol	75–80%	30 s
Hydrogen peroxide	3%	15–30 min
Mercuric chloride	0.01–0.1%	10–20 min
Benzalkonium chloride	0.01–0.1%	5–10 min

nucellar embryo (Jajoo, 2010), shoot tips, stem pieces and nodes with buds. The explants should be carefully selected, first washed with water, and then disinfected with an appropriate disinfectant solution followed by rinsing three times with sterilized double distilled water. The addition of 0.1% Tween 20 into sodium hypochlorite solution further enhances the effectiveness of the disinfectant. In most of the published literature, sodium hypochlorite and ethyl alcohol have been widely used as disinfectant for lemon and limes. Then, explants are cultured on nutrient medium and the culture room must be well equipped to keep it free of dust and microorganisms. Ultraviolet lights, an air conditioning system and a laminar flow hood with positive air pressure should be present in the room. Before using the flow cabinet, it should be thoroughly cleaned and sterilized with 70–80% ethanol. Different nutrient media, namely Murashige & Skoog (MS), White, Heller, B5 and Hildebrandt, Riker & Duggar are used but MS media is frequently used. The chemicals are dissolved in double distilled water for the preparation of media. The prepared media and the test tubes, jars or culture vessels are sterilized before inoculation.

In the media, some vitamins, myoinositol (100 mg/l), niacin (0.5 mg/l), pyridoxine HCl (0.5 mg/l), thiamine HCl (0.1 mg/l), glycine (2.0 mg/l), sucrose (30,000 mg/l), and agrose or phytagel (0.25%) are added for solidification of media for propagation of *C. limonia* Osbeck (Jajoo, 2010). Once the inoculation of explants is done, it results in the formation of enlarging shoot tips, rooted shoot tips or callus after 4–8 weeks. Culture should be free of any obvious infection; if some explants are not successful or got infected during the process, they should be immediately taken out of the growth room and disposed of.

Multiplication of propagula

The culture established in the first stage is divided and replanted in separate tube or jars on the nutrient media. The objective of this stage is rapid growth of organs that can lead to plants. At this stage, growing fragments can be divided again and further planted to increase the number of plants to get rapid propagation. Growing shoots have the ability to develop roots and then to become a complete plant. By changing the concentration of plant growth regulators (auxin and cytokinin), growing shoots develop roots. Jajoo (2010) found that the shoot growth from nucellar embryo of *C. limonia* Osbeck was maximum in the medium containing 2.22 μM 6-benzylaminopurine, whereas these shoots produced more roots in an MS medium containing indole-3-butyric acid (2.46 μM) and 6-benzylaminopurine (1.11 μM). Similarly, nodal segments of *Citrus limon* L. produced shoots on MS medium supplemented with 9 μM 6-benzylaminopurine, and sub-cultured several times without loss of vigour up to 48 months, and all the shoots produced roots successfully when cultured on half-strength MS medium supplemented with 27 μM NAA and 0.1% activated charcoal (Rathore *et al.*, 2007). In another study, nodal segments of *C. limon* L. produced the best shoots on MS medium with 2 mg/l 6-benzylaldenine and 2 mg/l GA while the highest rooting percentage was attained in MS media supplemented with 3 mg/l IBA (Perez-Tornero *et al.*, 2010). Keeping in view the result of previous studies described above, the appropriate concentration can be selected.

Preparation for re-establishment of plants in the soil

The success of the tissue culture techniques depends on the survival efficiency of the produced plants under field conditions. To prepare plants for transplanting in the field, they should be hardened and acclimatized in advance, so they can tolerate the outside environment. This can be done by shifting the small plants in a nutrient medium without hormones, and then to small pots with compost but still in the growth room. Some other approaches to hardening include increasing light intensity, reduction of humidity and use of growth retardants and antitranspirants during the process (Hazarika, 2003). Then they are transferred to relatively big containers and shifted to a greenhouse with partially controlled conditions, and finally transplanted to the field.

In commercial nurseries, the cost of production is an important consideration. It has been reported that *in vitro* rooting accounts for 35–75% of the cost of micropropagation (Debergh and Maene, 1981). Several studies have reported *ex vitro* rooting and acclimatization in walnut, peaches and citrus by using Soilrite and Soilrite with farmyard manure (Drivers and Suttle, 1987; Hazarika *et al.*, 1999). So, *ex vitro* rooting of lime plants could be a suitable option to reduce costs. However, the method still needs to be standardized.

Culture environment

The culture environment widely affects the success of propagation under *in vitro* conditions. Previous studies have suggested better results can be obtained by maintaining temperatures from 25–27°C, with a 16:8 h photoperiod, and fluorescent light intensity of 1000–1500 lux for the establishment and multiplication stages (Jajoo, 2010; Perez-Tornero *et al.*, 2010). However, more light intensity (2500–3000 or, even higher, 5000–10,000 lux) is required during the last stage (Murashige, 1974; Perez-Tornero *et al.*, 2010).

Tissue culture techniques can be linked with container-grown nurseries to further increase the quantity and quality of propagated plants. Micropropagation and shoot tip grafting provide suitable options for the production of self-rooted or grafted lime plants on a commercial scale. The seeds of rootstocks can be germinated on nutrient medium and at the initial stage of seedling development shoot tip grafting can be performed to prepare healthy and disease-free plants, while for self-rooted plants the seeds of the desired cultivar can be used. The seedless cultivars of lime and rootstock (if necessary) can be propagated by using shoot tips and nodal segments, and can be further used as required. The propagation of limes by different methods is described in Fig. 5.1.

Nursery Production Systems, Registration and Certification

Citrus plants are affected by a number of fungal, bacterial and viral diseases, and in many cases the nursery plants are infected, which becomes the main reason for orchard failure when they are transplanted to fields. Similarly, dooryard lime trees from local retail outlets and unregistered nurseries are sometimes mislabelled as to cultivar and/or rootstock. These situations demand improvement in production systems, and necessitate the need for registration and certification of lime nurseries.

Nursery production systems

There are two nursery production systems followed across the world, i.e. field nurseries and

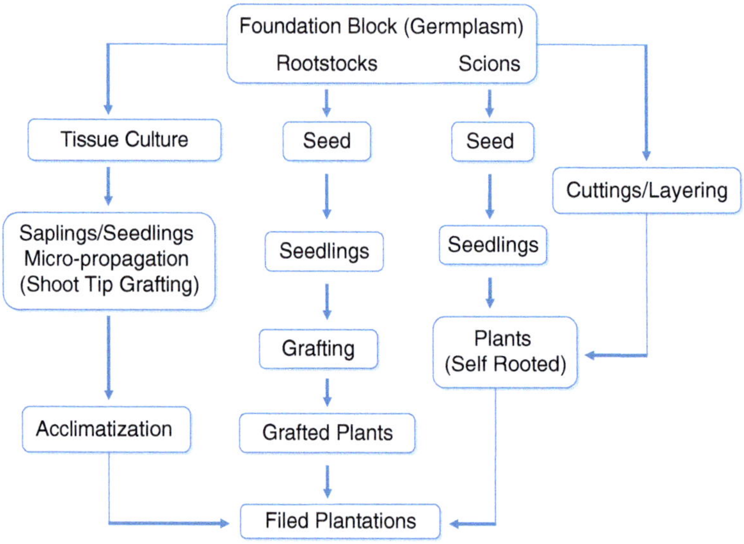

Fig. 5.1. Propagation summary of lime.

container nurseries. With the increasing demand for good quality, disease-free and certified plants, container nurseries have become more popular and the number of container nurseries is ever increasing. The container nurseries are better than field nurseries because the establishment of container nurseries requires less land area, they do not need to change sites again and again, and their plants grow better under controlled conditions. Moreover, control of diseases, insects, nematodes and weeds is easy, production time is reduced and finally the success rate after transplanting in the field is better because the root system of plants is not disturbed. However, establishment of container nurseries demands high initial capital cost for permanent structures and control equipment, there are intensive care and management requirements, and the plants require more frequent irrigation at the initial stage of orchard establishment, when transplanted to the field.

Foundation block

The foundation block (see Fig. 5.2) of each nursery is very important to obtain propagation materials used for the production of plants. Important rootstock and scion cultivars of lime are grown in foundation blocks. The foundation block should contain true to type, vigorous, healthy and disease-free plants. To establish a new foundation block, initially required plants can be procured from public and private sector research and development organizations involved in citrus and lime research. On the other hand, if a nursery does not have its own foundation block, care must be taken in the selection and procurement of the propagation materials (seed and budwood). The plants grown in the foundation block are kept under strict control, and grown in a screen house to restrict the intrusion of insects to minimize the chances of disease transfer. Similarly, irrigation water should be clean and free from disease. Plants are subjected to disease indexing on a regular basis and if any of the plants are found to be infected, they are immediately removed from the foundation block and properly disposed of.

Rootstock seedling production

Seeds are extracted from the rootstock plants of the foundation block or can be purchased from any other commercially available source. The seeds can be surface sterilized with warm water (52°C) for 10 min, and they should be treated with fungicide and sown in seed trays, seed tubes or cells. Pre-plant soaking of seeds in aerated water at 30°C for 24 h reduces germination time, and increases percentage germination and uniformity of germination. Germination time and seedling growth may further be managed by the use of heating beds (30°C), providing a photoperiod of 16:8 h and the use of plastic covering material to increase humidity (Sauls, 2008). Normally, commercial nurseries plant double the material that they actually need, as a lot of losses are observed, starting with the failure of some seeds to germinate and ranging up to the culling of grafted plants. In the production of self-rooted lime

Fig. 5.2. A view of a foundation block at the University of Agriculture, Faisalabad, Pakistan (2007). (Source: Courtesy of M.M. Khan.)

plants, all the processes are kept the same except grafting is not practised. A view of container grown rootstock seedlings is presented in Fig. 5.3. The management and care of seedlings is discussed earlier.

Sanitation practices

To avoid the spread of diseases during the propagation process, budding knives and the equipment used for pruning and removal of offshoots should be sterilized regularly. Before starting the operation, budding knives and secateurs should be washed with warm soapy water, followed by spraying with 10% chlorine bleach. When the workers change the variety, tools should be re-sterilized (Sauls, 2008). The nursery plants (seedlings, grafted plants) should be kept and grown in covered places (greenhouse, screen house) to restrict the movement of insects, birds, animals, etc. to reduce the chances of disease infection and transmission. If some plants are found to be infected, they should be separated from the main lot and kept aside for treatment or permanent culling, depending upon the nature of disease and extent of damage. Usually, fungal diseases can be treated, but the diseases caused by bacteria and virus lead to permanent rejection of the plants.

Propagation (grafting and microbudding)

Grafting has already been described earlier. As far as microbudding is concerned, rootstock seedlings are produced in tubes on a nutrient medium, and when they gain a height of 8–10 cm, the seedlings are decapitated and a tiny bud of scion is inserted in the rootstock with a vertical incision. The scion bud starts growing and after a few months the grafted seedling is shifted into a container for further growth and acclimatization; upon attaining reasonable size, it can be sold out. The grafted plants propagated by this technique might be infected with *Phytophthora* due to low budding height, when transplanted to field conditions.

Nursery registration and certification

With the discovery of attacks by viruses and virus-like diseases, the first mother plant testing and inspection service was started in 1936 in California, which led to the California Citrus Variety Improvement Program in 1957. The first virus-free plants were planted in the key foundation block in central California during 1961. This programme was taken and adopted as a model by many countries or regions including Spain, South Africa (1973), the Philippines, Cuba, Chinese Taiwan, Argentina, Uruguay, Indonesia (1987), Morocco (1997) and many other countries across the world (Passos *et al.*, 2000). With the development of shoot tip grafting, disease-free certified plants are being produced and planted. These clonal and variety improvement programmes were mainly initiated to control viruses (citrus tristeza virus) and bacterial diseases (citrus canker,

Fig. 5.3. Rootstock seedlings in container nursery (University of Agriculture, Faisalabad, Pakistan, 2007). (Source: Courtesy of M.M. Khan.)

greening). Developing countries are also trying to adopt certification on a limited scale and with the passage of time; the trend for nursery certification and registration is increasing. Without adopting these measures, the development of the citrus (lime) industry is impossible. The following model (Fig. 5.4) with appropriate modifications can be used for the registration and certification of lime nurseries.

1. The public sector research and development organization should conduct a survey in the respective country or state to collect basic data on lime nurseries (demographic figures, type of nursery, infrastructure, the process followed for production of plants, and strengths and weaknesses of nurseries).
2. The research organizations should develop the certified and disease-free foundation blocks of the economically important lime cultivars and rootstocks.
3. The nurseries should be encouraged to develop their own foundation blocks by acquiring high-quality, healthy and disease-free plants from respective research and development organizations.
4. The nursery manager and labourers should be trained to maintain the disease-free status of the foundation block and in the propagation of certified and disease-free plants.
5. Once they attain a level/meet the defined criteria, they should be registered with the government agriculture departments or another specific registration authority, and issued a registration certificate. During the initial phase of development, a period of 3–5 years may be given to nurseries owners to upgrade their status and follow the standards/regulations; and after that period if they do not follow the regulations, their nursery business may be banned until they upgrade their production system and meet the defined criteria.

In the production systems industry, traceability mechanisms are very important. Keeping in view the importance of certified nurseries, Italian scientists have developed and successfully implemented an integrated computer based information system (ICBIS) for maintaining traceability of certified plants, which can be followed in other major producing countries (Porto *et al.*, 2011).

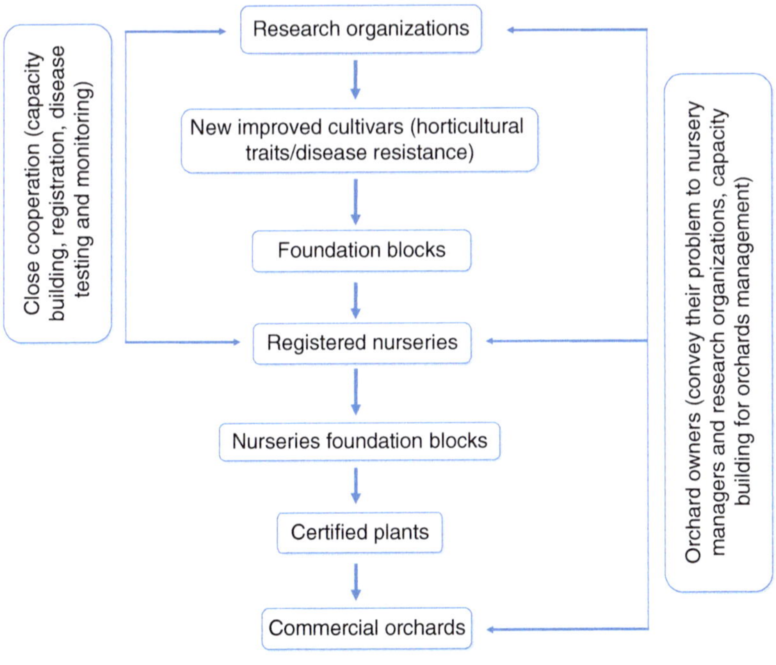

Fig. 5.4. Model for registration and certification of lime nurseries.

Recent Trends and Future Directions

Although seedless cultivars like Bearss lime/Tahiti lime/Persian lime exist, new cultivars with better fruit and plant characteristics (thin skin, higher percentage juice, seedless, longer shelf life, resistance to diseases, insects and abiotic stresses) are still required to make them acceptable for production and processing on a commercial scale, as well as for fresh use. The scientists of the Citrus Research and Education Center (CREC), University of Florida, USA, have recently released some seedless lime hybrids including C4-5-27 (a cross of Key lime and tetraploid lemon) and iced tea hybrid lemon with superior fruit and plant characteristics (Grosser *et al.*, 2015). Similarly the development of coloured Mexican lime (pigmented lime) has recently been reported through the introduction of transcription factor *Ruby* (Blood orange) and *VvmybA1* (Grapes) in Mexican lime (Koltunow *et al.*, 2000; Grosser *et al.*, 2015; Dutt *et al.*, 2016). It has opened up new avenues for the development of lime cultivars with different colours. Scion and stocks improvement programmes will continue forever to produce cultivars with better characteristics. In particular, multi-disease resistant compatible rootstock for limes still needs to be considered. To avoid soil-borne diseases, the trend for container-grown plants is increasing in developing countries and things are moving towards industrialized production of nursery plants. Similarly, the production systems for disease-free certified plants through various techniques such as tissue culture and shoot tip grafting are continuously increasing. During *in vitro* propagation, rooting involves considerable costs; so as has been done for some other plants (walnut, peaches), it is desired to standardize the techniques for *ex vitro* rooting of limes.

Nursery production and management is always a labour intensive task. Keeping in view the labour scarcity and costs involved, the use of machines will increase in the near future. Currently, in China, semi- or fully automated grafting robots have been invented by several agricultural machine industries for vegetable grafting with a working efficiency of 600 grafts/h and 90–95% success rate (Gu, 2006; Chu *et al.*, 2011). So, with the passage of time, the involvement of robots for lime propagation seems likely. Similarly, to reduce labour costs, the application of new materials such as photo-degradable tape will increase to facilitate the production process. The use of precise seeders, carrier vehicles, germination rooms and plant growth chambers, grafted seedling acclimatization facilities will increase with the passage of time, for the production of quality plants. The development of databases, software and crop models related to citrus and limes are likely to be developed by research and development organizations across the world. This will help the nursery managers and farming community in the selection of suitable cultivars of scion and rootstock, and will provide guidelines for management practices. In conclusion, we suggest researchers should consider undertaking research into limes, as there is limited information available on them compared with other fruit and citrus species.

References

Ahmed, S. (1994) Plant propagation. In. Malik, M.N. (ed.) *Horticulture.* National Book Foundation, Islamabad, Pakistan, pp. 184–213.

Ahmed, W., Pervez, M.A., Amjad, A., Khalid, M., Ayyub, C.M. and Nawaz, M.A. (2006) Effect of stionic combination on the growth and yield of Kinnow mandarin (*Citrus reticulata* Blanco). *Pakistan Journal of Botany* 38(3), 603–612.

Ahmed, W., Nawaz, M.A., Iqbal, M.A. and Khan, M.M. (2007) Effect of different rootstocks on plant nutrient status and yield in Kinnow mandarin (*Citrus reticulata* Blanco). *Pakistan Journal of Botany* 39(5), 1779–1786.

Albrigo, L.G. (1977) Rootstocks affect 'Valencia' orange fruit quality and water balance. *Proceedings of the International Society of Citriculture* 1, 62–65.

Bhatt, B.B. and Tomar, Y.K. (2010) Effects of IBA on rooting performance of *Citrus aurantifolia*Swingle (Kagzi lime) in different growing conditions. *Nature and Science* 8, 8–11.

Castle, W.S. (2010) A career perspective on citrus rootstocks, their development, and commercialization. *HortScience* 45(1), 11–15.

Castle, W.S. (2012) *Choose the Right Rootstock*. Fact Sheet HS-932, Horticultural Sciences Department, Florida Cooperative Extension Service, Institute of Food and Agricultural Sciences, University of Florida, Gainesville, Florida, pp. 1–3.

Castle, W.S., Youtest, C.O. and Hutchison, D.J. (1986) Rangpur lime × Troyer citrange, a hybrid citrus rootstock for closely spaced trees. *Proceedings of the Florida State Horticultural Society* 99, 33–35.

Chaudhary, M.I. (1994) Fruit crops. In: Malik, M.N. (ed.) *Horticulture*. National Book Foundation, Islamabad, Pakistan, pp. 437–487.

Chilembwe, E.H.C., Castle, W.S. and Cantliffe, D.J. (1992) Grading, hydrating, and osmotically priming seed of four citrus rootstock to increase germination rate and seedling uniformity. *Journal of American Society for Horticultural Sciences* 117, 368–372.

Chu, Q., Jiang, K., Liu, K. and Gu, S. (2011) Experimental study on 2JC-600 automatic grafting machine. *Journal of Agriculture and Mechanization Research* 1, 183–189.

Crane, J.H. (2013) *Key Lime Growing in the Florida Home Landscape*. Horticultural Sciences Department, UF/IFAS Extension, Gainesville, Florida, FC47.

Crane, J.H. and Osborne, J.L. (2013) *Growing 'Tahiti' Lime in the Home Landscape*. Horticultural Sciences Department, UF/IFAS Extension, Gainesville, Florida, HS8.

Debergh, P.C. and Maene, L.J. (1981) A scheme for commercial propagation of ornamental plants by tissue culture. *Scientia Horticulturae* 14, 335–345.

Dore, J. (1953) Seasonal variation in the regeneration of root cuttings. *Nature* 172, 1189.

Drivers, J.A. and Suttle, G.R.L. (1987) Nursery handling of propagules. In: Bonga, J.M. and Durzan, D.J. (eds) *Cell and Tissue Culture in Forestry*, Vol. 2. Martinus Nijhoff, Dordrecht, the Netherlands, pp. 320–335.

Dutt, M., Stanton, D. and Grosser, J.W. (2016) Ornacitrus: development of genetically modified Anthocyanin-expressing citrus with both ornamental and fresh fruit potential. *Journal of American Society for Horticultural Sciences* 141(1), 54–61.

Ferguson, J. and Young, M. (1985) The propagation of citrus rootstocks by stem cuttings. *Proceedings of the Florida State Horticultural Society* 98, 39–42.

Fishel, F.M. (2014) *Pesticide Toxicity Profile: Miscellaneous Organic Fungicides*. Agronomy Department, UF/IFAS Extension, Gainesville, Florida, PI-72.

Furlani, P.R., Zanetti, M. and Bataglia, O.C. (2009) Citrus nursery production in soilless culture. *Proceeding of the International Symposium on Soilless Culture and Hydroponics*. Lima, Peru.

Garner, R.J. (1976) Materials and methods. In: Garner, R.J. and Chaudhary, S.A. (eds) *The Propagation of Tropical Fruit Trees*, Part 1. Commonwealth Agricultural Beaureux, Slough, UK, pp. 4–222.

Grieneisen, V.A., Xu, J., Maree, A.F.M., Hogeweg, P. and Scheres, B. (2007) Auxin transport is sufficient to generate a maximum and gradient guiding root growth. *Nature* 449, 1008–1013.

Grosser, J., Viloria, Z. and Dutt, M. (2015) New acid citrus selections for Florida. *Citrus Industry*, July 2015, 30–33.

Gu, S. (2006) *Equipments and Techniques of Vegetable Industrial Production of Grafted Seedlings*. China Agricultural Press, Beijing, China.

Hartmann, H.T., Kester, D.E. and Davies, F.T. (1989) *Plants Propagation: Principles and Practices*, 5th edn. Prentice-Hall, Englewood Cliffs, New Jersey, pp. 199–255.

Hazarika, B.N. (2003) Acclimatization of tissue-cultured plants. *Current Science* 85, 1704–1712.

Hazarika, B.N., Nagaraju, V. and Parthasarathy, V.A. (1999) Acclimatization technique of *Citrus* plantlets from *in vitro*. *Advances in Plant Sciences* 12, 97–102.

Herrin, B.B and Carter, J. (1995) The response of stem girdling and covering material on air layering propagation of rubber plants. *HortScience* 30(4), 914.

Hodgson, R.W. (1967) Horticultural varieties of citrus. In: Reuther, W. (ed.) *The Citrus Industry*, Vol. I. Division of Agricultural Sciences, University of California, Riverside, California, pp. 431–592.

Hong, C.X. and Moorman, G.W. (2007) Plant pathogens in irrigation water: challenges and opportunities. *Critical Reviews in Plant Sciences* 24(3), 189–208.

Jajoo, A. (2010) *In vitro* propagation of *Citrus limonia* Osbeck through nucellar embryo culture. *Current Research Journal of Biological Sciences* 2(1), 6–8.

Jarvis, B.C. (1986) Endogenous control of adventitious rooting in non-woody cuttings. In: Jackson, M.B. (ed.) *New Root Formation in Plants and Cuttings*. Martinus Nijhoff, Dordrecht, the Netherlands, pp. 191–222.

Khan, M.M., Thompson, K., Usman, M. and Fatima, B. (2002) Role of moisture content and controlled atmosphere in citrus seed storage. *International Journal of Agriculture and Biology* 4, 259–263.

Klich, M.A., Arthur, K.S., Lax, A.R. and Bland, J.M. (1994) Iturin A: a potential new fungicide for stored grains. *Mycopathologia* 127, 123–127.

Koltunow, A.M., Brennan, P., Protopsaltis, S. and Nito, N. (2000) Regeneration of West Indian limes (*Citrus aurantifolia*) containing genes for decreased seed set. *First International Citrus Biotechnology Symposium, ISHS Acta Horticulturae* 535.

Larson, K.D., Schaffer, B. and Lara, S.B. (1991) Vegetative propagation of Spanish lime and Jaboticaba. *Proceedings of the Florida State Horticultural Society* 104, 54–56.

Mattos, D. Jr, Ramos, U.M., Quaggio, J.A. and Furlani, P.R. (2010) Nitrogen and copper for citrus nursery production on two different rootstocks. *Bragantia, Campinas* 69(1), 135–147.

Maust, B.E. and Williamson, J.G. (1994) Nitrogen nutrition of containerized citrus nursery plants. *Journal of American Society for Horticultural Sciences* 119(2), 95–201.

Melnyk, C.W. and Meyerowitz, E.M. (2015) Plant grafting. *Current Biology* 25, R183–R188.

Mobayen, R.G. (1980) Germination and emergence of citrus and tomato seeds in relation to temperature. *The Journal of Horticultural Science and Biotechnology* 55, 291–298.

Murashige, T. (1974) Plant propagation through tissue cultures. *Annual Review of Plant Physiology* 25, 135–166.

Newman, S.E. (2004) Disinfecting irrigation water for disease management. *20th Annual Conference on Pest Management on Ornamentals*. Society of American Florists, San Jose, California, pp. 1–10.

Passos, O.S., Sobrinho, A.P.C. and Filho, H.P.S. (2000) Citrus certification programs in Brazil. *Fourteenth IOCV Conference*, Short Communications, Riverside, California, pp. 391–399.

Perez-Tornero, O., Tallon, C.I. and Porras, I. (2010) An efficient protocol for micropropagation of lemon (*Citrus limon*) from mature nodal segments. *Plant Cell Tissue and Organ Culture* 100, 263–271.

Pierik, R.L.M. (1988) *In vitro* culture of higher plants as a tool in the propagation of horticultural crops. *Proceeding of the* International Symposium on Propagation of Ornamental Plants, Geisenheim, Germany.

Platt, R.G. and Optiz, K.W. (1973) The propagation of citrus. In. Reuther, W. (ed.) *The Citrus Industry*, Vol. III. Division of Agricultural Sciences, University of California, Riverside, California, pp. 1–45.

Porto, S.M.C., Arcidiacono, C. and Cascone, G. (2011) Developing integrated computer-based information systems for certified plant traceability: case study of Italian citrus-plant nursery chain. *Biosystems Engineering* 109, 120–129.

Rathore, J.S., Rathore, M.S., Singh, M., Singh, R.P. and Shekhawat, N.S. (2007) Micropropagation of mature tree of *Citrus limon*. *Indian Journal of Biotechnology* 6, 239–244.

Rogers, M.E., Stansly, P.A., Stelinski, L.L. and Burrow, J.D. (2014) *Quick Reference Guide to Citrus Insecticides and Miticides*. Department of Entomology, Florida Cooperative Extension Service, Institute of Food and Agricultural Sciences, University of Florida, Gainesville, Florida, ENY 854.

Rouse, R.E. and Sherrod, J.B. (1996) Optimum temperature for citrus seed germination. *Proceedings of the Florida State Horticultural Society* 109, 132–135.

Sagee, O., Raviv, M., Medina, S., Beckeff, D. and Cosse, A. (1992) Involvement of rooting factors and free IAA in the rootability of citrus species stem cuttings. *ScientiaHorticulturae* 51, 187–195.

Sauls, J.W. (2008) Citrus Nursery Production. Texas *Agri* Life Extension. Available at: http://aggie-horticulture.tamu.edu/citrus/nursery/L2305.htm (accessed 23 October 2016).

Schaffer, B. and Moon, P.A. (1991) Influence of rootstock on flood tolerance of Tahiti lime trees. *Proceedings of the Florida State Horticultural Society* 103, 318–321.

Seran, T.H and Umadevi, T. (2011) Influence of indole acetic acid (IAA) on the establishment of stem cuttings in lemon (*Citrus limon*L.). *Journal of Agriculture Research* 49(4), 517–524.

Shukla, M., Singh, J., Namdeo, K.N. and Ahirwar, M.K. (2012) Effect of growth regulators and fungicide on seed germination and seedling vigour of 'Kagzi lime'. *Annual Plant and Soil Research* 14(2), 175–176.

Singh, M. and Achhireddy, N.R. (1984) Tolerance of citrus rootstocks to preemergence herbicides. *Journal of Environmental Horticulture* 2(3), 73–76.

Siviero, A., Cristofani, M. and Machado, M.A. (2003) QTL mapping associated with rooting stem cuttings from *Citrus sunki vs. Poncirus trifoliata* hybrids. *Crop Breeding and Applied Biotechnology* 3, 83–88.

Stegemann, S. and Bock, R. (2009) Exchange of genetic material between cells in plant tissue grafts. *Science* 324, 649–651.

Sutton, N.E. (1954) Marcotting of Persian limes. *Proceedings of the Florida State Horticultural Society*, USA, 219–220.

Thomson, S.V. and Allen, R.M. (1974) Occurrence of *Phytophthora* species and other potential plant pathogens in recycled irrigation water. *Plant Disease Reporter* 58(10), 945–949.

Zekri, M. and Parsons, L.R. (1992) Salinity tolerance of citrus rootstocks: effects of salt on root and leaf mineral concentrations. *Plant and Soil* 147, 171–181.

Zhang, H. and Graves, W.R. (1995) Using subirrigation to root stem cuttings: comparison to mist and effects of nutrition. *HortScience* 30(4), 805.

6 Planning and Orchard Establishment

Muhammad Usman[1]* and M. Mumtaz Khan[2]
[1]*Institute of Horticultural Sciences, University of Agriculture, Faisalabad, Pakistan;*
[2]*Department of Crop Science, Sultan Qaboos University, Muscat, Oman*

The citrus industry has been rapidly expanding over time. However, land, water and input resources are becoming scarce and fierce competition has developed among plants for nutrients and other plant growth requirements. Hence, growers are keen to reduce their production costs and earn higher profits. This requires precise and efficient utilization of the precious available resources. A well established successful commercial orchard depends on better selection of plant material and successful initial establishment. Proper orchard establishment followed by better and timely cultural practices helps avoid many risks and reduces losses. Planning to establish a commercial lime orchard involves site selection, selection of rootstock and scion cultivars, selection of proper orchard design and layout, irrigation systems, selection and planting of the plant material, post-planting care and orchard hygiene.

Site Selection

Selection of a proper site for orchard establishment is the foundation of orchard planning and should be given due attention. This may reduce the operational cost, plant losses, and enhance plant growth and development. The site should be located in an area suitable for citrus cultivation with optimal climatic conditions and sufficient water availability for irrigation. Warm and humid areas with annual rainfall of 200–400 mm are highly suitable for lime cultivars (Chung *et al.*, 2006; Vand and Abdullah, 2012). Use of land with no previous history of citrus cultivation is best to avoid the incidence of soil-borne diseases caused by *Phytophthora*, *Pythium*, different bacteria and nematodes. Soil-borne fungi including *Pythium* and *Rhizoctonia* are known to cause damping off disease. In addition different weed seeds present in soils may also adversely affect plant growth and create competition for the available nutrients. In Punjab-Pakistan, soils under citrus are infected with various strains of *Phytophthora*. The soil and soil mixture at the planting site should be treated before planting to minimize the threat of soil-borne diseases. Environmentally friendly sterilization techniques like solarization and steam sterilization could be possible ways to minimize soil-borne infections at the nursery level to minimize disease spread (Usman *et al.*, 2014). Sanitized plants in containers should preferably be purchased from a reputable certified nursery before planting. Assess the orchard site for the risk of hail storms, frost injuries and wind storms. The following are important factors that should be considered before site selection.

*E-mail: m.usman@uaf.edu.pk

Soil requirements and reclamation practices

Before planning an orchard at a specific site, the complete soil profile should be analysed including basic soil properties like fertility status, soil moisture, texture, depth, soil pH and organic matter content, etc. Soil analysis of the samples taken from different locations and depths must be performed by soil scientists to gain a clear picture of the area. Generally sandy loam soils with good drainage capacity are better soils for the *Citrus* industry. The proportion of silt and clay should be 15–25% of soil. Soils containing >30% clay in the upper soil surface may block development of roots in plants necessitating drainage installations to prevent salt accumulation and high water table. There should be no hard pan, stones and calcareous substrata in the upper 1 m of soil to avoid any interference with the cultural practices. Soil pH has strong indirect effects on the availability of nutrients and beneficial microbial activity. Citrus plants grow well in soil pH ranging from 5–6, which can be regulated with additives like sulfur or by adding sulfuric or phosphoric acid to irrigation water (Davies and Albrigo, 1999). Hard and clayey soils are not suitable for lime growth. In sandy and low pH soils, Mexican lime (*Citrus aurantifolia* Swingle) grows well but produces few large fruit with thick peel and less juice content. However, regular addition of limestone to the soil to raise its pH could enhance plant growth and fruit produce. Soil additives may also disturb soil structure and texture, hence ammonium fertilizers may be used to decrease pH of alkaline soils, while calcium nitrate could be used for acidic soils. In very acidic soils, two-thirds of the recommended quantity of limestone must be spread in the planting area, mixed with top soil followed by deep ploughing about 9–12 months before actual planting. In the case of magnesium deficiency, dolomite could be applied to improve soil pH.

Land use history

It is highly important to have information about the past use of the land before site selection for an orchard. This includes knowledge of previous soil alterations, land levelling, soil surveys, areas with standing water (if any), irrigation systems used, source of water and its quality, cultivation practices, cropping schedule, diseases or insects on current vegetation, measures taken to check such problems including a record of pesticide use, any soil contamination and permanent weeds growing in the area. Such information is highly valuable for making future decisions, and a site free of such issues is highly preferred for orchard growing.

Temperature requirements

Lime trees are sensitive to both freezing, low and extremely high temperatures (>45–50°C). Hence it is important to collect information about the temporal history and fluctuations in the area and climatic trends in that region. However, in subtropical regions wide variations in freeze damage may exist in closely situated orchards. At extremely high temperatures, fresh growing flushes may be badly affected particularly in arid regions where leaf temperature may be as high as 8–10°C compared with the air temperature (Davies and Albrigo, 1999). Extreme temperatures associated with water stress could be highly damaging due to higher transpiration rates, closed stomata and decreased net photosynthesis (P_N).

Freeze damage to *Citrus* plants is a serious concern, particularly in subtropical areas. Freeze-related crop damage is rare in Brazil and other tropical regions. Frosts are common in Mediterranean regions. Young, succulent and expanding leaves in lime cultivars supercool at temperatures of −3°C to −5°C. Moderate day temperatures (20–25°C) and low night air and soil temperatures (<12°C) for 2–3 weeks help plants to attain higher freeze acclimation by quiescent growth. De-acclimation in *Citrus* is rapid and could be attained in a few days under favourable growth conditions. It is important to obtain long-term temperature data for a particular region to assess the chances of freeze damage. Mexican lime trees are more sensitive to cold than Persian lime, also known as Bearss or Tahiti lime (*Citrus latifolia* Tanaka), and should only be grown in frost-free areas with hot climatic conditions. Leaves of Persian lime trees could be damaged at temperatures below −2°C and plants may die at temperatures below −4°C.

Location, labour, transportation and other utilities

The orchard site should be located away from residential and other developmental areas. However, it should have easy access for all the stakeholders including labourers, customers, transportation and other services. Refrigerated transportation, storage and packing facilities should be available for the transportation of fruit to distant markets. Electricity, telephone and internet facilities should be available for ease of communication. When the site is developed, all roads should be paved for the passing of heavy traffic like trucks, tractors, etc. Parking areas should be evaluated and attention given to pedestrians. Vehicle flow and building must be taken into consideration.

Topography and land preparation

An orchard should be established on land that has remained fallow for some time, which should be ploughed deeply – up to 1–1.5 m – to break up hard layers of soil. These layers may block water and root penetration into the soil. This should be followed by disc ploughing to break up clods of dirt. Cover crops may be continuously grown for the first 2–3 years to prevent weeds and act as a mulch source. If an orchard is planned on bushy land, clearing of bushes is necessary before deep ploughing.

Land levelling must be done before actual layout for planting. A slight slope up to 2% could help with better surface drainage; however, a slope of >2% may cause soil erosion, translocation of soluble fertilizers and hindrance in the cultural practices. Moderately sloping areas can be modified by terracing; however, this may be expensive and expose less fertile subsoil affecting the crop yield. Uneven slopes may cause water accumulation and standing water can cause diseases and plant death by oxygen depletion in the root zone. On steep and sloping land with a 30–40% slope, contour trenching should be carried out in preference to deep ploughing. This should provide congenial tilth to the young plant roots for their healthy development. Land should be levelled properly providing a unidirectional gentle slope for ease of irrigation and drainage of excessive water during heavy rains. Trenches should be developed in soils with drainage problems to avoid water stagnation. Trees should be planted on mounds or in trenches where the land slope gradient is up to 10% to improve drainage and for better management. For steep slopes, terracing is recommended. Installation of irrigation systems is more difficult in areas with irregular topography as this may cause heavy leakage of irrigation water.

Water quality and quantity

Frequent and sufficient supply of water suitable for irrigation is the basic requirement for a commercial orchard. Irrigation water is supplied to reduce the damaging effects of water stress on plant growth, total yield and fruit quality. Hence the orchard site should be selected for better quality and quantity of water. Generally, mature *Citrus* trees require 1000–1600 mm water annually to replace the amount of water lost through evapotranspiration (ET). Sufficient irrigation plays a critical role in plant growth and development, fruit yield and reduces fruit drop, particularly June drop caused by higher summer temperatures in subtropical regions. Irrigation enhances juice content and reduces total soluble solids (TSS) and total acidity (TA) by dilution. Fruit quality may also be decreased by excessive irrigation. It may also enhance incidence of stylar end breakdown in Tahiti lime, which leads to fruit becoming unmarketable. Irrigation scheduling depends upon the daily water losses due to ET, deep percolation and surface runoff. In *Citrus*, plant ET is usually less (<80%) than potential ET measured by the USDA class A evaporation pan (Davies and Albrigo, 1999). Average pan evaporation (mm/day) can be calculated for each month of the year. Average tree canopy diameter data could be developed and water requirements can be calculated for each canopy diameter/month for the whole year. Irrigation scheduling could be planned on the basis of available soil moisture, which could be determined by soil tensiometers and other similar instruments. A series of soil water depletion data could be developed that may be helpful for planning irrigation. In young trees, root numbers are limited and they can store less water compared with mature trees, hence these should be irrigated more frequently. Measurements of plant characteristics like stomatal conductance, net CO_2 assimilation, leaf water potential and trunk growth could be helpful in regulating the irrigation schedule. However, much variation exists in these characteristics in the field and this information may be less useful for growers. Traits like diurnal trunk or fruit shrinkage could determine tree water status and could be more useful for growers but are less commonly used due to the difficulty in collecting these data. Therefore, growers have to depend on historical irrigation patterns, calendar system or a combination of both.

Plant water requirements at different vegetative and reproductive stages should be studied to ensure adequate and timely supply of irrigation water. Excessive irrigation may cause poor aeration and more accumulation of soluble salts in soils, hence it should be avoided. External irrigation sources should also be tested for any chemical contaminants like calcium, boron, etc. that may be introduced through soil or surface runoff. Therefore, every potential water source, ditches, lakes and streams for example, should be checked for mineral contents in the water. Saline irrigation water (1.3 dS/m) may be a major limitation to plant growth if applied regularly. Under such conditions use of micro irrigation systems is recommended as opposed to sprinkler and furrow irrigations systems. Contamination by weed seeds may result in unwanted vegetation in the orchard. Water-borne pathogens like *Phytophthora* may also infect plant root systems and foliage and cause diseases, in which case chemical treatment of water may be necessary.

Selection of Rootstock and Scion Cultivars

Rootstocks

Choice of a suitable rootstock for the orchard is vital in *Citrus* for better health and long productive life of the plants. Selecting a rootstock depends upon multiple factors including climate, locality, soil type, scion varieties and stress factors including fungi, viruses, nematodes, pH, salinity, drought, cold and frost (Ollitrault and Navarro, 2012; Snoussi *et al.*, 2012). Rootstocks offer great potential for acclimatization of the grafted scion cultivars across different agro-climatic conditions. Rootstocks indirectly manipulate plant phenotype by regulating tree growth pattern, fruit quantity, maturity and its quality (Warschefsky *et al.*, 2015). Selection of rootstock is more important in sites of replanting. Propagation of rootstocks through seeds is highly economical due to the presence of polyembryonic seeds and apomictic/nucellar embryos, which provide true to type plants. Use of rootstocks also reduces time to develop plants, manages plant canopy size and induces early flowering. Some important rootstocks for *Citrus* include Rough lemon, Sour orange, Trifoliate orange, Citranges, Swingle citrumello, Volkamer lime and Rangpur lime. All these have specific advantages and disadvantages. Growers are interested in early maturing cultivars with less fruit acidity. Use of rootstocks like Palestine sweet lime may reduce acidity in the fruit (Zekri, 2011).

In calcareous soils, rootstocks suitable for Mexican lime include Rough lemon (*Citrus jambhiri* L.), Alemow (*Citrus macrophylla*), Volkamer lemon (Volk) (*Citrus volkameriana*) and Rangpur lime (*Citrus limonia* L.). Rootstocks like Cleopatra mandarin provide tolerance against saline conditions, and problems using saline water may be reduced. Rough lemon is more tolerant to flooding conditions than Cleopatra mandarin, Sour orange and *Poncirus trifoliata*. Rootstocks like *P. trifoliata* and sour orange are more tolerant to foot rot than Rough lemon (Davies and Albrigo, 1999). Among *Citrus* rootstocks, Rangpur lime is sensitive to *Citrus* viroids (Wang *et al.*, 2009). In Brazil, Rangpur lime has been used as the main rootstock for Tahiti lime (*C. latifolia* Tanaka). After several years of establishment of the orchard, the greatest canopy development and higher yield were found in plants budded on C-13 citrange and African rough lemon, compared with plants budded on Trifoliate orange, Sunki and Cleopatra mandarins (Stenzel and Neves, 2004). Rootstocks such as Flying Dragon and FCAV trifoliate induced higher yield efficiency, early bearing and higher-quality fruit for foreign markets in Tahiti lime, with developing dwarf trees suitable for high-density plantations under rain-fed conditions (Cantuarias-Avilés *et al.*, 2012). Calculation of the area under the disease severity progress curve (AUDSPC) for citrus greening (HLB) severity was done on the basis of the numbers of qPCR positive replications. Rootstocks like Flying Dragon, Davis A trifoliate oranges and Swingle citrumelo showed lower values of AUDSPC compared with Morton citrange, Orlando tangelo, Rangpur lime and Volkamer lemon (Stuchi *et al.*, 2003). Thus, selection of rootstock for the orchard plays a pivotal role in orchard development, its productive life and yield.

Scion cultivars

A scion cultivar should be selected on the basis of the end product, market demand and the purpose of production, i.e. domestic or export. Market requirements could be determined by meeting with the local agents, fruit suppliers, processors and exporters before establishing the orchard. Healthy, true to type and disease-free plants should be selected from a reputable professional nursery. The plant growth habits of the scion cultivar to be planted should be known in order to develop a better orchard design. Acid lime (*C. aurantifolia* Swingle) cultivar Mexican lime is also known as Kaghzi lime, West Indian lime, Key lime and Omani lime, and has a relatively big plant with a large canopy size compared with Chinese lime, which is short in stature. Hence Mexican lime needs more space for proper tree growth and development. Mexican lime is highly sensitive to frost and cold conditions therefore it is only grown commercially in areas with a warm climate. It is more tolerant to drought conditions than other *Citrus* species. In Mexican lime, tree growth is vigorous and the plant is medium in size (2–4 m) with a bushy growth habit. The tree of Mexican lime blooms throughout the year in the tropics depending upon rainfall patterns; however, in subtropical areas it blooms once and the crop matures in the summer season. The fruit of Mexican lime is small (2–5 cm in diameter) and round. The fruit is more juicy, seedy and acidic than Tahiti lime. Juice TSS and acidity can be markedly enhanced if plants are raised under an organic farming system rather than conventional farming (de Castro *et al.*, 2014). Trees of Persian lime or Tahiti lime (*C. latifolia* Tanaka) are even larger than Mexican lime in their growth habit. Persian lime trees do not need pollination for fruit setting and normally bloom from February to April in areas with a warm climate. Sometimes flowering may be extended to the whole year in clusters of five to ten flowers. The fruit of Persian lime is large in size. The plant is more resistant to cold weather, drought conditions, diseases and pests. It is easier to grow and pick than Mexican lime since it has no thorns. Furthermore, the fruit are easy to ship to distant areas due to their thick skin. Normally fruit acidity decreases with rising temperature in *Citrus*; however, little or no decrease was found in lime cultivars (Zekri, 2011). Seedless fruit (5–7 cm in size) are produced in 90–120 days when planted away from other *Citrus* species. For propagation of *Citrus* plants it is important to select a proper budwood source and bear in mind its physiological status. In Tahiti lime and other *Citrus* species, trees propagated from buds taken from floral shoots produced more compared with trees propagated from vegetative mother shoots (Lovatt and Krueger, 2015).

Another type of acid lime is Yuzu ichandrin (*Citrus junos* Sieb. ex Tanaka), which is cultivated in China and Japan. Hybrids of Yuzu are Kabosu and Sudachi. Yuzu is used both as a fruit tree and rootstock and has more cold hardiness than satsuma mandarin. Yuzu is harvested from October to March and is commonly planted as a backyard tree in Japan. The sweet lime (*Citrus limmetoides* L.) plant is medium in size and needs more space. It is one of the most insipid tasting citrus fruit having only 0.1% citric acid, whereas Mexican lime and Persian lime have 6% acid content. The fruit of the sweet lime is very juicy, but its taste is not liked very much. Hence it is not grown on large acreages and little international trade is reported. However, it is very popular in Pakistan, India and other Middle East countries. Acid lime cultivars could be planted mixed with coconut palms, mangoes, papaya and guava.

Orchard Design

The primary objective of orchard design is to get higher yield and better fruit quality. This can be achieved by enhancing light interception by the tree canopy and minimizing the light penetration onto the bare land in between the rows. Furthermore, there must be enough space and ease for doing field operations like pruning, harvesting and insect pest management. Orientation of the tree rows can be determined in relation to slope, soil erosion, drainage and ease of movement of the operational equipment. Another important factor is the cultivar to be planted. Tree spacing is directly related to the cultivar being planted and controls the number of trees per acre.

Plant spacing and layout systems

Plant spacing mainly depends upon the row orientation. Generally there are two styles of row orientation: (i) planting at equal lengths in rows and columns, commonly known as the square system of plantation; and (ii) every alternate row is balanced by half-spacing the distance from the field border, commonly known as the hexagonal system or diamond formation according to its shape. The number of trees to be planted in a given space could be determined by using Table 6.1. To display the right number of trees per acre it is very important to find the distance between trees in rows and match this with the desired distance between each row in one of the two columns. Orchard planting design should be decided on the basis of a soil analysis report, land topography, climatic conditions of the area, plant requirements and orientation of the sun in the field. The rows would normally run from north to south.

Before planting the trees on the ground, the site layout should be staked with wooden sticks.

Table 6.1. Calculation of number of lime plants to be planted per hectare at different planting distances.

Distance between trees in the rows (m)	Distance between the rows (m)			
	3	4	5	6
2.5	*1333	1000	800	666
3	1111	833	666	555
3.5	952	714	571	476
4	833	625	500	416
5	666	500	400	333
6	555	416	333	277

Note: 10,000 m^2 per hectare divided by 7.5 m^2 (for a 3 by 2.5 m spacing) = *1333.33 trees per hectare given as round number 1333.

Selection of the layout system and planting distance can be decided according to needs and the cultivar to be planted. Layout on level land is relatively simple and a straight baseline is drawn next to the fence, walkway or roadway. Then lines are established from the baseline to both ends of the plot and in the middle. After drawing the boundary lines, division of the area into rectangles or squares becomes easier, keeping their sides equal to the planting distance. There are several layout systems available; however, square, rectangular, triangular, hexagonal, quincunx and contour systems are the most common. High-density plantations (HDPs) are laid out in a rectangular system using single hedge row or double hedge row systems.

Square and quincunx systems

In a square system, the available orchard space can be divided into equal squares. The trees are planted in straight rows at right angles to each other in all the four corners of each square (Fig. 6.1a). The plant to plant (P × P) and row to row (R × R) distances are kept equal. This system is one of the most desired systems of layout for establishing long lived fruit plant orchards under density plantations and is easy to lay out. Different types of cultural operations including cultivation, irrigation, spraying and harvesting can be carried out easily in the orchard in all directions. This system also provides the basis for the establishment of rectangular and quincunx systems of layout. The disadvantage of using this system is that there may be some unutilized space left in the centre of each square. This space could be utilized either by planting a filler plant or by reducing the P × P and R × R distances and establishing smaller squares. The planting method in the quincunx system is similar to the square or rectangular systems; however, a fifth plant is set in the centre of the square or rectangle (Fig. 6.1b). The central tree is termed as a filler tree – these are normally early bearing fruit trees for higher early return. The filler trees may be kept for a short period or till attainment of full canopy size by the main trees. This system provides about 75% more trees compared with the square system of planting. The filler trees may give early returns, which could be better than

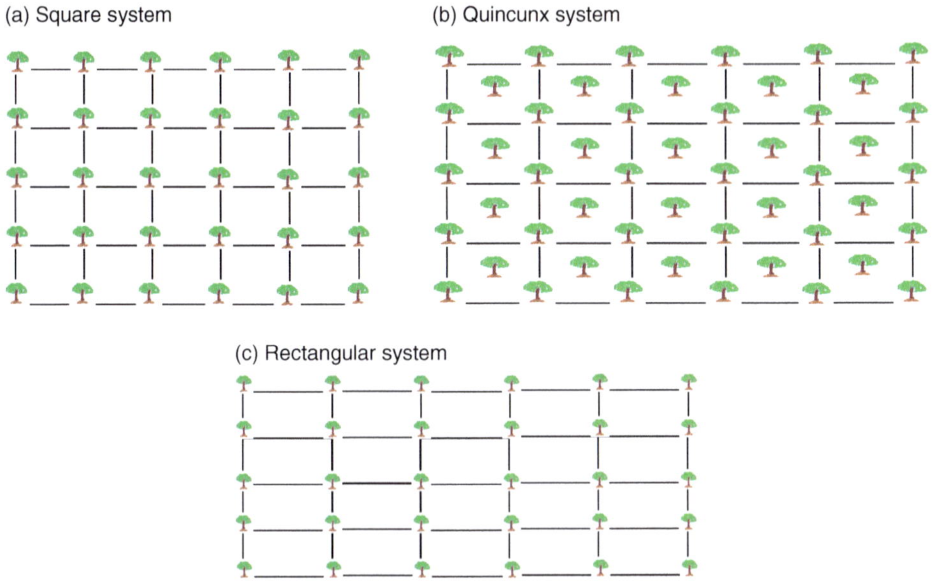

Fig. 6.1. Planting layout systems for commercial lime orchards.

practising intercropping. The filler trees may be headed back or removed when the main plants are fully developed or getting crowded.

Rectangular and triangular systems

In the rectangular system of layout, trees are planted in straight rows lined up at right angles to each other. The R × R distance is higher compared with P × P distance, making a rectangle. Trees are planted at the corners of the rectangle (Fig. 6.1c). This system has advantages similar to those of the square system of plantation, such as ease of layout and management. It is a more suitable system for medium to large sized trees. This layout is commonly adopted for establishing high-density plantations (HDPs) using single or double hedge row methods. Removal of alternate trees/fillers in each row will provide additional space for the trees to grow well and mature, and also space will be created for cultural practices at later stages. Use of additional trees/fillers helps to generate extra income in the early years. The rectangular system is not good for permanent tree plantations and causes crowding in two directions, which could be managed by regular pruning and canopy management. The triangular system is a modified form of the rectangular system in which a filler is planted in the centre of each rectangle. The planting distance between rows could be the same or more in a row leading to the establishment of isosceles triangles (two sides equal). This system is easy to lay out and provides 50–60% more plantation in a given area.

Single and double hedge row systems and high-density plantations (HDP)

These systems are also modifications of the rectangular system and plants could be placed closely in rows. The P × P distance in a row could be half to one-third of the R × R distance (Fig. 6.2a) and is used for HDPs in lime. Rows in rectangular plantations will convert into the hedge row system and should be oriented in a north to south direction for enhanced light interception in the field. In the double hedge row system, two closely spaced plant rows are planted with a wide middle space between the rows (Fig. 6.2b). Trees in a row should be planted at half of the recommended distance. Alternatively additional rows of the same crop or cultivar could be added

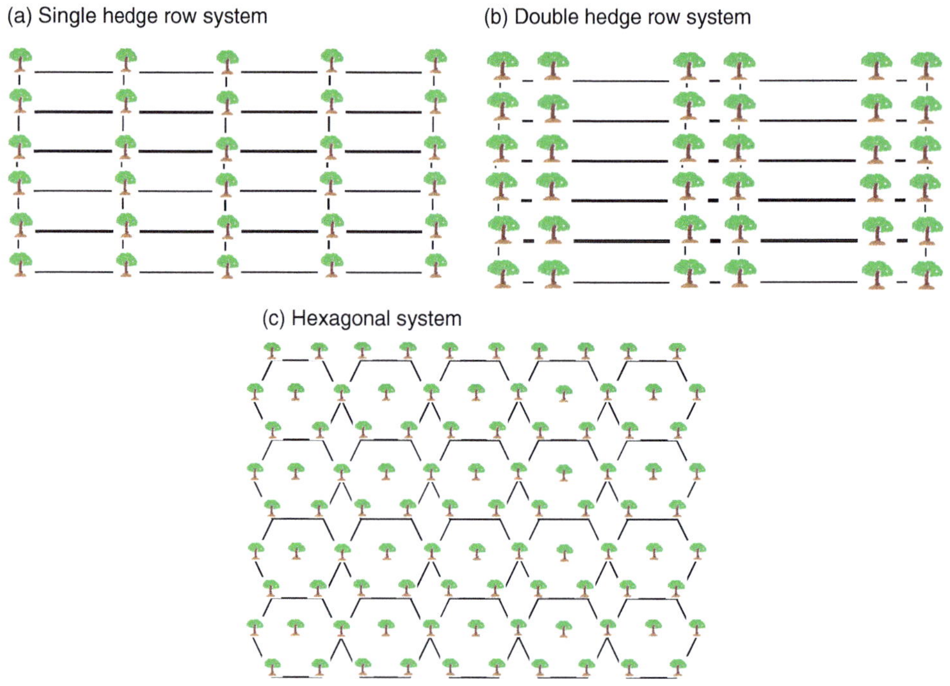

Fig. 6.2. Planting layout systems for commercial lime orchards.

in a square or rectangular system. Such commercial plantings are at the initial stages in the field as yield potential and longevity of the orchard life under such designs are yet to be determined. In hedge rows, radiant heat loss is reduced from the tree canopy and risk of frost and freeze damage is reduced. Penetration of pesticides and harvesting is difficult under these systems.

HDPs are categorized into low (<300), moderate (300–700), high (700–1500) and ultra-high (>1500 trees/ha) densities by Davies and Albrigo (1999). HDPs should give a better return during the early years of the plant growth; however, growers need to remove the extra plants to avoid orchard overcrowding. The orchard will attain its original square or rectangular design after removal of the extra plantings. Worldwide studies have suggested that properly managed moderate to high-density plantings have been more profitable and higher yielding compared with low- and ultra-high-density plantings (Koo and Muraro, 1982; Davies and Albrigo, 1999). Regular pruning, hedging and topping are key practices to be performed in HDPs to control light interception in plants and maintain optimal tree size. Orchards with HDPs make more efficient use of water and nutrients, and have higher root densities per hectare particularly during early developmental stages compared with low-density plantations (LDPs). It is more difficult to manage cultural practices in HDPs than LDPs.

Hexagonal system

This is one of the best layout systems as all the plants are set at an equal distance to each other and no unutilized space is available. The R × R distance will be less than the P × P distance. The layout is actually a grid of equidistant hexagons (Fig. 6.2c). Trees are planted at all vertexes and in the middle of each hexagon. This may also be considered as a grid of equidistant triangles where the length of each triangle is the desired planting distance. This system allows about 15% more plants per unit area compared with the square system and allows cultivation in three directions. However, the layout of this system requires technical skills.

Contour system

The contour planting design is used on slopes or hillsides where terracing may be required. This planting system allows fruit production from the waste land available in the sloping areas that are not normally utilized. There is a high chance of soil erosion caused by heavy rains or irrigation amongst the tree rows. Any of the above discussed planting systems could be used for making plantations on sloping land. In slopes with a higher gradient, terraces should be constructed with slightly higher edges against the hillside to retain irrigation water for longer. Drip irrigation or sod culture systems could be utilized to minimize soil erosion. At the time of planting, it is better to start marking tree points from the lowest level to the top. Rows should be laid out along the contour.

Irrigation Systems

The efficiency of any irrigation system is determined by its ability to balance or provide the amount of water transpired by the plants during the day. The amount of available water is normally measured by determination of soil moisture. However, adequate soil moisture alone sometimes does not ensure availability of water to the plant and may not protect plants from water stress. Soil moisture may be sufficient; however, the rate of stomatal conductance and net CO_2 assimilation may be low and young leaves can show wilting due to higher transpiration rates and less water absorption (Davies and Albrigo, 1999). These facts suggest that determination of soil moisture alone will not give the actual water status of the tree. In arid to semi-arid regions, tree roots grow in areas with water application. In tropical areas, root distribution is widespread due to heavy rainfall. Hence for optimum fruit yield, irrigation must be supplied to most (>50%) of the root volume.

Initially *Citrus* orchards were irrigated using rainwater basins in arid regions of the Middle East. This water was later diverted to the commercial orchards using ditches. Currently, orchards employ three different standard irrigation systems. The first system – flood irrigation – is the most common for old *Citrus* orchards and is being routinely used in the developing countries and in parts of Texas, Florida, northern México and central China. Trees may be planted on raised beds creating space between rows called furrows or ridges, or on levelled grounds with basins developed around the tree trunks (basin irrigation system). This basin may be connected with furrows (modified basin system). These systems are modified types of flood irrigation system, which utilize less water for irrigating *Citrus* fields compared with a conventional floor irrigation system. However, these basin and modified basin systems require more water than drip or sprinkler irrigation systems. It is best to calculate the plant water requirements at different growth stages as mentioned above and then the required amount of water may be applied through pumping.

In a drip irrigation system, water is steadily distributed in the effective root zone, which saturates the fibrous root zone area and standard drip emitters deliver water at 4–8 l/h. This system requires filtration to save the pumping system and lines from clogging. However, in areas with hot summers having temperatures round 40–45°C, dependence on drip irrigation systems could be a serious limitation in maintaining a healthy orchard (personal observations). Studies on the effect of different irrigation levels on plant growth, productivity and fruit quality in Tahiti lime trees grafted on Swingle citrumelo rootstock in Brazil revealed that different levels of irrigation did not show much influence on root distribution in depth and did not improve fruit quality. Fruit yield increased in all irrigated treatments; however, the most efficient yield per unit of water used was at 25% crop evapotranspiration (ET_c) treatment (Alves Junior *et al.*, 2011).

The third system is the micro irrigation system, which uses small sized sprinklers emitting water at 20–80 l/h. The target zone is the effective root zone. Citrus orchards have been irrigated using permanent sprinklers or travelling guns. The micro irrigation system was developed in the 1960s in South Africa and utilizes less water than flood irrigation without compromising fruit yield. This system also requires filtration; however, it is less sensitive to contaminants than the drip irrigation system. The pumping cost of the sprinkler irrigation system is less than the drip system as it can work at low psi values and requires less pressure. With rising water stress conditions and lack of quality irrigation water, micro irrigation systems are becoming widely used for *Citrus* orchard irrigation around the world. However, these systems require intensive management. Proper operation of a micro irrigation system requires good quality water, proper filtration and chlorination.

Under flooding conditions, citrus trees have less foliage and stunted growth with low yield. Trees usually don't die but production becomes marginal. Trees of Mexican lime do not tolerate over-irrigation and like well drained soils. Mostly trees are exposed to wet–dry cycles and conditions may be conducive for higher proliferation of soil-borne fungi like *Phytophthora* and *Pythium*. Under waterlogged conditions plants should be raised on beds or in artificial growing media. Yield reduction is a major physiological response of *Citrus* trees to flooding situations, and occurs before any phenotypic symptoms. Other physiological responses include reduced net CO_2 assimilation, transpiration and stomatal conductance.

Digging and Refilling of Pits

Holes or pits are dug in the soil to provide ideal conditions for better plant growth and development particularly at the initial establishment stage. The size of the pit should be 1 × 1 × 1 m diameter in light soils, while in heavy soils pit size may be increased. On hilly slopes, pit size should be maintained as 1 × 1 × 1 m along the contour. In sloped areas, pits should be formed as trenches by removing soil from the upper side of the slope and using it to level the lower side. Pits should remain open for 2–3 weeks in full sun to expose the dug soil to sunlight and higher temperatures to eliminate soil-borne diseases, insects and nematodes. Pits should be filled before the onset of the rainy season. The top soil should be mixed with suitable organic manure, compost, biofertilizer or *Trichoderma*-rich manures with silt and used to fill the hole at the bottom before planting. The top soil, compost or other supplements to the native soil should be mixed in a 50:50 ratio. The filled mixture will help the plant to establish a better root system in the lower part of the pit for improved plant development. When redigging the hole for planting, the hole should be wider than the root ball with equal depth. The container or pot should be removed and the roots visible on the sides of the root ball should be removed to avoid plant choking, which may lead to its death. Put the root section in the centre of the hole and ensure that the top portion of the root ball is 1–2 cm higher than the soil surface. Properly fill the hole with the soil mix and pat it down to remove air pockets in the soil. After planting, develop a small basin around the plant immediately for irrigation and thoroughly water the plants. Planting should be done in the autumn season (September–November) and spring season (March–April) in subtropical areas like Punjab-Pakistan. In other areas plantation is normally done in consistently warm seasons.

Selection and Planting of the Plant Material

Selection of the best available plant material provides the foundation for a healthy, long lived and productive orchard. Plant material should be purchased from a reputable registered and certified nursery to ensure true to type, disease-free and healthy plants. Growers may also produce their own plants; however, it may be difficult to ascertain all standard operating procedures (SOPs) for the development of a certified nursery. Furthermore, the *Citrus* nursery should be raised away from the orchard to ensure nursery sanitation. Healthy plants should have more mature leaves for higher photosynthesis efficiency. Dark green and glossy leaves depict better plant health. Deformed and discoloured leaves and compressed internodes with stunted plant growth may be due to malpractices in the nursery or insect pest or disease issues. The leaves of the selected plants should not be damaged and there should be no holes indicating any insect damage. Plant foliage should be inspected, particularly the lower side of the leaf, for any insect pest infestation. There should be few or no roots growing out of the holes on the bottom side of the pot or container. Root outgrowth indicates that the plant has been growing in this container for years, which is termed as a pot/root-bound plant, and is not suitable for transfer. It is better to select small plants growing in large containers compared with large plants growing in small containers. Select only vigorous plants, hardened by sufficient or reduced water application and growing under direct sunlight. Hardened plants survive better during field establishment. The top portion of the bare-rooted plants should be pruned before digging to avoid excessive transpirational losses. Trees should be planted at the same height as they were growing in the nursery. Although field-grown plants grow well during the initial 2–3 years compared with container-grown plants, later, container-grown plants produce yields similar to field-grown plants.

Post-planting Care

In less windy areas, it is optional to stake plants with a wooden or bamboo stake. Wire or nylon cord should not be used to tie trees to a stake as it may girdle and damage the tree trunk after growth. Cotton or natural fibre strings normally degrade slowly and could be more useful in this regard. Lime trees should be watered slowly but thoroughly so that moisture reaches deep into the soil. Irrigate the field once a week in hot summer seasons and fortnightly during winter. In the rainy season, irrigation may be reduced or stopped depending upon the frequency of rainfall. Use of a 5–10 cm layer of mulch over the soil surface can help to retain moisture content and prevent growth of weeds. Mulch should be kept 2–5 cm away from the bark of the tree to prevent diseases. The use of slow-release fertilizers with more nitrogen (with NPK ratios of 2:1:1 or 3:1:1) could be beneficial. Supplementation with nutrients like magnesium, iron, zinc and manganese is also very important and these may be applied as two or three foliar sprays per year, particularly for Mexican lime trees growing on calcareous soils. In neutral to acidic soils these nutrients may be applied to soil or foliage. Poor drainage or poor fertilization may cause yellowing of leaves. Like other *Citrus* species, Mexican lime trees need limited pruning to shape trees, for removal of dead wood and to limit tree size (1.5–2.5 m high and 3–3.5 m wide). It is difficult to take care of and maintain larger trees, and they are also more vulnerable to wind damage.

Use of wind breaks, tree covers and clean cultivation are passive means of protection against freeze damage. A wind break reduces heat loss from the orchard thus providing freeze protection. Natural wind breaks planted near *Citrus* trees could reduce tree vigour and yield due to shade and competition for nutrition. The use of tree covers is cost effective for young plants only and in large trees it is difficult to manage. Clean cultivation combined with irrigation is quite effective in minimizing freeze damage. Other active methods of freeze protection include orchard heating, using wind machines, different irrigation methods like micro sprinklers and flood irrigation. Irrigation water is normally warm (15–25°C), provides heat to the orchard and protects against freeze. Pest problems in Mexican lime include leaf miner, Citrus psylla and snow scale. Snow scale may appear during prolonged drought conditions. Common diseases in lime cultivars include Citrus canker, wither tip, lime anthracnose, algal disease, collar rot and *Sphaeropsis tumefaciens*.

Orchard Hygiene

The hygiene status of the site plays a pivotal role in the establishment and maintenance of a good orchard. Therefore considerable attention should be paid to hygiene at different stages, and following these practices can minimize disease incidence (Tucker *et al.*, 1993). Regular cleaning and disinfection of orchard tools including pruning scissors, cutters, hedge cutters, etc. with 10–15% household bleach (NaClO) is necessary, and contaminated or used tools should not be moved from infected areas to clean areas. Commonly diseases spread from one plant to other via these tools. Rinse sterilized tools thoroughly in tap water and dry well because bleach solution is caustic to metals. At the end of all operations apply a light coating of oil. Removal of diseased and declining plants and their residues from the orchard is important to check disease spread in the whole area. Storage or packing areas should be free of any sort of debris and should be treated regularly with appropriate fungicides to check disease spread.

References

Alves Junior, J., Folegatti, M.V., Silva, C.R.D., da Silva, T.J. and Evangelista, A.W. (2011) Response of young 'Tahiti' lime trees to different irrigation levels. *Engenharia Agrícola* 31(2), 303–314.

Cantuarias-Avilés, T., Mourão Filho, F.D.A.A., Stuchi, E.S., da Silva, S.R., Espinoza-Núñez, E. and Neto, H.B. (2012) Rootstocks for high fruit yield and quality of 'Tahiti' lime under rain-fed conditions. *Scientia Horticulturae* 142, 105–111.

Chung, K.R., Khan, I.A. and Brlansky, R.H. (2006) *Citrus Diseases Exotic to Florida: Witches' Broom Disease of Lime (WBDL)*. EDIS Publications, Institute of Food and Agricultural Sciences Citrus Extension, University of Florida, Gainesville, Florida, pp. 228.

Davies, F.S. and Albrigo, L.G. (1999) *Citrus*. CAB International, Wallingford, UK.

de Castro, D.D.S.B., de Souza Rosa, L., da Silva Menezes, E.M. and Teodoro, A.J. (2014) Comparative evaluation of organic and conventional farming on chemical quality parameters and antioxidant activity in fruits. *African Journal of Biotechnology* 13(18), 1883–1890.

Koo, R.C.J. and Muraro, R.P. (1982) Effect of tree spacing on fruit production and net returns of 'Pineapple' oranges. *Proceedings of Florida State Horticultural Society*, Florida, 95, 29–33.

Lovatt, C.J. and Krueger, R.R. (2015) Morphological and yield characteristics of 'Washington' navel orange and 'Tahiti' lime trees produced with buds from 'Floral' versus 'Vegetative' mother shoots. *Acta Horticulturae* 1065, 1831–1837.

Ollitrault, P. and Navarro, L. (2012) Citrus. In: Badenes, M.L. and Byrne, D.H. (eds) *Fruit Breeding*. Springer US, New York, pp. 623–662.

Snoussi, H., Duval, M.F., Garcia-Lor, A., Belfalah, Z., Froelicher, Y., Risterucci, A.M., Perrier, X., Jacquemoud-Collet, J.P., Navarro, L., Harrabi, M. and Ollitrault, P. (2012) Assessment of the genetic diversity of the Tunisian citrus rootstock germplasm. *BMC Genetics* 13(1), 1.

Stenzel, N.M.C. and Neves, C.S.V.J. (2004) Rootstocks for 'Tahiti' lime. *Scientia Agricola* 61(2), 151–155.

Stuchi, E.S., Donadio, L.C. and Sempionato, O.R. (2003) Performance of Tahiti lime on *Poncirus trifoliata* var. monstrosa flying dragon in four densities. *Fruits* 58(1), 13–17.

Tucker, D.P.H., Hearn, C.J. and Youtsey, C.O. (1993) *Florida citrus varieties*. SP 102. University of Florida, Institute of Food and Agricultural Sciences, Gainesville, Florida, pp. 57.

Usman, M., Shah, M.H., Badar, A., Fatima, B., Sabir, M. and Zaman, Q. (2014) Media steaming and coco-coir enhance growth of rough lemon (*Citrus jambhiri* L.) stock. *Pakistan Journal of Agricultural Sciences* 51(3), 615–623.

Vand, S.H. and Abdullah, T.L. (2012) Identification and introduction of thornless lime (*Citrus aurantifolia*) in Hormozgan, Iran. *Indian Journal of Science and Technology* 5(11), 3670–3673.

Wang, X., Zhou, C., Tang, K., Zhou, Y. and Li, Z. (2009) A rapid one-step multiplex RT-PCR assay for the simultaneous detection of five citrus viroids in China. *European Journal of Plant Pathology* 124(1), 175–180.

Warschefsky, E.J., Klein, L.L., Frank, M.H., Chitwood, D.H., Londo, J.P., von Wettberg, E.J. and Miller, A.J. (2015) Rootstocks: diversity, domestication, and impacts on shoot phenotypes. *Trends in Plant Science* 21(5), 418–437.

Zekri, M. (2011) Factors affecting citrus production and quality. *Citrus industry, Citrus Extension Trade Journal*, Institute of Food and Agricultural Sciences Citrus Extension, University of Florida, Gainesville, Florida, 92(12), 6–9.

7 Irrigation and Fertilization Management in Lime Trees

Francisco García-Sánchez[1]*, Silvia Simón-Grao[1], Vicente Gimeno[2], Juan José Martínez-Nicolás[3] and A.K. Srivastava[4]

[1]Department of Plant Nutrition, CEBAS-CSIC, Espinardo, Murcia, Spain; [2]Department of Agronomy, ISA University, Santiago, Dominican Republic; [3]Department of Vegetal Production and Microbiology, EPSO-UMH, Orihuela, Alicante, Spain; [4]Indian Council of Agricultural Research - Central Citrus Research Institute, Maharashtra, India

Introduction

The genetic origin of limes and lemons reveals that limes derive from crosses between papeda as the female parent and *Citrus medica* as the male parent, whereas lemons are a close relative of citron or sour orange. The lime group is known by different names worldwide: lime in English and French, lima in Italian and Spanish, limoon in Arabic-speaking counties and usually nimboo in India. Australia has a range of true citrus native limes. Limes, lemons and other citrus fruits are enormously valued by consumers for their flavour and nutritional properties in the human diet. The flavours provided by limes and lemons are enjoyed throughout the world. Besides being a good source of Vitamin C, limes and lemons also contain an impressive list of other essential nutrients including sugars, fibre, potassium, foliate, calcium, thiamin, niacin, vitamins, phosphorus, magnesium, copper and a variety of phytochemicals. Antioxidant compounds such as polyphenols, and the acids p-coumaric, caffeic, ferutic, sinapic, etc. are richer in lemons than in limes. Normally, acid limes are classified by varieties that give small-sized fruit such as West Indian lime, Mexican lime and Kaghzi lime, or large fruit such as Tahiti and Bearrs. These varieties are seedless. Limes can also be classified into varieties that are sweet or bitter according to the sweetness of the juice. Sweet limes are used above all as an ornamental plant in gardens. Bitter limes contain citric acid, which gives them an acidic and bitter taste. Sweet limes, as far as their smell and aspect, are similar to the acidic ones, but the content of citric acid is relatively low, and the pulp is sweet and edible (Srivastava and Singh, 2002).

For optimum production, appropriate amounts of water are especially important during the period when the tree blooms and fruit set. Production decreases when there is a water deficit, caused by watering at 33% evapotranspiration coefficient (ET_c) during flowering. In addition to irrigation management, citrus trees need to be fertilized. These trees demand large amounts of mineral nutrients, and adequate fertilization is essential in order to obtain good vegetative growth and high production. The amount of fertilizers applied in their cultivation mainly depends on the size and production levels of the trees and the soil where they are grown.

*E-mail: fgs@cebas.csic.es; aksrivas2007@gmail.com

As with other woody crops, citrus trees need an adequate balance of nitrogen, potassium and phosphorus, as well as micronutrients such as iron, manganese, boron and copper.

Bearing this in mind, citrus growers have to establish irrigation and fertilization strategies that ensure the economic viability of the crop. This chapter will outline irrigation and fertilizer management in lime trees, exploring the water requirement of lime trees, the different irrigation systems used currently and how these strategies have to be modified according to the different soil, climatic and water resource characteristics. In the fertilization section we will outline the role of the different mineral nutrients in the physiological processes of this crop, and the most commonly used fertilizers will be discussed, along with how to reach an optimum nutritional range in the trees, thereby avoiding nutritional imbalance.

Irrigation Management

Depending on the environmental conditions and plant cover, citrus trees require anything from 800–1200 mm of water per year; grapefruits require more water than oranges, lemon or limes. The amount of water given to a crop for irrigation (ET_c) depends on two factors: the water transpired by the plants and the water lost through evaporation from the soil. When the water lost to evapotranspiration is greater than the water contributed by precipitation, watering of the citrus trees in general, and limes in particular, is required. Therefore, irrigation plays an important part in the production and quality of the crop. Water status is very important for lime trees, as water is a major component of the organs, and almost all the physiological processes in the plant can be limited if the water content in the tissues is inadequate. However, the fact that the trees are suffering from a water deficit does not necessarily mean that production will be damaged. In citrus trees, there are some phenological stages that are more sensitive to water stress than others. For example, if the trees suffer from water deficit during the initial growth stages or fruit maturation and during the post-harvest period, fruit production will not be harmed. But if the lack of water is during flowering and fruit set, then there will be a loss of production. This is why production and fruit quality sometimes do not reach the levels desired. Therefore, irrigation scheduling is very important in order to obtain good outputs and to improve the size and quality of the fruit. To schedule irrigation, three parameters must be established: (i) the amount of water needed by the tree; (ii) the time when water should be applied to the tree; and (iii) the manner in which water should be applied.

The amount of water that should be applied to the crops is usually based on a parameter known as the evapotranspiration potential of the crop (ET_c), which is calculated starting with meteorological data or by measuring the humidity of the soil. In general, it is calculated as $ET_c = ET_o \times K_c$, where ET_o is the reference evapotranspiration and K_c is the crop's coefficient. Currently, the Food and Agriculture Organization of the United Nations (FAO) recommends the Penman–Monteith method for calculating this parameter. The FAO Penman–Monteith method requires the measurement of several environmental parameters including wind speed, atmospheric humidity, air temperature and solar radiation. The crop coefficient K_c includes the crop's characteristics, such as genotype, age and size of the plants, and the effects of soil evaporation. Allen *et al.* (1998) proposed that ET_c should also contain the parameter K_s, which is equivalent to the soil depletion coefficient. Thus ET_c is calculated as $ET_o \times K_c \times K_s$, where ET_c and ET_o are expressed in units of mm/day. In citrus, for large trees K_s ranged between 0.6 and 1.2 mm/day from winter until the summer period (Martin *et al.*, 1997; Fares and Alva, 1999; Morgan *et al.*, 2006; Jia *et al.*, 2007). As well as the above-mentioned factors, the fraction of the water that will not be easily accessible by the plant should be taken into account. Allen *et al.* (1998) suggested that there is a range of water content between the wilting point (*WP*) and the field capacity (*FC*) in most soils, where the water is easily available to the plants. To take this into account when calculating the amount of water applied to the crop, K_s is used, the value of which is '1' when the availability of the water in the soil is not a limiting factor, but if this is not the case, it should be calculated with the following formula (Morgan *et al.*, 2010):

$$K_s = \frac{\theta - \theta_{WP}}{\theta_t - \theta_{WP}} \quad (7.1)$$

where K_s = soil water depletion coefficient ($K_s \leq 1$); θ_{WP} = permanent wilting point soil water content (cm^3/cm^3); θ = soil water content (cm^3/cm^3); and θ_t = minimum soil water content with no reduction in plant water uptake (cm^3/cm^3). Once ET_c is known the crop water requirement (WR) can be calculated following the next steps:

$$WR = N_n / Ef_t \quad (7.2)$$

where N_n (net needs) is calculated as $ET_c - PE$ (simplified model). PE is effective precipitation and Ef_t is calculated as the total efficiency obtained as the lesser product of both equations: $Ef_p \times Ef_u$ or $Ef_u \times \cdot Ef_s$, where Ef_u is efficiency of irrigation system uniformity (0.9 in drip irrigation, 0.5 in gravity irrigation) (Keller and Karmeli, 1974). Ef_p is soil leaching efficiency (clay = 1, silt = 0.95, sandy = 0.9) (Hoare *et al.*, 1974) and Ef_s is salt leaching efficiency. $Ef_s = 1 - LR$; $LR = EC_{ar}/2EC_{max}$. LR is leaching requirements, EC_{iw} is electrical conductivity of irrigation water, and EC_{max} is the maximum soil EC wanted in the rootzone (Ayers and Wescot, 1987; Martín de Santa Olalla and De Juan, 1993).

The water requirement varies according to the age of the Nagpur mandarin or acid lime tree. It has been observed that acid lime has a lower water requirement than Nagpur mandarin (Table 7.1). For citrus irrigation ET_c is currently being used, however, in some cases calculation of ET_c is not easy since it requires the recording of data such as soil type, stage of crop growth and climatic conditions. Presently, as an alternative to meet the water needs of this crop, sensors are being utilized that measure the soil water content, leaf temperature, the flow of evaporative heat, etc. For example, in an acid lime (*Citrus latifolia* Tanaka) orchard, researchers Marín and Angelocci (2011) scheduled irrigation based on the ET_c calculated according to evaporative flow using an aerodynamic method that takes into account the mass and energy between the crop and the atmosphere. For this, they installed a copper–constantan thermocouple psychrometer at different heights from 2.5–6.5 m above the ground between two trees, and they also measured wind speed at these heights with a Met-One anemometer. With the resulting data, evapotranspiration (ET_c) and the crop's transpiration (T) were measured. They then compared these data with the reference evapotranspiration data (ET_o) obtained using the Penman–Monteith equation. The irrigation needs were expressed starting with K_c and K_{cb} calculated as $K_c = ET_c/ET_o$ and $K_{cb} = T/ET_o$. With this experiment the authors suggested that K_{cb} could be a very useful parameter to schedule irrigation, but it should not be forgotten that it depends on ET_o. K_{cb} decreased by almost 40% when ET_o increased from 3 mm/day to 5 mm/day due to the high sensitivity that the lime tree stomas have to changes in vapour pressure deficit, solar radiation and air temperature. Sepaskhash and Kashefipour (1995) linked evapotranspiration (ET) and the crop coefficient (K_c) of sweet lime trees (*Citrus limetta* (*Citrus limettioides*)) with the difference in temperature between the tree's canopy (T_c) and the air (T_a) through the formula ET (mm/day) = $3.02 - 0.94 \times (T_c - T_a)$. Also, they observed that the difference in evapotranspiration between trees that suffer drought and those that were well watered could be adjusted with the formula $-0.43 \times (T_c - T_c^*)$, where T_c^* is the temperature of well irrigated tree canopies.

Growers can use sensors to measure the soil's water content, thereby maintaining adequate humidity for the crops. The sensors that are most commonly used are tensiometers.

Table 7.1. Comparative studies on the water requirement of Nagpur mandarin versus acid lime (l/day/plant) during monsoon months (June–October) under the subhumid tropical climate of central India.

	1–3 years of age	4–7 years of age	8 years and above
Nagpur mandarin			
June	20	64	135
July	16	53	110
August	12	40	80
September	13	42	90
October	15	50	110
Acid lime			
June	18	55	88
July	12	49	84
August	9	36	60
September	8	38	65
October	10	44	68

Source: Shirgure *et al.*, 2002c; Shirgure *et al.*, 2004b; Shirgure and Srivastava, 2012

These sensors act as if they were artificial roots, measuring the force needed to extract water from the soil. In order to obtain reliable data, these should be well maintained and a direct contact should be maintained between the ceramic plate, which is installed on the tip of the sensor, and the soil. Normally, these work better in thick-textured soils, and do not work well with clay soils that break up easily. These sensors measure pressure up to −85 kPa, and can be placed at different depths, although they are typically placed at 30, 45 and 60 cm depths close to the root zone. As the soil becomes dry, the sensor reading increases, advising that in the case of citrus, the values should be maintained between −10 and −60 kPa, depending on the soil texture and the crop conditions.

Other sensors used to measure the water content of soils are neutron probes (NPs) and multisensor capacitance probes (MCPs). NP measurements are based on the dispersion of neutrons, and they can measure water content to a great depth (0–25 m) instantly. However, their main inconvenience is that the data cannot be monitored in real time; single measurements can only be done at set time points (Fig. 7.1). However, MCPs, which work by measuring the changes in the dielectric constant of the medium through frequency domain reflectometry (FDR), have simplified soil humidity measurements to temporal and spatial scales, as acceptably precise data can be automatically acquired.

The continuous recording of micro-variations in the size of the trunk and the branches

Fig. 7.1. (A), Tensiometer; (B), Irrometer sensor; (C), the neutron probe (NP); (D), multi-sensor capacitance probe.

is another technique that provides valuable data on the water status of the plant, allowing for the optimization of the irrigation management of the crops. The measurements of the tree trunk are carried out by sensors known as 'dendrometers'. Most of the continuous measurement sensors used today are based on linear variable differential transformer (LVDT) technology, the principle of which is to transform the movement or deformation of a sensor into an electrical signal whose voltage varies according to the position or deformation of the sensor. In this way, the contraction of the trunk's circumference is measured from the morning until the afternoon, and this has been linked to the storage of water in the tree. At the end of the afternoon and at night, the trunk again expands, as the water in the plant is replenished. In this cycle, when the level of contraction is large, the tree is considered to be under water stress. Many experiments on citrus have shown promising results for scheduling or irrigation using values of maximum daily shrinkage (MDS) of the trunk. On adult 'Fino' lemon trees, Ortuño *et al.* (2009) compared two irrigation treatments, one based on conventional watering using ET_c values, and another using the MDS. When the intensity of the MDS signal was lower than the intensity threshold value for 2 to 3 consecutive days, watering was done with 10% less water than the ET_c required. On the other hand, when the MDS value was higher than the threshold, watering was done with 10% more water than for ET_c. The results showed that the total water used through MDS measurements was only 9% more than that estimated by the crop's evapotranspiration, suggesting that this technique could be successful for the growing of citrus, with water only being given as needed by the tree.

Despite the benefits irrigation brings to citrus, information on how to schedule irrigation is scarce. Excessive use of water in irrigation can lead to low-quality fruit, loss of nutrients due to leaching and could cause pathophysiology in the roots. On the other hand, prolonged periods of drought could cause damage to production and fruit quality. The results obtained by Alves *et al.* (2011) on lime trees indicate that in this crop it is necessary to create a good irrigation schedule to avoid unwanted problems. The authors of this research used acid lime Tahiti trees in Piracicaba (Brazil), and applied the following irrigation treatments, based on ET_c: without irrigation, 25%, 50%, 75% and 100% for three specific periods during the year. The data showed that the maximum production was achieved with the 25% treatment, and from this point on, significant differences among the rest of the treatments were not observed. These types of experiments show that it is feasible to use irrigation techniques that optimize the use of water, independently of the ET_c values (Boller *et al.*, 2004). The increase in efficiency of irrigation through the reduction of water that is not directly exploited by the crop is an important way to save water without having an effect on productivity. The water saved could be used to increase the area irrigated. Pereira and Villa Nova (2009) conducted an experiment on Tahiti lime trees grafted onto 'Swingle' to characterize, out of all the parameters that could be measured in the soil, those that could be useful for conducting precision irrigation. The authors arrived at the conclusion that because the soil acts as a reservoir of water, and due to the fact that the energy needed to use this water increases as the soil dries, it is necessary to take into account the readily available water (RAW), establishing this value as the threshold, after which the production could be affected – with this parameter being a better measure than the total available water (AW) that accumulates in the root zone. Da Silva *et al.* (2005) also observed that in lime trees, the RAW is a good parameter that could be used to activate irrigation. These authors established the threshold of 60% RAW, after which negative physiological effects are produced, although this parameter has to be modified if for some reason the water potential of a few lower leaves reaches −0.62 MPa, or if the soil's water potential at a depth of 0.6 m reaches −48.8 kPa.

In citrus orchards, there are three main methods that are utilized to water the trees: surface irrigation, sprinkler irrigation and drip irrigation (Fig. 7.2). Each system has its advantages and disadvantages. The physical factors that influence the selection of these systems are the soil, weather, topography, quality and availability of water, groundwater depth, size of the orchard and performance, system maintenance and repair. In surface irrigation, the water is distributed around the plot through the action of gravity, freely circulating according to the gradient. The entire site is therefore watered, and the

Fig. 7.2. (A), Drip irrigation; (B), sprinkle irrigation; (C), surface irrigation; (D), sub-irrigation.

distribution of water through physical modifications of the terrain is necessary, such as by using ditches or framing to control its distribution. This is recommended in areas where labour is cheap, where water is abundant or where the distribution of precipitation matches the water needs of the trees, with irrigation only needed every once in a while (Singh *et al.*, 2012). In sprinkler irrigation, the water is moved by pressure, and when it arrives at the emitters (sprinklers) they produce drops that wet the entire area, similar to rain. Therefore, this system needs a water source, a pump, a system to handle pressure (water pipe network) and a set of nozzles to spray the water into the air and water the trees. The use of this system is recommended in freeze-prone areas or in areas with low humidity, in order to counteract the detrimental effects of this type of environment (Singh *et al.*, 2012). Drip irrigation is a localized system where water is applied drop by drop in a specific area of the plot. The objective is to make a small contribution of water in a continuous and frequent manner in a place near the plant, wetting only part of the soil's volume. This system distributes water through polyethylene tubing at low pressure, where drippers are placed at regular intervals. From the water use point of view, the main advantage of localized irrigation is the possibility of total control of the water supplied to the plants. This allows for provoking stress or guaranteeing optimal humidity at the moments of the growing cycle desired, saving water as compared with other irrigation systems. These savings derive from two aspects, the first being the elimination of losses during water transport, as it arrives straight to the plant through tubing, and the second is the reduction of direct evaporation from the soil, as only a specific part of the soil is wetted. Well designed irrigation systems allow for greater uniformity and efficiency of irrigation. Without a doubt, today drip irrigation is considered to be the best system for irrigating woody crops.

Among all the irrigation methods, surface irrigation is the most commonly used in acid lime orchards in Brazil, India and México where the main production areas are situated. However, in recent years micro irrigation has been adopted due to its many advantages in the citrus orchards. Micro irrigation leads to savings in labour, water and energy since it allows one to apply irrigation uniformity when the crops need it, leads to favourable root growth, enhances yield and fruit quality, and increases water and fertilizer use efficiency. In developed countries, under-tree sprinkler irrigation systems are used in citrus orchards for efficient use of water. Studies carried out in lemon orchards found that trees under sprinkler and micro irrigation gave the best pomological results. Shirgure and Panchariya (2012b) carried out an experiment during 1995–1998 in acid lime trees at Nagpur applying different treatments consisting of dripper (8 l/h), microjet 300° (Rayjet 1/plant), 180° (Ejet, 2/plant) and surface irrigation methods. They observed that trees watered with microjet 300° had the greatest plant height and canopy volume, followed by trees under dripper and microjet 180°. Therefore, these results suggest that micro irrigation systems are superior to the conventional methods of basin irrigation.

In citrus-growing areas in the Mediterranean regions, the arid or semi-arid climate is conducive to the crops suffering from drought. In order to solve this problem, growers use low-quality water for irrigation. This comes from wells that have an elevated soluble salt concentration, mainly sulfate, chloride, sodium and calcium. Normally, these waters have an electrical conductivity above 3 dS/m, which is the critical value when citrus crops start to decrease their production (Garcia-Sánchez *et al.*, 2003). Citrus, as compared with other crops, tend to be more sensitive to salinity (Maas, 1993; Storey and Walker, 1999), as the high concentrations of salts in irrigation water or the soil lead to considerable damage to production and fruit quality (Grieve *et al.*, 2007; Prior *et al.*, 2007). Under these conditions, irrigation management has to take into account that watering should be above 100% ET_c in order to leach the salts and eliminate them from the soil profile where the roots are found. García-Sánchez *et al.* (2003) reported the agronomical and mineral nutrition responses of plants under different irrigation management treatments in a lemon Fino '49' orchard of 6-year-old trees watered with water containing low (CE = 1 dS/m), moderate (CE = 2.5 dS/m) and high (CE = 4 dS/m) NaCl concentrations. For each salinity level, two different amounts of water were applied: 100 and 125% ET_c. As was expected, fruit yield decreased with moderate and high salt concentration, but this decrease was lower for trees treated with 125% ET_c relative to those treated with the 100% ET_c. This differential response was due to the fact that the salinized trees under 125% ET_c had a lower leaf chloride concentration, as increasing the amount of applied water by 25% decreased the chloride concentration in the rootzone.

As well as the use of saline water in agricultural areas with drought problems, the re-utilization of residual agricultural water is an important management strategy in areas with limited good water resources. This strategy is important due to the economic and environmental benefits that are produced. One of the advantages is that in addition to re-utilizing the water, the mineral elements contained in the water are also taken advantage of, evading environmental contamination problems. However, depending on the source and degree of treatment, residual waters can also contain a high concentration of soluble salts, heavy metals, viruses and bacteria, etc. Irrigation with low-quality residual water can create undesirable effects on soils and plants, or could be a potential threat to the consumers' health, as if the water is not treated well previously, it can contain pathogens that are transmitted to the food items. Pedrero *et al.* (2012) conducted an experiment on Fino '49' trees grafted onto *Citrus macrophylla* in Murcia, Spain. These authors employed different types of water for irrigation, including residual waters purified with a secondary sewage treatment (conventional activated sludge; STW), and residual water from a tertiary treatment plant (conventional activated sludge with tertiary ultraviolet treatment; TTW). The qualities of the two sources were different in terms of sodium, chloride and boron concentrations as well as electrical conductivity (EC), total dissolved solid (TDS) and pathogen contents (thermotolerant coliforms, *E. coli* and helminth eggs), which were higher in the irrigation water coming from STW. This experiment, with a total experimentation period of 3 years, showed that total production was 15% higher in trees watered with the TTW treatment; however, the fruit quality was better for trees under the STW treatment. The most important conclusion was that in both wastewater treatments there was an absence of microbiological contamination on fruit, and that water coming from secondary treatment could have adverse effects when applied over a long period.

Micro Irrigation-based Fertigation

Various micro irrigation systems have proved to be very useful in meeting the water requirements of citrus trees (Azzena *et al.*, 1988; Madrid *et al.*, 1995; Kanber *et al.*, 1996). The 'fertigation' term refers to the application of fertilizers through irrigation systems. Usually, the fertilizers are liquids or solids with a high solubility. This technique allows the provision of the precise amount of nutrients that the trees need, which notably improves plant nutritional status, and therefore increases growth and fruit yield (Haynes, 1988; Bowman, 1996; Shirgure *et al.*, 2001b; Shirgure *et al.*, 2002c; Shirgure and Srivastava, 2014). Basin irrigation for citrus trees is usually used in countries like India, Pakistan, Thailand, etc. in South Asia (Ghosh and Singh, 1993), Argentina (Castel *et al.*,1989), Australia (Simpson, 1978), Turkey (Tuzucu *et al.*, 1997), Italy (Capra and Nicosia,1987) and South Africa (Plessis, 1985). When basin irrigation is used in the loam soils of north-west and the clay soils of central India, a temporary excess soil moisture condition occurs as well as the leaching of applied nutrients below the effective root zone (Castel *et al.*, 1994; Shirgure *et al.*, 2004b). A strategy that allows judicious use of water as well as nutrients in concurrence with plant demand is likely to impart an improvement in citrus production as well as fruit quality.

Irrigation scheduling based on pan evaporation through drip irrigation systems using four levels of open pan evaporation-based irrigations was carried out (0.6, 0.7, 0.8 and 0.9 of open pan evaporation), and the incremental growth, leaf nutrient status, yield and fruit quality were recorded. The water used in acid lime scheduled at 0.8 of pan evaporation ranged from 16–73 l (1998–1999), 20–76 l (1999–2000) and 20–80 l (2000–2001), respectively. The incremental increases in plant height (0.63 m), stock girth (5.63 cm) and canopy volume (7.07 m^3) were greater with the irrigation scheduled at 0.8 of open pan evaporation. The average fruit yield (14.08 kg/tree), fruit weight (37.9 g), the total soluble solids (7.24°Brix), juice percentages (45.58%) and acidity (6.16%) of the lime were higher with drip irrigation scheduled at 0.8 of open pan evaporation (Shirgure *et al.*, 2004a). Another study using four levels of open pan evaporation-based irrigation schedules (0.6, 0.7, 0.8 and 0.9 evaporation) with respect to irrigation requirement, growth, yield and fruit quality was carried out by Shirgure *et al.* (2004b). The study indicated that the irrigation requirement using a drip irrigation system in acid lime scheduled at 0.8 of pan evaporation ranged from 15–73 l and 20–76 l per day during 1998–1999 and 1999–2000, respectively. The annual incremental increase in plant height (0.89 m) and girth (6.2 cm) was greater with irrigation scheduled at 0.8 of open pan evaporation. A significant difference in canopy volume increase (8.02 m^3) in irrigation scheduled at 0.8 of open pan evaporation was observed. As it was the optimum water regime the irrigation scheduled at 0.8 of open pan evaporation showed better physiological growth of acid lime. The average fruit yield recorded was 9.45 kg/tree (37.8 q/ha) with the 0.8 of pan evaporation irrigation schedule. The average fruit weight in the treatment irrigated at 0.8 of open pan evaporation was found to be 37.6 g. The total soluble solid in fruit was greater with the irrigation schedule at 0.7 of open pan evaporation treatment. The juice percentage of the irrigation schedule 0.8 of open pan evaporation was found to be greater (44.3%) in comparison with the other irrigation schedules. Lime acidity was increased (6.18 and 6.16%) with the irrigation schedules 0.8 and 0.9 of open pan evaporation treatments. The study indicated that the growth, yield and fruit quality of acid lime improved with 0.8 of open pan evaporation irrigation scheduling through a drip irrigation system (Shirgure *et al.*, 2004b).

The comparative efficiency of micro irrigation systems (dripper 8 l/h, microjet 300° and microjet 180°) versus basin irrigation in acid lime is indicated as the greatest plant height and canopy volume is found in trees under microjet 300°, followed by the dripper and microjet 180° irrigation systems and the basin method of irrigation. The year-round changing pattern of available N showed a reduction with all the treatments. However, the basin method of irrigation recorded 42.2 mg/kg available N compared with 6.9–11.4 mg/kg using micro irrigation systems. The available soil P improved in the dripper and microjet 300° treatments, but indicated a depletion pattern with the microjet 180° and basin irrigation methods. The available soil K improved with the dripper and microjet 300° treatments, but decreased with microjet 180° and basin irrigation. However, changes in the available nutrients in the soil were not reflected by corresponding changes in leaf nutrient composition. The results revealed that the growth and available soil nutrient build-up in the rhizosphere of acid lime was superior using micro irrigation systems compared with the conventional method of basin irrigation (Shirgure *et al.*, 2001b).

Fertigation is becoming an increasingly popular method of fertilizing citrus trees. According to Lekchiri (1983), the phosphorus and potassium requirements of citrus trees are relatively high. But soil conditions and restricted root colonization may limit the availability and uptake of soil nutrients. To overcome these difficulties, two alternatives can be adapted. These are: (i) using a micro irrigation system, fertilizer application using fertigation or by placement in a furrow parallel to the dripping ramp where the soil is moist, thereby improving the mobilities of P and K and enriching the soil where roots are concentrated to improve fertilizer uptake efficiency; and (ii) application of fertilizers by placement in the zone receiving water, to improve the mobility of P and especially of K up to a depth of 60 cm. Besides nutrient mobility, fertigation has another technical advantage over granular fertilizer application (Willis *et al.*, 1990; Willis *et al.*, 1991): it provides nourishment at the appropriate time and place (Ferguson and Davies, 1989), favouring uniform root distribution, which increases better use of fertilizer by the plants (Zhang *et al.*, 1996). Fouche and Bester (1987) assayed various fertigation strategies, combining it with different granular fertilizers in a navel orange orchard. Fertigation was supplemented with additional granular fertilizers in the following ways: (i) soluble fertilizer 'Trisol' (3:1:5) plus 350 g urea by broadcast; (ii) crystalline fertilizers of N and K with broadcast of single superphosphate; and (iii) N P K granular fertilizers through broadcast applications. The highest yield was obtained with the first and third strategies and no significant differences were observed with regard to fruit quality parameters, such as fruit fresh weight, amount of juice per fruit, titratable acidity and total soluble solids in the juices when compared with other treatments.

Another field experiment on pre-bearing acid lime plants evaluating the efficacy of differential fertigation (60, 80 and 100% of recommended dose of fertilizers, RDF) and the band placement method of fertilizer application on leaf–soil nutrient build-up as well as incremental growth of the plant showed that the amount of soil available N built up was greater in fertigation with 80% N treatment during 1995–1997 followed by fertigation with 100% N band placement and fertigation with 60% N. The available P build-up in the soil was highest in the band placement treatment, followed by fertigation with 100% N and fertigation with 80% N. The available P in the soil was decreased in fertigation with 60% N. The available K built up in the soil was greater in fertigation with 100% N, followed by band placement, fertigation with 80% N and fertigation with 60% N. The available leaf N concentration was highest in fertigation with 80% N followed by fertigation with 100% N, fertigation with 60% N and band placement. The leaf P concentration was decreased in all the treatments. The highest depletion was recorded in fertigation with 60% N, followed by fertigation with 80% N, band placement and fertigation with 100% N. The available K build-up in the acid lime leaves was greater in fertigation with 80% N, followed by fertigation with 100% N, fertigation with 60% N and band placement. The incremental plant height was greater in fertigation with 80% N, followed by fertigation with 60% N, band placement and fertigation with 100% N (Table 7.2). The canopy volume of acid lime plants was highest in fertigation with 80% N, followed by 4.9 m^3 in fertigation with 60% N and band placement, and 4.4 m^3 in fertigation with 100% N in decreasing order. The results revealed that the soil–leaf nutrient build-up and incremental plant growth were superior in fertigation plants compared with the band

Table 7.2. Response of acid lime to varying levels of irrigation versus band placement with reference to growth, yield and leaf nutrient composition.

Treatments	Growth response			Leaf nutrient concentration (%)		
	Plant height (cm)	Stock girth (cm)	Canopy volume (m³)	N	P	K
Fertigation with 60% N of recommended dose	59.77	11.54	2.89	2.02 (0.34)	0.158 (−0.019)	1.57 (0.37)
Fertigation with 80% N of recommended dose	66.67	11.72	3.04	2.33 (0.515)	0.153 (−0.02)	1.62 (0.41)
Fertigation with 100% N of recommended dose	62.43	12.33	2.80	2.30 (0.48)	0.159 (0.003)	1.58 (0.33)
Band placement with 100% N of recommended dose	58.67	11.62	2.887	2.04 (0.18)	0.150 (−0.02)	1.52 (0.25)

Source: Shirgure *et al.*, 2001a

placement method of fertilizer application (Shirgure *et al.*, 2001a).

Bowman (1996) compared the effect of applying granular fertilizers by broadcasting with a fertilization strategy based on combining fertigation with broadcast application. This successful assay was carried out on mature 'Ruby Red' grapefruit trees in Florida (USA). Conventional fertilization consisted of three broadcast applications during a year (winter, spring and autumn), while trees under fertigation/broadcast treatment received 33% of their annual N and K in winter by broadcast, and the rest by fertigation every 2 weeks beginning in spring. In this latter treatment, the fruit had higher total soluble solids (cumulative of 4 years; 10.9 t/ha) than those from conventional fertilization (10.1 t/ha).

A field experiment on integrated irrigation and fertigation through a drip irrigation system in acid lime conducted by using four levels of irrigation schedules (10, 20, 30 and 40% depletion of AWC) and three levels of fertigation (600:200:100, 500:140:70 and 400:80:40 g NPK per plant) with respect to irrigation depth, growth, yield and fruit quality was carried out. Two years of study revealed that the depth of irrigation in acid lime ranged from 620–862 mm and 889–1428 mm during the first year and second year of the experiment, respectively. The incremental growth of plant height (0.60 m) and girth (4.26 cm) were greater in irrigation scheduled at 30% depletion of AWC. There was a significant difference in canopy volume increase in irrigation scheduled at 30% depletion of AWC with 500:140:70 g N:P:K/plant. The combined effect of irrigation at 30% depletion of AWC and 500:140:70 g N:P:K/plant showed better growth of acid lime. The average fruit yield recorded was 14.93 kg/tree and fruit weight was found to be 28.84 g at 30% depletion of AWC. The total soluble solids, juice percentage and acidity in fruit were greater with the irrigation schedule at 30% depletion of AWC and 500:140:70 g N:P:K/plant in comparison with the other treatments (Table 7.3). The study hence indicated that growth, yield and fruit quality improved with the integrated use of water and nutrients (Shirgure *et al.*, 2004b).

In another study, irrigation at 30% depletion of AWC produced maximum increase in canopy volume, an average that was significantly superior to canopy volume with irrigation at 20% depletion of AWC and with irrigation at 40% depletion of AWC. The best treatment, irrigation at 30% depletion of AWC along with 500:140:70 fertilizer doses through fertigation, produced the highest growth canopy volume rate with regard to previous years and the rest of the assayed treatments. Fruit yield was also significantly affected by irrigation and fertilizer levels applied individually or in combination (Table 7.4). The above work demonstrated the greater efficacy of both nutrients and water with combined use of nutrient and water (fertigation

Table 7.3. Effect of irrigation and fertigation on incremental growth and leaf nutrient status of pre-bearing acid lime.

Treatments: Irrigation	Treatments: Fertigation $N:P_2O_5:K_2O$	Plant height (m)	Stock girth (m)	Canopy volume (m^3)	Leaf nutrient composition (%): N	P	K
Irrigation at 10% depletion of AWC (I_1)	F_1 600:200:100	0.39	3.10	6.00	2.18	0.16	1.98
	F_2 500:140:70	0.41	3.52	6.42	2.00	0.14	2.04
	F_3 400: 80:40	0.32	3.11	6.03	1.92	0.12	1.81
Irrigation at 20% depletion of AWC (I_2)	F_1 600:200:100	0.45	3.57	6.42	2.14	0.14	1.99
	F_2 500:140:70	0.46	3.17	6.02	2.12	0.14	1.88
	F_3 400:80:40	0.44	3.50	6.03	2.07	0.10	1.79
Irrigation at 30% depletion of AWC (I_3)	F_1 600:200:100	0.53	3.64	6.75	2.19	0.18	2.17
	F_2 500:140:70	0.60	4.26	6.93	2.38	0.18	2.22
	F_3 400:80:40	0.48	3.98	6.79	2.22	0.10	1.95
Irrigation at 40% depletion of AWC (I_4)	F_1 600:200:100	0.39	2.97	5.82	2.17	0.13	1.91
	F_2 500:140:70	0.41	2.98	5.74	2.13	0.12	1.75
	F_3 400:80: 40	0.32	2.56	5.53	2.06	0.08	1.59
CD($P = 0.05$)	Irrigation (I)	0.08	NS	0.28	0.10	0.03	0.12
	Fertigation (F)	0.05	NS	0.18	0.07	0.04	0.07
	Interaction (I × F)	0.12	NS	0.42	0.14	0.04	0.15

Source: Shirgure *et al.*, 2004b

Table 7.4. Response to different drip irrigation and fertilizer treatments (fertigation) in terms of growth yield and quality of acid lime (*Citrus aurantifolia* Swingle) in Typic Ustrochrept.

Treatment	Canopy volume (m^3)	Yield (kg/tree)	Fruit quality (%): TSS	Acidity	Juice	Leaf nutrients (%): N	P	K
Irrigation at 10% depletion of AWC (I_1)	4.72	8.26	7.94	6.21	35.93	2.08	0.13	1.80
	4.99	9.26	7.92	6.22	36.56	2.06	0.12	1.91
	4.73	10.07	8.02	6.15	36.55	1.87	0.11	1.83
Irrigation at 20% depletion of AWC (I_2)	4.97	12.2	8.20	6.13	37.34	2.14	0.10	1.83
	4.90	9.70	8.26	6.40	39.94	2.03	0.14	1.96
	4.83	10.75	8.12	6.37	37.96	2.03	0.18	1.91
Irrigation at 30% depletion of AWC (I_3)	5.80	12.94	8.16	6.70	39.79	2.16	0.14	2.02
	6.40	15.38	8.36	7.01	38.60	2.34	0.16	2.20
	5.25	9.73	8.24	6.77	41.09	2.18	0.11	1.91
Irrigation at 40% depletion of AWC (I_4)	4.90	9.66	7.98	6.28	37.55	2.01	0.12	1.89
	4.73	10.87	7.76	6.65	37.55	2.03	0.11	1.77
	4.50	9.50	7.97	6.38	38.57	1.95	0.10	1.71
LSD ($P = 0.05$) I	0.29	1.43	NS	0.13	1.61	0.12	0.78	0.15
F	0.18	0.72	NS	0.06	0.50	0.09	0.25	0.09
I × F	0.45	2.22	NS	0.20	2.22	0.18	1.18	0.19

Note: AWC stands for available water capacity of soil. Canopy volume is expressed in terms of the increase compared with the previous year.
Source: Shirgure *et al.*, 2002a, Shirgure *et al.*, 2002b

through micro irrigation): they were more effective than either of the two when used in separation. Fine-tuning the fertilizer requirement through site-specific nutrient management and introducing microbially loaded substrate will further add the required dynamism within the rhizosphere to assure nutrient acquisition vis-à-vis sustained quality production.

Diagnosis and Management of Nutrient Constraints

Diagnosis and management of nutrient constraints are the two important pillars of citrus fertilization programmes. The success of citrus fertigation depends on how precisely the nutrient constraints in the field are identified (Srivastava and Singh, 2008d; Srivastava, 2012b).

Deficiency symptoms as markers of diagnosis

Renewed efforts are being made consistently to develop and refine nutrient constraint diagnostic techniques. Currently, in intensive citriculture there are a variety of nutrient deficiencies, which need to be corrected to reach the maximum agronomic potential of the trees. Various diagnostic techniques like visual symptomology, leaf analysis, soil test methods and use of biochemical markers are continuously under critical scrutiny, testing and recurrent usage (Srivastava *et al.*, 2008). In the absence of suitable diagnostic reference for different nutrients, often nutritional problems of citrus orchards are erroneously diagnosed. Another core problem in effective nutrient management programmes is the determination of fertilizer requirement based on the nutrient constraints identified in the field. Visual deficiency symptoms should be regarded as just one of several kinds of evidence of deficiency of a given nutrient element. Several factors combine to introduce considerable uncertainty in diagnoses based on symptomology alone. Therefore, examining visual deficiency symptoms is often regarded as 'post-mortem diagnosis' (Srivastava, 2013a).

Nitrogen

ROLE IN PLANT METABOLISM. The many functions of nitrogen in the plant should be considered on the basis of its role as a component of a great number of organic compounds that are essential for the plant's metabolism. Besides being part of the structure of all the proteins and important molecules such as purines and pyrimidines, it is a component of nucleic acids (DNA and RNA), which are essential for protein synthesis. Nitrogen is also found in chlorophylls and enzymes from the cytochrome groups (nuclear porphyrin), which are indispensable for photosynthesis and respiration; and in various coenzymes, such as pyridoxal phosphate and nicotinamide adenine dinucleotide (NAD and NADP). Many phosphatides, alkaloids, glucosides, etc. are composed of nitrogen, and are abundant and important in plants.

VISUAL SYMPTOMS. Citrus N deficiency is characterized by a reduction in leaf size and generalized yellowing of the leaves, which is more evident on the veins. These symptoms are more intense in the leaf shoots with fruit (Fig. 7.3). This, however, should not be confused with chlorosis due to other causes, such as root alterations produced by excess of water, lesions of mechanical or pathological origin, viruses, phytotoxicities produced by herbicides, etc., all of which can lead to similar symptoms. In N-deficient trees, fruit set tends to be deficient, leading to fruit with reduced colour of peel, which tends to be pale and smooth (Srivastava and Singh, 1998; Srivastava 2013a).

Fig. 7.3. Nitrogen deficiency: yellowing of periphery of tree canopy. K deficiency: undersized fruit coupled with longitudinal fruit growth. P deficiency: thick peel and bigger hollow central core. Zn deficiency: reduction in size of leaves, typical interveinal chlorosis, trenching, rossetting, etc. are common features. Mg deficiency: formation of inverted V-shaped structure in the chlorotic background. (From: Srivastava, 2012a; Srivastava and Singh, 2008a; Srivastava and Singh, 1998.)

Phosphorus

ROLE IN PLANT METABOLISM. P forms part of organic compounds that are of great importance in the plant's metabolic processes. P forms part of phosphoglucosides, phospholipids, nucleic acids and nucleotides, co-enzymes, etc. This element plays a vital role in all the processes that require a transfer of energy. Phosphates form part of the chemical structure of ADP and ATP, which are the sources of energy of a multitude of chemical reactions in plants. The transfer of energy from ADP and ATP to other molecules (a process named phosphorylation) unchains a great number of essential processes in the plant. Among these processes, we find photosynthesis, carbohydrate metabolism, gene transfer, nutrient transport – processes that affect the development of roots, seed maturation, resistance to abiotic and biotic stresses, etc. (Srivastava *et al.*, 2006).

VISUAL SYMPTOMS. The symptoms first appear in the older leaves, which with time lose their intense green. The younger leaves stay small and thin, and have a dull bronze or purple colour (Fig. 7.3). If the deficiency persists, necrotic areas can appear. The shoots are softer and break easily. In roots, growth is arrested, and the branched shape is lost. Flowering is more scarce, and the fruit are large but do not contain as much juice, the ring is thicker and less robust, with the wedges separating from the central part of the fruit (Srivastava *et al.*, 2008).

Potassium

ROLE IN PLANT METABOLISM. Potassium is an essential nutrient that participates as a co-factor for more than 40 enzymes. It is notable because it is mobile in plants (Srivastava and Singh, 2003a). Some of its most important functions are: (i) regulation of stomatal movement; (ii) maintaining the charge balance of cells, (iii) participating in vital biochemical processes including carbohydrate metabolism, synthesis and regulation of proteins, and cellular division, among others; (iv) it is fundamental in processes of acclimation of plants against freezes, salt stress and drought; (v) regulating water balance in plants, among other processes (Srivastava, 2012a).

VISUAL SYMPTOMS. Symptoms begin with yellowing of tips and margins; the yellow area then gets broader, with necrotic areas and spotting first appearing on older leaves. Early symptoms often consist of stunted growth, sparse foliage and somewhat bronzed and lustreless appearance of leaves with more acute deficiency. Leaves often wrinkle and twist, and only weak new lateral shoots emerge due to a lack of mechanical strength. There is a reduction in fruit size with very thin smooth peel coupled with premature shedding of fruit possessing very low acidity. Excess K, on the other hand, produces very coarse textured fruit with reduced juice content (Fig. 7.3) (Srivastava and Singh, 2003a).

Magnesium

ROLE IN PLANT METABOLISM. Magnesium performs important and essential functions in plants. Without this element, life on Earth would not be possible, as it is an integral part of the composition of green pigments, use of solar energy and the synthesis of organic constituents that are indispensable for plant and animal life. It also acts as a specific co-enzyme for many enzymes that are necessary for life in all its forms. In summary, this element is involved in many biochemical processes such as (i) transformations of glycolysis; (ii) the tricarboxylic acid cycle; (iii) synthesis of nucleic acids; and (iv) transport of P by the plant (Srivastava and Singh, 1998).

VISUAL SYMPTOMS. In citrus, deficiency symptoms of this nutrient first appear in younger leaves, and it has a specific symptomology for this crop. Chlorosis in the leaf starts with yellowing in the main vein, progressing in a manner so that only the tip and the base remain green (Fig. 7.3). A non-chlorotic base appears in the shape of an inverted V, or a triangle with its base on the petiole. However, in severe cases, the leaf can become uniformly yellow (Srivastava, 2013d). On the trees, this deficiency leads to the dropping of fruit and leaves, and the fruit that remain are small in size (Srivastava, 2013b).

Iron

ROLE IN PLANT METABOLISM. Iron is an essential element in many metabolic processes of plants. This nutrient is involved in processes such as photosynthesis, respiration, nitrogen fixation, DNA synthesis, chlorophyll formation, hormone production and nitrate reduction. Iron in the cell is localized principally in the chloroplasts, which are responsible for photosynthesis, where it is involved in redox reactions. The rest of the iron is distributed in the cytoplasm and other organelles in forming haem and/or sulfide groups (Srivastava and Singh, 2008a). Iron is a co-factor in over 139 enzymes that catalyse biochemical reactions.

VISUAL SYMPTOMS. Interveinal white chlorosis appears first on younger leaves. In some cases leaves may be completely bleached; margins and tips remain scorched. Leaves take on the appearance of paper when visualized against sunlight. In acute cases, the leaves are reduced in size, fragile and very thin, and are shed early. Trees die back severely on the periphery, and especially at the top. Often trees with dead tops are seen with the lower limbs carrying almost normal foliage (Srivastava and Singh, 1998).

Manganese

ROLE IN PLANT METABOLISM. Although many of the functions of manganese are still unknown, it is known that it intervenes in many metabolic processes that take place in plants. In these, its chemical behaviour is similar in certain aspects to calcium and magnesium, and in other cases to the trace elements Fe and Zn. Manganese intervenes in processes such as photosynthesis, auxin metabolism, nitrogen metabolism, and synthesis of organic acids, proteins and nucleic acids (Srivastava and Singh, 2008b).

VISUAL SYMPTOMS. In citrus, the symptoms also appear visually as a thin web of green veins over a lighter background. In this crop, which is very sensitive to this deficiency, a certain degree of resemblance is found to Fe and Zn deficiencies, but there are clear differences between them. The common denominator to all three, Mn, Fe and Zn, is a decrease in chlorophylls, as they are elements that are not very mobile in plants. But Mn deficiency only appears in interveinal areas, with a green stripe remaining between the veins. In the case of Fe, the nerves are always green, while the rest of the leaf is an intense yellow. Zn deficiency could be confused with Mn deficiency, but it leads to a decrease in leaf size, with a shape that is thinner and pointier (Srivastava and Singh, 2003a).

Copper

ROLE IN PLANT METABOLISM. Copper plays the following roles in plant metabolism: it is an essential component of ascorbic acid oxidase, phenolase, laccase, diamine oxidase, urease, cytochrome oxidase and galactose oxidase; it has a role in carbohydrate metabolism and chlorophyll formation; and, showing an indirect role, there is reduced water movement in instances of Cu deficiency, primarily due to collapse of xylem vessels (Srivastava *et al.*, 2008).

VISUAL SYMPTOMS. Cu deficiency affects reproductive growth (formation of grains, seeds and fruit) more than vegetative growth. In the flowers of plants with adequate Cu status, anthers (which containing pollen) and ovaries demand large amounts of this nutrient, but pollen from plants with Cu deficiency is not feasible. Typical symptoms of Cu deficiency are chlorosis, necrosis, leaf dystrophy and dieback. Symptoms usually appear in tissue buds, which is indicative of poor Cu distribution in the deficient plant. Brown stained areas on the fruit will darken and turn black later. In severe cases, twigs will be covered with reddish brown droplets of gums, with likely fruit splitting (Srivastava and Singh, 2003b; Srivastava *et al.*, 2008).

Zinc

ROLE IN PLANT METABOLISM. The functions performed by zinc in plants are varied, but not all are well known. Most of these are due to its role in the formation and functioning of diverse enzymatic systems that take part in vital processes of plants. Some of these are: auxin biosynthesis, nitrogen metabolism, glycolysis and transformation of phosphorylated hexoses, as well as playing a role as a co-enzyme of the enzymes glutamic dehydrogenase, lactic dehydrogenase and carbonic anhydrase (Srivastava and Singh, 2004).

VISUAL SYMPTOMS. Zn deficiency is characterized by the formation of yellow areas around the secondary veins of the leaf, which are prominent over a green background, becoming more intense as the intensity of the deficiency increases. In severe states of deficiency, the leaves are smaller in size, they become thin and pointy and fall prematurely. These symptoms are found in young leaves, due to the scarce translocation of this element in plants. In citrus grown in Mediterranean areas, Zn deficiency is usually found in spring. The more acute deficiencies reduce the development of shoots, which decrease their internodes, thin out their stems and cause a generalized defoliation. Also, harvest is decreased, and the fruit are smaller in size, having a thinner rind, dense pulp, little juice and a low concentration of soluble solids (Srivastava and Singh, 2003a; Srivastava and Singh 2003b; Srivastava and Singh 2008b).

Boron

ROLE IN PLANT METABOLISM. Boron plays many important roles in plant metabolism such as: (i) a role in translocation of sugars from leaves, an important step towards enhanced photosynthesis – the compounds would more easily traverse cellular membrane than would the highly polar sugar molecules themselves; (ii) it is believed that B could be involved in carbohydrate distribution in the whole plant by forming certain boron–carbohydrate complexes, but direct evidence has not been observed; and (iii) this element is important in flower formation including pollen quality and growth of the pollen tube, N metabolism, hormone activity and the maintenance of Ca in soluble form (Srivastava and Singh, 1998; Srivastava, 2013b).

VISUAL SYMPTOMS. Apical meristems blacken and die, with breakdown of meristematic tissue. Terminal leaves turn necrotic and are shed prematurely, the internodes of terminal shoots remaining in shortened rosette form. Thickened leaves have a tendency to curl downward. Younger leaves show small water soaked spots or flecks, becoming translucent as the leaves mature. As mentioned above B deficiencies cause damage in flower production, and usually the fruit have lower numbers of seeds. Fruit symptoms are more reliable indicators and include hard dry fruit with lump formation in the rind caused by gum impregnations. Often brown pigmented spots are seen in the white albedo portion of fruit (Srivastava and Singh, 2008a).

Molybdenum

ROLE IN PLANT METABOLISM. Molybdenum is a co-factor of enzymes such as nitrate reductase and xanthine oxidase; participates in the synthesis of vitamin, proteins, starch and amino acid; and it is important in nitrogen metabolism (Srivastava *et al.*, 2007).

VISUAL SYMPTOMS. Leaf blades fail to expand further with the appearance of light yellow chlorosis of the leaves. As the leaves mature, yellow spots become deposits of brown gum on the lower leaf surface, which turn black. Under severe deficiency, symptoms are found on fruit: large, irregular, brown spots surrounded by yellow discolouration may develop on the peel without affecting the albedo (Srivastava and Singh, 2008b).

Leaf analysis as a diagnostic tool

Leaf mineral status in plants depends on, mainly, the availability of nutrients in the soils and nutrient plant uptake. These two factors are influenced by edaphoclimatic conditions and plant genotypes. A leaf analysis report shows the mineral status of the plants at the time of sampling, and can be used to readjust the fertilization programme.

Sampling index leaves

An experiment was carried out in the Vertic Ustochrept soil type during 1993–1994 in a 10-year-old acid lime orchard established in medium-deep black soil at Sangam Kheri village, Nagpur, with medium fertility status. The same experiment was also carried out in the Typic Haplustert soil type in a 10-year-old acid lime orchard during 1995–1996 at Sahuli village, Nagpur (Srivastava and Singh, 2003b).

Soil type: Vertic Ustochrept

Analysis of leaf samples during the entire growth period revealed that the leaf N content was stable from 3 October to 5th month (December) at a concentration of 1.9% during 1993–1994. Thereafter, it showed irregular variation till the end of the season. In comparison, leaf K concentration reached a maximum of 1.04% in

October and, thereafter, no distinct variation was observed between 0.71 and 0.64% during November to February (Fig. 7.4A, C). Leaf Ca content was recorded at a concentration of 3.5% during October–November, with no significant variation, and in December, a significantly higher concentration of 7.4% was found. The leaf calcium content was at its maximum (7.9%) in June towards the end of the season. On the other hand, leaf Mg status was low, varying from 0.29 to 0.27% during October to December and thereafter reduced to its lowest (0.12%) in March. In April, it again reached the maximum concentration of 0.52% and in the following months it regularly declined from 0.37 to 0.12% during May to July. The leaf S content continued to increase from 0.43 to 0.87% from August to October, and further up to December no considerable change was noticed (Fig. 7.4).

Leaf Fe and Mn content also showed minimal variation in concentration from 83.2 to 88.7 ppm and from 32.5 to 36.2 ppm during October to December, respectively. Thereafter, Fe content indicated irregular variation till the end of season, whereas leaf Mn content showed irregular variation up to May and, in the following months, it reduced to 25.0 ppm in July, the end of the season (Fig. 7.4E). Nair and Mukherjee (1970a) observed that seasonal variation in leaf Zn content was dependent on seasonal variation in root growth in the first 30 cm of soil, which contained most of the available Zn; and changes in demand for the nutrients in different growth flushes and different seasons. However, absorption, translocation and utilization of Zn from one season to another depended on the presence of Cu in the root zone (Nair and Mukherjee, 1970b).

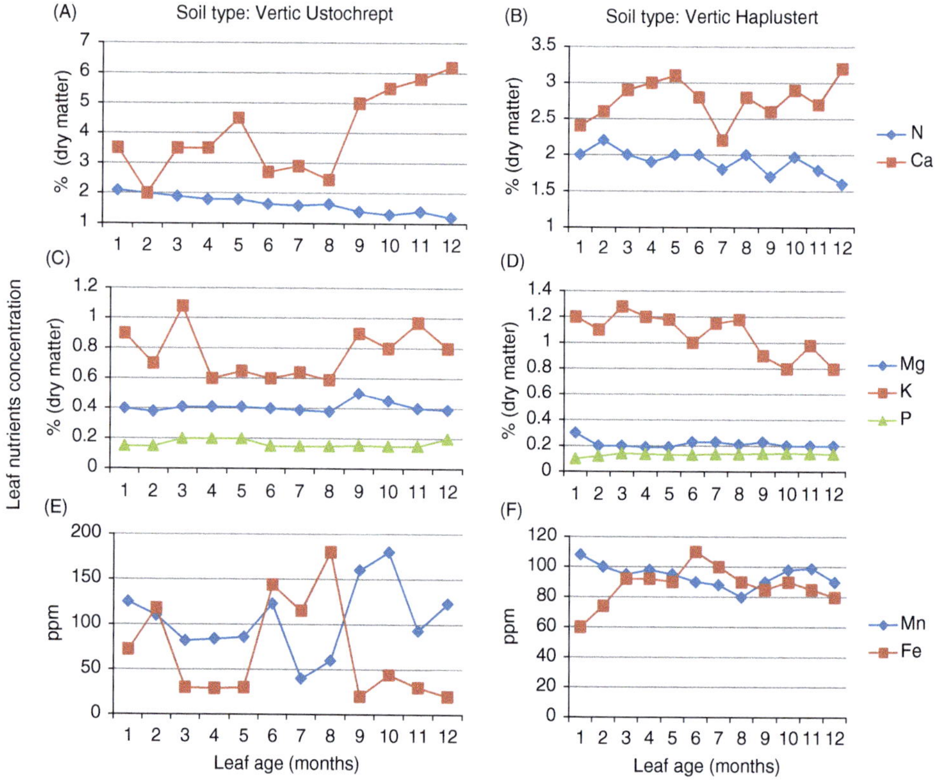

Fig. 7.4. Seasonal variation in leaf N, P, K, Ca, Mg, Fe and Mn contents in acid lime trees in soil type Vertic Ustochrept (A, C, E) and Typic Haplustert (B, D, F). (From: Srivastava and Singh, 1998; Srivastava and Singh, 2003a.)

Soil type: Typic Haplustert

The analyses of collected samples indicated variation in leaf N, P and K content from 1.6–2.2%, 0.08–0.14% and from 1.00–1.28%, respectively (Fig. 7.4). The minimum variation in leaf N (1.0–2.0%), P (0.13–0.14%) and K (1.18–1.28%) was observed when leaves were 3–5 months old. Similarly, leaf Ca and Mg concentrations were observed to vary from 2.9–3.1% and from 0.20–0.24%, respectively, at 3–5 months of leaf age, and at later stages the magnitude of variation was comparatively much higher (Fig. 7.4B, D).

The leaf Fe, Mn, Zn and Cu concentrations in 3–5-month-old leaves varied from 94.6–98.6 ppm, 94.2–100.4 ppm, 22.0–28.0 ppm and from 8.2–9.1 ppm, respectively (Fig. 7.4D, F). During the rest of the period, Fe, Mn, Zn and Cu varied from 62.0–112.4 ppm, 81.6–108.6 ppm, 18.0–26.4 ppm and from 5.2–16.0 ppm, respectively. These observations indicated minimum variation in leaf nutrient composition when leaves were 3–5 months old irrespective of soil type. Foliar analysis of *Citrus nobilis*, *Citrus sinensis*, *Citrus tangerine* and *Citrus reticulata* by Liu *et al.* (1985) showed that leaf N content was highest at the stage of flower bud physiological differentiation (September–October) followed by the early development stage (May) with the lowest N content at the stage of flower bud morphological differentiation (January). Amounts of P and K were highest in May and decreased with leaf development. Leaf Ca content increased with leaf age until September. Leaf N content during flower bud differentiation was significantly correlated with fruit yield. Socalo and Guzman (1986) observed that leaf N, P and K decreased during the flowering periods, namely, September–November and April–June in *C. aurantifolia* budded on *Citrus jambhiri* on the coast of Peru. The average lowest values of N, P and K were observed as 2.7, 0.16 and 0.70%, respectively.

Studies carried out on 4-year-old lemon trees by Sema *et al.* (1999) to investigate the effects of flushes (spring, rainy season and winter), leaf age and shoot type on foliar macronutrient levels under the low hill agro-climatic conditions of Nagaland, India, showed that the spring flush exhibited higher levels of Fe, Zn and Mn, but lower Cu levels than the winter flush. Leaf age had a significant influence only on Fe, Zn and Mn, whereas Cu concentration was more or less steady throughout the season. Fe levels increased with leaf age. The stable period (leaf age in months) was 5–6 for Fe, 2–3 for Zn and 7–8 for Mn. Lateral shoots indicated higher foliar contents. Interaction between growth flushes and leaf age produced a marked effect on foliar micronutrient concentration.

Leaf position

A study carried out by Srivastava and Singh (1998) pointed out that leaf mineral composition hardly changed with leaf position on a tree branch. These authors did not observe significant differences in the N, P, K, Ca, Mg, S, Fe, Mn, Cu and Zn concentrations in leaves when collected at positions of second, third and fourth on a shoot in 6–8-month-old leaves of acid lime (Table 7.5).

Variation in leaf N, P and K content in relation to all three leaf positions likewise was observed to be non-significant. Leaf N, P and K content varied from 2.08–2.32%, 0.08–0.12% and from 0.89–1.36%, respectively. Similarly, the variation in leaf Ca and Mg content was also noted to be non-significant (Table 7.5) during 1996–1997. The Ca and Mg content showed variations from 2.08–2.41% and from 0.29–0.42%, respectively, during the entire leaf sampling period of 1996–1997, indicating the equal effectiveness of all the three leaf positions. Leaf micronutrient content of elements such as Fe, Mn, Cu and Zn also showed non-significant variation with reference to leaf position, though varied from 898.2–110.0 ppm, 40.2–62.5 ppm, 8.2–12.2 ppm and from 20.2–28.2 ppm, respectively, during 1996–1997. These observations, hence, showed that any leaf in second, third or fourth leaf positions could serve as an index leaf for leaf analysis in acid lime.

Leaf nutrient norms

Leaf analysis as a method to estimate the crop nutrients needed assumes that the doses of the nutrient supplied, leaf nutrient content and yield are positively related. The Diagnosis and Recommendation Integrated System (DRIS) indices estimate the optimum value of different nutrients and what the relationship should be between these nutrients.

Table 7.5. Changes in leaf nutrient composition related to leaf position in acid lime (*Citrus aurantifolia* Swingle).

	Second leaf				Third leaf				Fourth leaf			
	Age in months			Critical difference ($P = 0.05$)	Age in months			Critical difference ($P = 0.05$)	Age in months			Critical difference ($P = 0.05$)
Nutrient	3	4	5		3	4	5		3	4	5	
N (%)	2.14	2.17	2.22	NS	2.22	2.14	2.16	NS	2.17	2.36	2.11	NS
P (%)	0.10	0.11	0.11	NS	0.09	0.11	0.09	NS	0.11	0.12	0.09	NS
K (%)	0.99	0.61	0.63	NS	1.28	1.03	0.70	NS	1.20	1.21	1.10	NS
Ca (%)	2.19	2.26	2.28	NS	2.04	2.10	2.28	NS	2.16	2.07	2.37	NS
Mg (%)	0.32	0.29	0.27	NS	0.27	0.28	0.33	NS	0.29	0.23	0.29	NS
Fe (ppm)	98.4	90.8	93.0	NS	100.2	88.8	92.1	NS	94.5	98.8	87.7	NS
Mn (ppm)	53.8	55.9	54.4	NS	51.8	53.0	41.3	NS	46.2	48.5	47.2	NS
Cu (ppm)	10.3	10.2	8.2	NS	10.0	8.7	8.4	NS	9.5	9.4	8.7	NS
Zn (ppm)	22.8	20.8	23.6	NS	26.6	22.3	24.6	NS	25.4	22.6	22.3	NS

Note: Non-significant at 5% level of significance.
Source: Srivastava and Singh, 1998

Some examples are: 1.80–2.12% N, 0.09–0.13% P, 1.79–1.43% K, 2.04–3.12% Ca, 0.28–0.46% Mg, 38.4–98.3 ppm Fe, 28.1–58.4 ppm Mn, 6.1–9.9 ppm Cu and 16.9–21.4 ppm Zn in relation to fruit yield of 22.8–41.2 kg/tree (Table 7.6). In another study, Varalakshmi and Bhargava (1998) suggested optimum leaf nutrient concentration for acid lime to be: 1.53–2.10% N, 0.10–0.15% P, 0.96–1.66% K, 3.05–3.42% Ca, 0.40–0.60% Mg, 0.25–0.29% S, 117.0–194.0 ppm Fe, 21.0–63.0 Mn, 8.68–14.8 ppm Cu and 25.0–50.0 ppm Zn for a fruit yield of 15.7–19.4 kg/tree.

Soil fertility norms

DRIS-based soil nutrient norms predict optimum values of: $KMnO_4$-N 106.3–118.2 mg/kg, Olsen-P 9.22–14.6 mg/kg, NH_4OAc-K 102.4–146.6 mg/kg, NH_4OAc-Ca 210.3–318.7 mg/kg, NH_4OAc-Mg 89.6–106.3 mg/kg, DTPA-Fe 4.6–12.3 mg/kg, DTPA-Mn 3.2–10.1 mg/kg, DTPA-Cu 0.80–1.40 mg/kg and DTPA-Zn 0.78–0.89 mg/kg in relation to optimum fruit yield of 22.0–41.2 kg/tree (Table 7.7).

Optimum fertilizer requirement

Optimum nutrient dose is the one that sustains the optimum productivity without affecting quality production or causing any nutrient depletion in soil (Srivastava and Singh 2002; Srivastava and Singh 2009). This aspect over the

Table 7.6. Leaf nutrient norms in acid lime (*Citrus aurantifolia* Swingle).

	DRIS norms				
Leaf nutrients	Deficient	Low	Optimum	High	Excess
Nitrogen (%)	<1.41	1.41–1.80	1.80–2.12	2.12–2.62	>2.62
Phosphorus (%)	<0.06	0.06–0.09	0.09–0.13	0.13–0.16	>0.16
Potassium (%)	<0.52	0.52–0.79	0.79–1.43	1.43–1.92	>1.92
Calcium (%)	<2.12	2.12–2.84	2.04–3.12	3.12–3.52	>3.52
Magnesium (%)	<0.12	0.12–0.28	0.28–0.46	0.46–0.82	>0.82
Iron (ppm)	<24.3	24.3–38.4	38.4–98.3	98.3–129.2	>129.2
Manganese (ppm)	<19.6	19.6–28.1	28.1–58.4	58.4–74.6	>79.6
Copper (ppm)	<3.8	3.8–6.1	6.1–9.9	9.9–12.9	>12.9
Zinc (ppm)	<14.2	14.2–16.9	16.9–21.4	21.4–29.3	>29.3
Fruit yield (kg/tree)	<12.1*	12.1–22.8	22.8–41.2	41.2–58.3	>58.3**

Note: *and ** stand for very low and very high, respectively.
Source: Srivastava, 2013b; Srivastava and Patil, 2014

Table 7.7. Soil fertility norms in acid lime (*Citrus aurantifolia* Swingle).

	DRIS norms				
Available nutrients	Deficient	Low	Optimum	High	Excess
$KMnO_4$-N (mg/kg)	<98.2	98.2–106.3	106.3–118.2	118.2–134.1	>134.1
Olsen-P (mg/kg)	<3.8	3.8–9.2	9.2–14.6	14.6–20.2	>20.2
NH_4OAc-K (mg/kg)	<78.3	78.3–102.4	102.4–146.6	146.6–191.6	>191.6
NH_4OAc-Ca (mg/kg)	<151.6	151.6–210.3	210.3–318.7	318.7–462.8	>462.8
NH_4OAc-Mg (mg/kg)	<52.1	52.1–89.6	89.6–106.3	106.3–142.3	>142.3
DTPA-Fe (mg/kg)	<1.2	1.2–4.6	4.6–12.3	12.3–20.2	>20.2
DTPA-Mn (mg/kg)	<1.12	1.12–3.2	3.2–10.1	10.1–14.6	>14.6
DTPA-Cu (mg/kg)	<0.20	0.20–0.80	0.80–1.40	1.40–2.10	>2.10
DIPA-Zn (mg/kg)	<0.45	0.45–0.78	0.78–0.89	0.89–1.06	>1.06
Fruit yield (kg/tree)	<12.1	12.1–22.8	22.0–41.2	41.2–58.3	>58.1

Note: * and ** stand for very low and very high, respectively.
Source: Srivastava and Patil, 2014

Table 7.8. Optimum nutrient requirement (soil application and foliar spray) for different cultivars in acid lime.

Dose	Crop/*Citrus* spp.	Reference
Macronutrients (Soil application)		
750 g N – 200 g P_2O_5 – 500 g K_2O/tree	Egyptian Balady lime	Ahmed *et al.* (1988)
1500 g N – 400 g P_2O_5 – 750 g K_2O/tree	Egyptian Balady lime	Maatouk *et al.* (1988)
Micronutrients (Soil application)		
$ZnSO_4$-K_2SO_4(0.5% foliar spray) – K_2O as K_2SO_4 (210 g/tree soil application)	Kaghzi lime	Singh *et al.* (1989)
$ZnSO_4$ (810 g/tree soil application) – $MnSO_4$ (630 g/100 gallon foliar spray)	Lemon	Embleton *et al.* (1966)
Micronutrients (foliar spray)		
Fe-polyflavonoid (1%)	*Citrus limon*	Fernandez-Lopez *et al.* (1993)
$MnSO_4$ (378 g/l) – $ZnSO_4$ (378 g/l)	Lemon	Alcaraz *et al.* (1986)
$ZnSO_4$ (0.6%) – 20 ppm 2,4-D	Kaghzi lime	Singh and Misra (1986)
$ZnSO_4$ (0.5%) – K_2SO_4 (4%)	Kaghzi lime	Singh *et al.* (1989)
$ZnSO_4$(0.5%) – $FeSO_4$(0.5%)	Kaghzi lime	Ingle *et al.* (2002)
Zn-EDTA – Mn-EDTA (0.10%)	Lemon	Rawash *et al.* (1983)
$ZnSO_4$ (0.5%) – urea (1.5%)	Kaghzi lime	Rathore and Chandra (2001)

years has undergone a sea change from conventional fertilizer response experiments to site-specific nutrient management, transforming the rhizosphere environment through substrate addition, etc. (Srivastava and Shirgure, 2014).

Long-term fertilizer response studies have shown a significant increase in growth and yield of acid lime with applications of 400 g and 800 g N with 55.52 kg, 73.66 kg and 81.81 kg fruit yield per plant, respectively, in the second year. The leaf nitrogen content also increased with increasing doses of nitrogen (Huchche *et al.*, 1996). Other field responses of both macro- and micronutrients (Table 7.8) showed a large variation in optimum nutrient requirement depending on soil, climate, initial soil fertility level, age of the orchard, etc. But such attempts do provide some idea about the nutrient requirement of limes versus mandarin or sweet oranges.

Lastly, precision agriculture must increase fertilizer use efficiency by knowing the spatial variability of the soil fertility (Srivastava and Singh, 2008c; Srivastava, 2013c), the nutrients and doses required for all phenological stages of the crops and understanding the soil's ability to supply those needed nutrients (Srivastava, 2013e). Even with precise diagnosis of soil fertility constraints, we still see undesirable nutrient use inefficiency due to non-redressal of spatial variability in soil fertility leading to sub-optimum fertilizer use. The emergence of site-specific nutrient management exploiting the spatial variability in soil fertility, the indigenous nutrient supply of soil and crop phenology-based nutrient demand has started sensitizing researchers as well as practitioners to tailor fertilizer requirements based either on soil test value in annual crops or on plant canopy size in perennial crops. The success of site-specific nutrient management over the past 10 years has been prominently realized for a number of perennial crops including citrus.

Another very effective nutrient management practice is integrated nutrient management involving three basic components—inorganic fertilizers, organic manures and microbes (single or multiple inoculation). This has displayed excellent success (Srivastava, 2009; Srivastava 2012b; Srivastava and Ngullie, 2009) in citrus.

References

Ahmed, F.F., El-Sayed, M.A. and Maatouk, M.A. (1988) Effect of nitrogen, potassium and phosphorus fertilization on yield and quality of Egyptian Balady lime trees (*Citrus aurantifolia*) II. Yield and fruit quality. *Annals of Agricultural Science* 33, 1249–1268.

Alcaraz, C.F., Martinez-Sánchez, F., Sevilla, F. and Hellin, E. (1986) Influence of ferrodoxin levels on nitrate reductase activity in iron deficient lemon trees. *Journal of American Society of Horticultural Science* 111, 4265–4270.

Allen, R.G., Pereira, L.S., Raes, D. and Smith, M. (1998) Crop evapotranspiration – guidelines for computing crop water requirements. *FAO Irrigation and Drainage Paper No. 56*. Food and Agriculture Organization of the United Nations, Rome.

Alves, J.J, Folegatti, M.V., Da Silva, C.R., Da Silva, T.J.A. and Evangelista, A.W.P. (2011) Response of young 'Tahiti' lime trees to different irrigation levels. *Engenharia Agricola Jaboticabal* 31, 303–304.

Ayers, R.S. and Wescot, D.W. (1987) Calidad del agua para la agricultura. FAO: Riego y drenaje n° 29. Rome.

Azzena, M., Deidda, P. and Dettori, S. (1988) Drip and micro-sprinkler irrigation for young Valencia orange trees. *Proceedings of the Sixth International Citrus Congress, Tel Aviv, Israel*, vol. 2, pp. 747–751.

Boller, E.F., Avilla, J., Jörg, E., Malavolta, C., Wijnands, F. and Esbjerg, P. (2004) Integrated production: principles and technical guidelines. *IOBC/WPRS Bulletin* 27(2). Available at: https://www.iobc-wprs.org/ip_ipm/iobc_bas.pdf (accessed 2 November 2016).

Bowman, B.J. (1996) Fertigation versus conventional fertilization of flatwoods grapefruit. *Fertilizer Research* 44, 123–128.

Capra, A. and Nicosia, O.U.D. (1987) Irrigation management in citrus orchards. *Irrigazione* 34, 3–15.

Castel, J.R., Buj, A. and Ramos, C. (1989) Comparison of drip and border irrigation of mature Salustina orange trees. *Investigation Agraria, Producción y Protection Vegetables* 4, 393–412.

Castel, J.R., Lidon, A.L., Ginestar, C. and Ramos, C. (1994) Yield, growth and nitrate leaching in drip irrigated citrus tree under different fertigation treatments. *Paper presented at 17th ICIW European Regional Conference on Irrigation and Drainage, Varna, Bulgaria*, 16–22 May 1994.

Da Silva, C.R., Vinícius, M., Araújo, T.J., Alves, J., Fonseca, C. and Vasconcelos, R. (2005) Water relations and photosynthesis as criteria for adequate irrigation management in 'Tahiti' lime trees. *Scientia Agricola* 62, 415–422.

Embleton, T.W., Wallihan, E.F. and Goodall, G.E. (1966) Coastal lemon. Zinc and manganese fertilization. *Citrog* 51, 131, 140, 142–143, bibl.2.

Fares, A. and Alva, A.K. (1999) Estimation of citrus evapotranspiration by soil water mass balance. *Soil Science* 164, 302–310.

Ferguson, J.J. and Davies, F.S. (1989) *Fertilization of Young Citrus. Florida Cooperation Extension Service. Fact Sheet* FC-79, University of Florida, Gainesville, Florida.

Fernandez-Lopez, J.A., Lopez-Roca, J.M. and Almela, L. (1993) Mineral composition of iron chlorotic *Citrus limon* L. leaves. *Journal of Plant Nutrition* 16, 1395–1407.

Fouche, P.S. and Bester, D.H. (1987) The influence of water soluble fertilizer on nutrition and productivity of Navel orange trees under micro-jet irrigation. *Citrus and Sub-Tropical Fruits Journal* 62, 8–12.

García-Sánchez, F., Carvajal, M., Porras, I. and Martínez, V. (2003) Effects of salinity and rate of irrigation on yield, fruit quality and mineral composition of 'Fino 49' lemon. *European Journal of Agronomy* 19, 427–437.

Ghosh, S.P. and Singh, R.B. (1993) *Citrus in South Asia*. RAPA Publication, FAO, Regional Office, Bangkok, p. 21.

Grieve, A.M., Prior, L.D. and Bevington, K.B. (2007) Long-term effects of saline irrigation water on growth, yield, and fruit quality of Valencia orange trees. *Australian Journal of Agricultural Research* 58, 342–348.

Haynes, R.J. (1988) Comparison of fertigation with broadcast applications of Urea-N on levels of available soil nutrients and on growth and yield of trickle irrigated peppers. *Scientia Horticulturae* 35, 189–198.

Hoare, E., Garzoli, K. and Blackwell, J. (1974) Plant water requirements as related trickle irrigation. *Proceedings of II International Drip Irrigation Congress, California*, pp. 323–328.

Huchche, A.D., Dass, H.C., Lallan Ram, Srivastava, A.K. and Kohli, R.R. (1996) Response of acid lime (*Citrus aurantifolia* Linn.) to nitrogen fertilization. *Indian Journal of Horticulture* 53, 14–18.

Ingle, H.V., Kokate, S.S., Athawale, R.B. and Katole, S.R. (2002) Effect of foliar application of zinc and iron on growth, yield and quality of acid lime. *Indian Journal of Citriculture* 1, 43–45.

Jia, X., Swancar, A., Jacobs, J.M., Dukes, M.D. and Morgan, K.T. (2007) Comparison of evapotranspiration rates for flatwoods and ridge citrus. *Transactions of the American Society of Agricultural Engineers* 50, 83–94.

Kanber, R., Koksal, H., Ouder, S. and Eyten, M. (1996) Effect of different irrigation methods of yield, evapotranspiration and root development of young orange trees. *Turkish Journal of Agriculture and Forestry* 20, 163–172.

Keller, J. and Karmeli, D. (1974) Trickle irrigation design parameters. *Transactions of the American Society of Agricultural Engineers* 17, 678–684.
Lekchiri, A. (1983) *Interactions between water and phosphate and potash fertilizers, nutrient balances and the need for the fertilizers in semi-arid and arid regions*. International Potash Institute, Bern, Switzerland, pp. 305–314.
Liu, X.H., Zheng, J.J., Zhao, Y.Q. and Liao, J.S. (1985) A study on the variation of nutrient element contents in citrus leaves in different phenological stages. *Journal of Fujian Agricultural College* 14, 31–36.
Maas, E.V. (1993) Salinity and citriculture. *Tree Physiology* 12, 195–216.
Maatouk, M.A, Ahmed, F.F. and El-Sayed, M.A. (1988) Effect of nitrogen, potassium and phosphorus fertilization on yield and quality of Egyptian Balady lime trees (*Citrus aurantifolia*). I. Vegetative growth and chemical composition of leaves. *Annals of Agricultural Science* 33, 1233–1247.
Madrid, R., Lacomba, R.F., Egea, C. and Amo, M.L. (1995) Effect of irrigation systems on fruit quality of orange cv (Valencia late): evaluation of chemical parameter. *ITEA Production Vegetable* 91, 197–206.
Marin, F.R. and Angelocci, L.R. (2011) Irrigation requirements and transpiration coupling to the atmosphere of a citrus orchard in Southern Brazil. *Agricultural Water Management* 98, 1091–1096.
Martín de Santa Olalla, F. and De Juan, J.A. (1993) *Agronomía del riego*. Mundi-Prensa, Madrid.
Martin, E.C., Hla, A.K., Waller, P.M. and Slack, D.C. (1997) Heat unit-based crop coefficient for grapefruit trees. *Applied Engineering in Agriculture* 13, 485–489.
Morgan, K.T., Obreza, T.A., Scholberg, J.M.S., Parsons, L.R. and Wheaton, T.A. (2006) Citrus water uptake dynamics on a sandy Florida Entisol. *Soil Science Society of America Journal* 70, 90–97.
Morgan, K.T., Zotarelli, L. and Dukes, M.D. (2010) Use of irrigation technologies for citrus trees in Florida. *HortTechnology* 20, 74–81.
Nair, P.C.S. and Mukherjee, S.K. (1970a) Studies of zinc deficiency in citrus. I. Causes of seasonal occurrence of zinc deficiency. *Horticultural Research* 10, 1–13.
Nair, P.C.S. and Mukherjee, S.K. (1970b) Effect of zinc sprays on checking chlorosis and dieback symptoms on citrus. *Indian Journal of Agricultural Sciences* 40, 379–388.
Ortuño, M.F., Brito, J.J., Conejero, W., García-Orellana, Y. and Torrecillas, A. (2009) Using continuously recorded trunk diameter fluctuations for estimating water requirements of lemon trees. *Irrigation Science* 27, 271–276.
Pedrero, F., Allende, A., Gil, M.I. and Alarcón, J.J. (2012) Soil chemical properties, leaf mineral status and crop production in a lemon tree orchard irrigated with two types of wastewater. *Agricultural Water Management* 109, 54–60.
Pereira, A.R. and Villa Nova, N.A. (2009) Transpiration of irrigated Tahiti acid lime trees with different canopy leaf areas, reference evapotranspiration and water balance for precision irrigation schemes. *Tree and Forestry Science and Biotechnology* 3, 87–91.
Plessis, S.F.D. (1985) Irrigation of citrus. *Citrus and Sub Tropical Fruit Journal* 614, 12.
Prior, L.D., Grieve, A.M., Bevington, K.B. and Slavich, P.G. (2007) Long-term effects of saline irrigation water on Valencia orange trees: relationships between growth and yield, and salt levels in soil and leaves. *Australian Journal of Agricultural Research* 58, 349–358.
Rathore, R.S. and Chandra, A. (2001) Effect of application of nitrogen in combination with zinc sulphate on nutrients content, quality and yield of acid lime (*Citrus aurantifolia* Swingle) cv. Kagzi lime. *Current Agriculture* 25, 107–110.
Rawash, M.A., Bondok, A. and El-shazly, S. (1983) Leaf mineral content of lime trees affected by foliar application of chelated zinc and manganese. *Annals of Agricultural Science* 28, 1021–1031.
Sema, A., Kar, P.L. and Sanyal, D. (1999) Influence of flushes, age of leaf and types of shoot on the fluctuation of foliar micronutrient levels in Assam lemon. *Horticulture Journal* 12, 13–19.
Sepaskhash, A.R. and Kashefipour, S.M. (1995) Evaporation and crop coefficient of sweet lime under drip irrigation. *Agricultural Water Management* 27, 331–340.
Shirgure, P.S. and Panchariya, G.S. (2012a) Auto-drip irrigation scheduling with e-6 hybrid station controller and production of Nagpur mandarin. *The Third National Conference on Agro-Informatic and Precision*. Allied publishers.
Shirgure, P.S. and Panchariya, G.S. (2012b) Automation of micro-jet irrigation systems and production of Nagpur mandarin (*Citrus Reticulata* Blanco). *Proceeding of AIPA*, pp. 190–193.
Shirgure, P.S. and Srivastava, A.K. (2012) Fertigation in citrus. Technical Bulletin No. 16, National Research Centre for Citrus, Nagpur, Maharashtra, India, pp. 1–90.
Shirgure, P.S. and Srivastava, A.K. (2014) Fertigation in perennial fruit crops: major concerns. *Agrotechnology* 3, 1 Available at: http://dx.doi.org/10.4172/2168-9881.1000e109.

Shirgure, P.S., Srivastava, A.K. and Singh, S. (2001a) Effect of nitrogen fertigation and band placement fertilizer application on soil-leaf nutrient build-up and incremental growth of acid lime. *Journal of Soil and Water Conservation* 45 (3 and 4), 176–181.

Shirgure, P.S., Srivastava, A.K. and Singh, S. (2001b) Effect of drip, microjets and basin irrigation method on growth, soil and leaf nutrient change in acid lime. *Indian Journal of Soil Conservation* 29, 229–234.

Shirgure, P.S., Srivastava, A.K. and Singh, S. (2002a) Economics of drip and fertigation in acid lime orchards. *Journal of Soil and Water Conservation* 46, 56–60.

Shirgure, P.S., Srivastava, A.K. and Singh, S. (2002b) Effect of micro-irrigation systems and basin irrigation on growth and yield of acid lime (*Citrus aurantifolia* Swingle). *Indian Journal of Citriculture* 1, 154–161.

Shirgure, P.S., Srivastava, A.K. and Singh, S. (2002c) Water management in citrus. Technical Bulletin No. 6, National Research Centre for Citrus, Nagpur, Maharashtra, India, pp.1–170.

Shirgure, P.S., Srivastava, A.K. and Singh, S. (2004a) Growth, yield and quality of acid lime under pan evaporation based drip irrigation scheduling. *Indian Journal of Soil Conservation* 32, 32–35.

Shirgure, P.S., Srivastava, A.K. and Singh, S. (2004b) Integrated water and nutrient management in acid lime. *Indian Journal of Soil Conservation* 32, 148–151.

Simpson, G.H. (1978) Developments in under tree irrigation systems in the Murray valley. *Proceedings of the International Society of Citriculture* 1, 234–235.

Singh, S.B. and Misra, R.S. (1986) Effect of plant growth regulators and micronutrients on fruit drop, size and quality of Kinnow orange. *Progress in Horticulture* 18, 260–264.

Singh, A.R., Maurya, V.N., Pande, N.C. and Rajput, R.S. (1989) Role of potash and zinc on the biochemical parameters of Kagzi lime (*Citrus aurantifolia* Swingle). *Haryana Journal of Horticultural Science* 18, 46–50.

Singh, G., Nath, V., Pandey, S.D., Ray, P.K. and Singh, H.S. (2012) *The Litchi. Water Requirements and Irrigation*. Food and Agriculture Organization of the United Nations, New Delhi, India.

Socalo, V. and Guzman, L. (1986) Annual variations in the nitrogen, phosphorus, and potassium content of leaves from a lime (*Citrus aurantifolia*) on the coast of Peru. *Turrialba* 36, 25–31.

Srivastava, A.K. (2009) Integrated nutrient management: concept and application in citrus. In: Tennant, P. and Beakebhia, N. (eds) *Citrus II. Tree and Forestry Science and Biotechnology*. 3, pp. 32–58.

Srivastava, A.K. (2012a) Nutritional disorders in citrus: morphological descriptors. Extension Bulletin No. 39. National Research Centre for Citrus, Amravati Road, Nagpur, Maharashtra, India, pp.1–32.

Srivastava, A.K. (2012b) Integrated nutrient management in citrus. In: Srivastava, A.K. (ed.) *Advances in Citrus Nutrition*. Springer, the Netherlands, pp. 369–390.

Srivastava, A.K. (2013a) Nutrient deficiency symptomology in citrus: an effective diagnostic tool or just an aid for post–mortem analysis. *Agricultural Advances* 2, 177–194.

Srivastava, A.K. (2013b) Recent developments in diagnosis and management of nutrient constraints in acid lime. *Agricultural Advances* 2, 86–96.

Srivastava, A.K. (2013c) Site specific nutrient management in citrus. *Agricultural Advances* 2, 1–15.

Srivastava, A.K. (2013d) Early warning system for plant nutrient deficiency: future Toolbox. *Agrotechnology* 2, 3. Available at: http://dx.doi.org/10.4172/2168-9881.1000e106 (Editorial).

Srivastava, A.K. (2013e) Nutrient diagnostics in citrus: are they applicable to current season crop. *Agrotechnology* 2, 104–105.

Srivastava, A.K. and Ngullie, E. (2009) Integrated nutrient management: theory and practice. *Dynamic Soil, Dynamic Plant* 3, 1–30.

Srivastava, A.K. and Patil, P. (2014) Soil fertility indexing for acid lime growing smectite soil. *Annals of Plant and Soil Research* 16, 25–28.

Srivastava, A.K. and Shirgure, P.S. (2014) Nutrient use efficiency in fruit crops: future strategy. In: Chadha, K.L., Kalia, P. and Singh, S.K. (ed.) *Horticulture for Inclusive Growth. The Horticultural Society of India*. Westville Publishing House, New Delhi, pp. 292–313.

Srivastava, A.K. and Singh, S. (1998) Fertilizer use efficiency in citrus. Technical Bulletin No. 1, National Research Centre for Citrus, Nagpur, Maharashtra, India, pp. 1–66.

Srivastava, A.K. and Singh, S. (2002) *Citrus: Climate and Soil*. International Book Distribution Company, Lucknow, Uttar Pradesh, India, pp. 1–559.

Srivastava, A.K. and Singh, S. (2003a) *Citrus Nutrition*. International Book Distribution Company, Lucknow, Uttar Pradesh, India, pp. 1–960.

Srivastava, A.K. and Singh, S. (2003b) Diagnosis of nutrient constraints in citrus. Manual No. 2. National Research Centre for Citrus, Nagpur, Maharashtra, India, pp. 1–70.

Srivastava, A.K. and Singh, S. (2004) Zinc nutrition, a global concern for sustainable citrus production. *Journal of Sustainable Agriculture* 25, 5–42.

Srivastava, A.K. and Singh, S. (2008a) Nutrient diagnostic and management in citrus. Technical Bulletin No. 8, National Research Centre for Citrus, Nagpur, Maharashtra, India, pp. 1–130.
Srivastava, A.K. and Singh, S. (2008b) Micronutrients in citrus. Technical Bulletin No. 13, National Research Centre for Citrus, Nagpur, Maharashtra, India, pp. 1–172.
Srivastava, A.K. and Singh, S. (2008c) Site specific nutrient management in Citrus. Technical Bulletin No. 10, National Research Centre for Citrus, Nagpur, Maharashtra, India, pp. 1–130.
Srivastava, A.K. and Singh, S. (2008d) Citrus nutrition research in India: problems and prospects. *Indian Journal of Agricultural Sciences* 78, 3–16.
Srivastava, A.K. and Singh, S. (2009) Citrus decline: soil fertility and plant nutrition. *Journal of Plant Nutrition* 32, 197–245.
Srivastava, A.K., Singh, S. and Tiwari, K.N. (2006) *Citrus: A Practical Guide to Nutrient Management.* National Research Centre for Citrus, Nagpur and Potash and Phosphate Institute of Canada India Programme, Haryana, India, pp. 1–24.
Srivastava, A.K., Singh, S. and Tiwari, K.N. (2007) Diagnostic tools for citrus: their use and implications in India. *Better Crops – India* 1, 26–29.
Srivastava, A.K., Singh, S. and Albrigo, L.G. (2008) Diagnosis and remediation of nutrient constraints in citrus. *Horticultural Reviews* 34, 277–63.
Storey, R. and Walker, R.R. (1999) Citrus and salinity. *Scientia Horticulturae* 8, 39–81.
Tuzucu, O., Cevik, B., Kaplankiran, M., Yesiloglu, T. and Eebeci, Z. (1997) The effect of different irrigation methods on root distribution of Kudi-ken lemon cv in Adana, Turkey. *Proceedings of the 5th World Congress of the International Society of Citrus Nurseryman, Montepellier, France*, 5–8 March 1997.
Varalakshmi, L.R. and Bhargava, B.S. (1998) Plant nutrient status and leaf diagnostic norms for acid lime (*Citrus aurantifolia*). *Indian Journal of Agricultural Sciences* 68, 375–376.
Willis, L.E., Davies, F.S. and Graetz, D.A. (1990) Fertilization, nitrogen leaching, and growth of young Hamlin orange trees on two rootstocks. *Proceedings of the Florida State Horticultural Society* 103, 30–37.
Willis, L.E., Davies, F.S. and Graetz, D.A. (1991) Fertigation and growth of young Hamlin orange trees in Florida. *Scientia Horticulturae* 26, 106–109.
Zhang, M., Alva, A.K., Li, Y.C. and Calvert, D.V. (1996) Root distribution of grapefruit trees under dry granular broadcast versus fertigation method. *Plants and Soil* 183, 79–84.

8 Cultural Practices

Dalmo Lopes de Siqueira*, Luiz Carlos Chamhum Salomão and César Fernandes Aquino
Departamento de Fitotecnia, Universidade Federal de Viçosa, Minas Gerais, Brazil

Tahiti lime (*Citrus latifolia* (Yu. Tanaka) Tanaka) is a widely cultivated species among all lime species, therefore, the information presented in this chapter is mainly focused on this species. Brazil was the fifth largest producer of lemons and limes in 2012, with a production of 1.2 million tonnes in 47 thousand hectares of harvested area (FAOSTAT, 2014). The average yield is 25.5 t/ha, and the state of São Paulo accounts for 76% of the total production of Tahiti lime.

Tahiti lime trees are medium to large sized, present vigorous growth and few thorns. In tropical regions, such as the Brazilian semi-arid region, the growth occurs in continuous flushes with several blooms and several annual crops. Owing to these characteristics, practices aimed at containment of the growth of the tree, as well as intensive and balanced supply of mineral nutrients are essential to achieve high yield and fruit quality, whether for the local market or export.

Planting Density

Fruit production generally occupies large areas of land because of the large spacing required by trees. To compensate for this characteristic, it is necessary to increase the productivity and profitability of orchards. Denser orchards generally provide higher profitability than the less dense ones, to the extent that the increase in revenue is greater than the increase in production costs (Paes and Esperancini, 2006).

Selecting the correct spacing used in the establishment of citrus orchards is important due to the perennial nature of the trees and also particularly for trees propagated by grafting (Teófilo Sobrinho *et al.*, 2012), whose size depends on the rootstock used.

Most orchards of Tahiti lime in Brazil have Rangpur lime (*Citrus limonia* Osbeck) as their rootstock, which induces the canopy to reach a large volume in a few years after planting, with canopies overlapping in the planting rows (Neves *et al.*, 2004). Trees with this combination of scion and rootstock can reach over 4 m in height, with a canopy diameter larger than 6 m and a canopy volume that exceeds 120 m^3 (Fig. 8.1). Stuchi *et al.* (2002) found that Tahiti lime tree at 9 years old budded on 'Cleopatra' mandarin rootstock (*Citrus reshni* hort. ex Tanaka) reached 4.5 m in height, 6.0 m in canopy diameter, with a volume and area of canopy projection of 87.9 m^3 and 29.2 m^2, respectively. This height reduces the incidence of solar radiation on the branches and leaves below the canopy, resulting in the reduction in fruit yield on these branches and concentrating the production of fruit at the

*E-mail: siqueira@ufv.br

Fig. 8.1. Tahiti lime trees budded on Rangpur lime, presenting canopy overlapping in the planting line due to the vigorous growth.

top of the tree (Whitney and Wheaton, 1984), increasing the time and cost of harvesting, since the fruit on the highest parts require the use of ladders to be harvested.

Under these conditions, trees require large spacing for development and production. Therefore, 8 × 5 m or 8 × 6 m spacings are recommended. With the use of mechanized pruning, it is possible to reduce the spacing between planting rows to 7 m or less (Figueiredo *et al.*, 2002; Medina and Silva, 2003; Stenzel and Neves, 2004), facilitating the management of orchards. In the semi-arid climatic conditions of Brazil, clones of Tahiti lime, IAC-5 and Quebra-galho at 4 years old, spaced 6 × 5 m, both budded on Rangpur lime, reached 4.5 and 2.8 m height, respectively. Despite the broad spacing, the canopies of the trees already reached each other towards the planting line (Fig. 8.1(A)–(C)), resulting in the need for pruning.

The use of dwarfing rootstocks such as 'Flying Dragon' (*Poncirus trifoliata* (L.) Ralf. var. *Monstrosa*) allows a higher planting density, which facilitates management and harvesting. In this case, the recommended spacing ranges from 6 × 1 m to 6 × 3 m.

Increasing the planting density of lime orchards has increased productivity, giving a faster return on invested capital, easier harvests, better use of fertilizers and sanitary treatments of the trees, earlier stabilization of production from the orchard, and rendering it unnecessary to replace dead trees. However, some of the factors identified as disadvantages of high planting densities include: the high cost of nursery trees for orchard planting, the need for irrigation in years of drought and of pruning when there is excessive tree growth, the increased competition for water and nutrients and the lack of knowledge of the ideal combination of rootstock and scion for high density planting (Donadio and Stuchi, 2001; Teófilo Sobrinho *et al.*, 2012).

The use of high density planting in citrus provides considerable increases in yield (Donadio and Stuchi, 2001; Teófilo Sobrinho *et al.*, 2002; Grizotto *et al.*, 2012; Teófilo Sobrinho *et al.*, 2012; Machado, 2014). Although fruit production per tree is higher in less dense plantings, due to the lower number of trees per area, yield per area is lower. The opposite occurs at higher density. Furthermore, because of the lower yield per plant, the longevity of trees at reduced spacing tends to be greater (Teófilo Sobrinho *et al.*, 2012; Machado, 2014) with fewer issues related to alternate bearing.

It seems that increasing the number of trees per area does not affect fruit quality, i.e. the fruit produced in less dense orchards are of similar quality to those produced in denser ones (Donadio and Stuchi, 2001; Grizotto *et al.*, 2012; Teófilo Sobrinho *et al.*, 2012).

In northern Minas Gerais state, in a semi-arid climate region, Machado (2014) evaluated different planting densities for Tahiti lime, IAC-5 clone, budded on *P. trifoliata* (L.) Ralf. var. *Monstrosa* Flying Dragon using the following spacing: (i) 6×1.0 m (1666 trees/ha); (ii) 6 × 1.5 m (1111 trees/ha); (iii) 6 × 2.0 m (833 trees/ha); (iv) 6 × 2.5 m (666 trees/ha); and (v) 6 × 3.0 m (555 trees/ha). Trees were drip irrigated. Five years after planting, the different spacing used did not alter the size of the trees. Trees spaced at 6 × 1 m and 6 × 1.5 m showed yields 75 and 79% higher, respectively, than those spaced at 6 × 3.0 m, resulting in increased yield per unit area (Table 8.1). No changes were found in the size of the fruit, in the levels of citric acid, in the content of soluble solids, in the juice yield and in the Vitamin C levels.

In the State of São Paulo, south-eastern Brazil, Donadio and Stuchi (2001) also evaluated planting densities for Tahiti lime budded on *P. trifoliata* (L.) Ralf. var. *Monstrosa* Flying dragon, using the following spacings: (i) 4 × 1.0 m (2500 trees/ha); (ii) 4 × 1.5 m (1666 trees/ha); (iii) 4 × 2.0 m (1250 trees/ha); and (iv) 4 × 2.5 m (1000 trees/ha). Sixty-six months after planting, trees spaced by 4 × 1.0 m reached the greatest height, the largest canopy diameter and the highest yield (2.54 m; 2.75 m and 22 t/ha, respectively). On the other hand, yield was about 41% lower in trees spaced by 4 × 2.5 m. The different spacing did not affect the content of soluble solids and fruit weight.

For other citrus species, the results are similar to those presented for Tahiti lime in high density tree orchards (Teófilo Sobrinho *et al.*, 2002; Grizotto *et al.*, 2012; Teófilo Sobrinho *et al.*, 2012).

Pruning

Studies on the effect of pruning on Tahiti lime are scarce; however, information is available in

Table 8.1. Mean values of tree height (H) (m), canopy diameter (CD) (m), canopy volume (CV) (m^3), production per tree (PT) (kg/tree), fruit weight (FW) (g), number of fruit per tree (NFT), yield (Y) (t/ha/year) in 5-year-old trees of Tahiti lime tree budded on *P. trifoliata* (L.) Ralf. var. *Monstrosa* Flying Dragon in semi-arid climate conditions in Minas Gerais, Brazil.

Spacing	H	CD	CV	PT	FW	NFT	Y
6 × 1.0 m	2.34 a	2.66 a	10.89 a	13.22 b	80.32 a	165 b	22.03 a
6 × 1.5 m	2.44 a	2.79 a	12.31 a	18.06 b	78.09 a	231 a	20.07 ab
6 × 2.0 m	2.39 a	2.55 a	10.18 a	22.17 a	81.94 a	272 a	18.48 b
6 × 2.5 m	2.48 a	2.79 a	12.55 a	20.84 a	79.21 a	263 a	13.89 c
6 × 3.0 m	2.33 a	2.70 a	11.91 a	21.11 a	81.99 a	258 a	11.73 c

Note: Means followed by the same letter in the column do not differ from each other according to the Tukey test at 5% of probability.
Source: Adapted from Machado, 2014

the literature for other citrus species, such as sweet orange and mandarin, which can be taken into account in the management of Tahiti lime pruning.

Types of pruning

Tree training

In Brazil, nursery trees of Tahiti lime can be commercialized in single stems (Fig. 8.2(A)) or with three to four mature branches, which will form the base of the tree structure (Fig. 8.2(B)). In the latter type, pruning is carried out in the nursery and the main stem is reduced to 0.30–0.60 m in height, and three to four shoots are selected on the 0.20 m distal end. In the single-stem nursery tree, pruning is carried out in the nursery at 0.30–0.60 m height (Normative Instruction 48, 2013). All shoots that arise in the rootstock are eliminated in both types of nursery trees when the buds are still tender, facilitating its removal.

After planting in the field, the single-stem nursery trees emit various sprouts above the budding point. When these sprouts reach approximately 0.10–0.15 m in length, they start thinning, leaving three to four sprouts radially opposed and equally spaced about 0.05 m from each other to form the canopy.

The most vigorous branches must be selected to avoid tipping and breakage. These branches must be very well constructed since they will ensure the maximum capture of solar radiation, they will bear the entire aerial part, including fruit, and withstand winds during the cultivation cycle.

Sprouts in the rootstock are vigorous and develop rapidly using large amounts of nutrients and water, without contributing to the production capacity of the tree; it is necessary to eliminate them periodically when they are still tender.

Cleaning pruning

The objective of the cleaning pruning is to improve fruit quality, to minimize alternate bearing, to control vegetative growth, to stimulate the formation of new fruiting branches, and to increase luminosity and ventilation within the canopy (Oliveira *et al.*, 2011; Santarosa *et al.*, 2013).

In orchards with high densities of trees, issues related to productivity, especially when the fruit are intended for fresh consumption, will arise over time. With the vegetative development of the tree, overlapping canopy and tree competition for sunlight will happen, causing reduction in the yield of the orchards, caused by a reduction in the photosynthetic rate in the inner canopy (Santarosa *et al.*, 2010; Yıldirim *et al.*, 2010).

In trees where sunlight penetration is reduced, production occurs almost exclusively on the outer branches of the tree. Furthermore, the canopy remains for long periods without adequate ventilation, creating conditions for the development of diseases and pests, and hindering the entry of predators and parasitoids, which act to reduce the population of insect pests. Another inconvenience is the difficulty in performing

Fig. 8.2. Tahiti lime nursery trees commercialized in Brazil. (A) single-stem nursery tree, and (B) nursery tree with three ripe stems arranged radially.

phytosanitary treatments, demanding higher frequency of sprays and spray volumes for tree wetting (Andrade *et al.*, 2013; Van Zyl *et al.*, 2014.). This condition will have negative effects on the productivity, and will raise production costs as well.

Cleaning pruning consists of removing all branches that are dry, diseased, broken, damaged by pests or badly positioned in the canopy, as well as the excessively vigorous or unproductive shoots or sprouts arising in the rootstock. Badly located and unproductive branches contribute to the weakening of the tree and they may reduce production as they are strong drains due to excessive vigour and they can shade branches with productive potential.

Dry branches and twigs should be cut close to the trunk to facilitate fast tissue healing. However, sick or plagued branches should be cut to the healthy part of the tree, below the infected part (Carvalho *et al.*, 2005b).

This type of pruning should be performed annually, preferably in winter. However, branches severely damaged by diseases or pests should be eliminated at any time of the year. Pruning should be carried out carefully, especially in young trees because drastic pruning in young citrus trees with a vigorous growing habit stimulates excessive growth of branches, therefore, extending the juvenile phase and delaying fruiting (Intrigliolo and Roccuzzo, 2011).

Tree size control

The main objective of pruning is to reduce the height and diameter of the canopy, to fit it to the spacing adopted. Topping aims to reduce tree height. It removes the branches located in the upper part of the trees and is performed before the trees become excessively high (Fig. 8.3). Hedging is to reduce the diameter of the trees along the rows to prevent or reduce overlapping of the canopy and to maximize the utilization of solar radiation. The more vigorous the rootstock and the denser the planting, the greater the need for frequent pruning. If not carried out at the correct time and if there is a complete overlap of canopies, the removal of most of the branches of the trees will be required, reducing the production of the current and the next year (Tucker *et al.*, 1994).

Topping and hedging are usually performed with circular saws mounted on an adjustable

Fig. 8.3. Tahiti limes trees taller than 4 m and overlapping canopies inter-rows.

mechanical arm coupled to a tractor (Fig. 8.4); however, this does not preclude manual complementation to remove excess new sprouts that will arise after the cut, as well as suckers and dried branches, maintaining adequate lighting for the internal part of the tree (Medina and Silva, 2003).

Pruning should be carried out with saws inclined in such a manner that the canopy acquires the shape of frustum; that is, narrower at the top than at the bottom of the tree, allowing higher incidence of solar radiation on the lower portion of the trees. Topping must be made at angles ranging from 15–30° to the horizontal, and hedging at angles ranging from 5–15° relative to the vertical tree axis (Tucker *et al.*, 1994; Azevedo *et al.*, 2013). Tree size control can be performed every 2 or more years, always after fruit harvest (Oliveira *et al.*, 2011).

Maintenance of smaller trees can reduce production costs by facilitating the harvesting operation and the removal of the fruit of the orchard, and reducing costs of phytosanitary management due to the reduced number of applied sprays and the spray volume (Van Zyl *et al.*, 2014).

Rejuvenation pruning

Rejuvenation pruning is performed to recover trees with low production, or those that have suffered the action of strong winds, severe attack of pests and/or diseases (Loussert, 1992). In such cases, pruning is required to force the tree to emit new productive branches. It ranges in intensity, from mild thinning of the canopy to the complete removal of the tree canopy.

The thinning of the canopy or partial rejuvenation of the branches is performed on the oldest or weakest branches, unable to bear fruit. The removal of these branches allows greater light penetration into the canopy, promotes bud sprouting and favours new growth. The selection of these new branches should be performed only about 6 months after the onset of sprouting, when the branches are partially lignified, so

Fig. 8.4. Top pruning in Tahiti lime trees.

that they will not break because of the wind (Oliveira *et al.*, 2011). Regarding complete removal of the canopy, only the trunk and about 0.5 m of each primary branch are left, eliminating all foliage of the canopy. This should be done in winter to minimize sun burns on the trunk and on the remaining branches. Furthermore, after pruning, the trunk and branches must be protected by painting them with hydrated lime.

Operations to rejuvenate the trees have lower costs and shorter time intervals for production than the planting of new orchards. However, the phytosanitary condition of the trees to be pruned has to be evaluated because this pruning is only advantageous if there is potential for full recovery of the trees, otherwise the recommendation is to form a new orchard.

The rejuvenation pruning has variable success, depending on the extent of the decline of the tree and on the quality of other cultural practices employed in the orchard (Tucker *et al.*, 1994), such as soil correction, fertilization, irrigation and phytosanitary control to promote the balance between the aerial parts and root system, helping the recovery of the trees.

To perform the pruning, it is necessary to use appropriate tools, which should be disinfected with a solution of sodium hypochlorite at 20 g/l when going from one tree to another, to avoid the transmission of diseases. Furthermore, in all pruning performed in the trees, the protection of the cuts is very important, to avoid the entry of pathogens. Protection is made through the brushing of the cuts with a copper-based solution (Medina and Silva, 2003).

Weed Management

Unlike annual crops, in which the short cycle and the short height imply a high sensitivity to weeds, perennial citrus trees are relatively tolerant to the presence of other species in the same area, after the establishment of the orchard (Theisen and Theisen, 2011). However, caution is needed to diagnose the time when weeds are causing negative interference in the target trees.

Weeds compete for water, light, nutrients and CO_2 (Carvalho *et al.*, 1993; Neves *et al.*, 1998; Carvalho *et al.*, 2002). In orchards where proper weed management is not carried out at the correct time, weeds will develop. The weeds grow extensively in the soil portion where fertilizers are distributed to compete for the nutrients (Durigan and Timossi, 2002). The degree of this competition depends on the species, incidence and time of year (Carvalho *et al.*, 1993). In addition, another major negative aspect of some weed species, such as climbing ones, is that if they are not controlled properly, they can delay fruit harvest, which results in the reduction of yield.

Due to the large spacing used in citrus orchards, an extensive surface of soil may be vegetation-free when weeds are completely eliminated from the area. The absence of vegetation leaves the soil with no protection, exposing it to the direct action of the rain, wind and sun, leading to considerable loss of soil by erosion.

In addition, soils with no vegetation cover, machinery traffic and soil management practices contribute to increase compaction and degradation, reducing macro porosity and water availability. This environment restricts the development of the root system, directly or indirectly reducing the absorption of nutrients by trees (Neves *et al.*, 1998; Carvalho *et al.*, 2002; Auler *et al.*, 2008), which is one of the main causes of drop in the yield and the short longevity of orchards (Carvalho *et al.*, 2005b; Auler *et al.*, 2008).

The presence of weeds in citrus orchards should not be regarded as a purely negative factor. If handled properly and not totally eradicated from the orchard, they indirectly benefit the citrus trees, with improvement in the chemical and biological characteristics of the soil, contributing to good development of the root system of the trees (Auler *et al.*, 2008). The presence of plant cover or dead plants in the orchard protects the soil surface from erosion, preventing the breakdown of particles, improving water infiltration, while retaining moisture and avoiding sudden temperature changes (Medina and Silva, 2003; Theisen and Theisen, 2011). After mowing, there is a partial increase in fertility of soil due to the presence of weed residues in the orchard and after decomposition this promotes a partial return of the nutrients to the soil (Durigan and Timossi, 2002). In addition, they can also attract and maintain natural enemies of pests in the orchard (Silva *et al.*, 2010). Minatel *et al.* (2006) recommend direct sowing of green manure (*Crotalaria spectabilis*) on inter-rows of Valencia orange orchards to mitigate the harmful effects of soil compaction.

Carvalho *et al.* (1999) compared two systems for weed control in an orange Pera (*Citrus sinensis* L. Osbeck) orchard, the first ever used by the farmer. The first system consists of three hoeings on the row and three harrowing on inter-rows per year and the second system proposed the use of glyphosate in rows and inter-rows, planting *Canavalia ensiformis* L. as a plant cover and subsoiling at 0.55 m depth when required. The improvement of soil structure caused by the *C. ensiformis* management in the rows, associated with subsoiling, and control of weeds with glyphosate resulted in higher and better root growth, with 19.3% of the roots found between 0 and 0.2 m of soil depth, while this percentage was 69.9% in the system based on hoeing and harrowing. However, 24.8% of the roots were found between 1.0 and 1.2 m of soil depth in the second system, whereas only 7.69% were found in the first system, showing a lack of a deeper root system for this treatment. Furthermore, an increase of 40% was found in fruit yield in the second system.

These results show that the root systems of citrus species, in proper conditions of soil management, are able to drill down into the soil profile, absorbing water and nutrients, reducing the effects caused by competition with weeds.

Competition critical period

The period immediately after planting the Tahiti nursery trees in the field is one of the most critical for competition of weeds with lime trees, and can delay the start of production. It is during this phase that loss of some trees in the orchard occurs due to the competition with weeds for water, nutrients and light (Fig. 8.5), especially in a period with low water availability because the root system of the Tahiti nursery trees still occupies a small range of land and the size of the trees is relatively small, with reduced photosynthetic surface (Theisen and Theisen, 2011).

Fig. 8.5. High weed infestation in a young orchard of Tahiti lime trees.

Excessive shading on the Tahiti nursery trees due to the accelerated growth of weeds is another limiting factor. In addition, many pests that attack young Tahiti nursery trees can take shelter in the weeds.

The competition for growth factors (water, CO_2, nutrients and light) between weeds and Tahiti lime trees occurs mostly during periods when one or more of these factors is scarce and coincides with critical stages of the crop. Generally, the greatest limitation is water for non-irrigated orchards in the dry season. Thus, the most appropriate management is to remove weeds during the period with little rain (when not irrigated) to reduce competition for moisture. On the other hand, in the rainy season, the grass should be kept in inter-rows, forming a cover to avoid erosion. In the dry season, the weeds must be kept, preferentially desiccated with herbicide or mowing (Durigan and Timossi, 2002).

In north-eastern Brazil, in semi-arid climate conditions, an orchard of Pera orange (*C. sinensis* L. Osbeck) budded on Rangpur lime was evaluated for 4 years. Carvalho *et al.* (1993) demonstrated that when the availability of water in the soil decreases, and water deficiency starts to increase, the orchard should be maintained with no interference with the weeds. This situation should be maintained until the water storage in the soil increases. The authors found an increase in productivity when weeds were managed in the dry season only, without increasing the production costs.

Weed management types

Mechanical management

Mechanical methods to control weeds are a very helpful tool for farmers in areas where orchard production is most restricted in terms of the use of pesticides. The main objective in mechanical management is to reduce the population of weeds by hoeing and mowing. Nowadays, it is recommended not to eradicate them because some species, when well-managed, are important allies for the management of soil and pests, since they shelter their natural enemies (Durigan and Timossi, 2002).

Hoeing is costly and a low operating income method, therefore it is only used in orchards in training or on small farms. It is very useful to keep the surroundings weed-free after the first year of planting. Alternatively the weeds should be mowed at a low height in a radius of 0.5–1 m as the young trees are very sensitive to certain herbicides.

During hoeing, one should be careful not to injure the trunks of young trees exposing them to the attack of pathogens such as *Phytophthora* spp. When hoeing is complete, it is recommended that the weed debris should be left in the soil, for protection against erosion, water loss and new infestations of weeds in the area of canopy projection (Durigan and Timossi, 2002; Medina and Silva, 2003). The best results for the control of weeds by hoeing are achieved when the weather is dry.

Mechanical weeding should be performed to reduce production costs and can be performed with a central or lateral mounted rotary cutter, depending on the extent of the orchard, the conditions and ground slope, and on the weed species present in the area.

Orchard mowing should be performed before the critical period for Tahiti lime trees, particularly in relation to water availability. Theisen and Theisen (2011) indicate that mowing should be carried out in the spring and summer, to reduce the biomass of weeds that stand out. However, in the summer mowing should be more frequent (Medina and Silva, 2003) to avoid the production of seeds, as they may germinate or remain dormant in the soil to re-infect the area.

Central mounted rotary cutters can be used for the management of weeds and green manure between the rows of orchards, leaving crop residues to form mulch. Lateral mounted rotary cutters can be used in parallel to the trunks of the trees under the canopy, and the cut material should be kept on the soil surface, forming an organic matter layer, which will act in weed management and improvement of the physical, chemical and biological characteristics of the soil (Durigan and Timossi, 2002; Matheis *et al.*, 2006). On the other hand, the continuous use of this equipment may lead to the selection of some weed species, mainly those with perennial cycles and, in particular, grasses with asexual propagation and prostrate growth habits (Durigan and Timossi, 2002). In some cases, integration with other forms of control, such as herbicides will be needed.

Green manure and mulching

Green manure is the incorporation of undecomposed plant matter, locally produced or not, in the soil. This operation results in desirable changes in the soil, such as the supply of mineral nutrients, prevention of erosion and water retention in the soil. Moreover, it can be a tool for integrated weed management (Neves *et al.*, 1998; Severino and Christoffoleti, 2001; Fidalski and Stenzel, 2006; Matheis *et al.*, 2006).

The planting of grasses and legumes or the use of plant species present in the planting rows of Tahiti lime trees can be carried out as part of weed management. The planting of fast-growing species will cover the soil to inhibit seed germination of the weeds (Sediyama *et al.*, 2010). In addition, it suppresses growth by mulching and by competition for light, water and mineral nutrients, and the release of chemicals that may affect their germination or development (Matheis *et al.*, 2006).

Carvalho *et al.* (2001) argued that the use of cover crops without using harrows, the planting of legumes in inter-rows and the replacement of hoeing by herbicide glyphosate, along with subsoiling when needed, is the best management for weed control in citrus.

The planting of the legumes *Crotalaria juncea* L., *Crotalaria spectabilis* Roth, *Cajanus cajan* L. Millsp, *Mucuna aterrima*, *Mucuna deeringiana*, *Dolichos labe-labe* L. and *Canavalia ensiformis* L. DC, besides acting in the management of the weeds in the orchard area, supplied the soil with 6.55, 1.23, 3.42, 1.78, 1.75, 1.61 and 3.03 t/ha of dry matter for the soil, respectively, representing the incorporation of considerable amounts of macro- and micronutrients.

Fidalski *et al.* (2006) found in sweet orange Pera, in 4 years of study, that the fruit production was not altered by the management of the orchard inter-rows with permanent coverage of *Paspalum notatum* plus mechanical mowing, or with *Arachis pintoi* without mechanical mowing or weed control with glyphosate. Weed control in the rows was done with glyphosate and hoeing.

However, the management of *P. notatum* in the rows ensured greater soil moisture and a higher rate of photosynthesis in the orange tree when compared with the legume *A. pintoi*. The results in terms of fruit production support the recommendation to maintain permanent vegetation with grass in the inter-rows of orange orchards. Moreover, the maintenance of the grass *P. notatum* as permanent vegetation in inter-rows improves water storage capacity and soil aeration, compared with the management using *A. pintoi* and the traditional post-emergent herbicide.

Neves *et al.* (1998), after 9 years evaluating perennial legumes *Indigofera campestris* Benth, *Arachis prostrata* Bong. ex Benth. and annual legume (*Stizolobium pruriens*) in the rainy season, with alternate use of a harrow in the dry season and three mowings in the rainy season and hoeing when necessary, in a Poncã mandarin orchard (*C. reticulata* Blanco), concluded that the production, fruit quality and nutrition of the trees were not significantly affected by the treatments. Among the soil chemical properties, only the organic matter was affected, increasing with *A. prostrata* and harrow/mowing treatments.

Regarding the presence of beneficial insects in orchards, Silva *et al.* (2010) concluded that increasing the diversity of plant species covering the soil in the orchard significantly increased the presence of arthropods and parasitoid wasps in the tree canopy compared with the maintenance of uncovered soil.

Chemical control

In any of the weed control methods, the farmer must evaluate the costs and efficiency of the desired method. In orchards where weeds are allowed only in inter-rows, keeping the coverage area of the roots free of competition will cause no damage to Tahiti lime trees (Fig. 8.6). Moreover,

Fig. 8.6. Invasive *Brachiaria* sp. in the inter-rows of citrus orchard, with the area of higher concentration of Tahiti lime roots free of competition.

many of the weeds present have an annual cycle, they senesce naturally in the dry season, not competing with the target trees in the critical period of water scarcity, so there is no need for intervention with the use of herbicides, which will surely generate a reduction in production costs.

Weed control with herbicides in citrus orchards is widely used because of the convenience and the good results achieved. However, as with other control techniques, there are some aspects that should be noted. Knowledge of the predominant weed species in the orchard, the best time to control them, the selectivity of the herbicides, the mode of action, Restricted Entry Interval (REI) and the method of herbicide application is fundamental to the farmer to decide on the best option in each case, ensuring the greatest efficiency and the lowest cost.

The application of herbicides should be conducted under conditions that ensure the efficiency of the product, prevent the poisoning of operators and avoid damage due to droplet drift, which can reach new sprouts of trees.

The ideal temperature for herbicide application is from 15–30°C. Lower temperatures may delay the effect of herbicides, while higher temperatures contribute to the rapid evaporation of the spray volume, to the closure of stomata and production of wax on the leaves, thus reducing the absorption of herbicides. The relative humidity of the air should be above 55%. Below that, the stomata will close and waxes will be produced, with effects similar to those of high temperature (Theisen and Theisen, 2011).

Weeds should be in the active stage of their metabolism so that herbicides exert their effects, especially those with systemic action. Application in conditions of water stress is usually ineffective and should be avoided. Wind speed from 1.5–8 km/h is suitable for application of herbicides. Above this range, negative effects caused by drift may occur (Matuo, 1990; Shiratsuchi and Fontes, 2002; Theisen and Theisen, 2011).

The low efficiency of herbicides and the need for reapplication are due, in large part, to the use of inappropriate techniques and/or application on weeds in advanced growth stages. Depending on the herbicide used, the younger the weed, the smaller the dose required, and possibly the chemical control will be more efficient, particularly of species that are difficult to control (Matuo, 1990; Theisen and Theisen, 2011).

Glyphosate (N-(phosphonomethyl) glycine) is the most widely used herbicide in the Brazilian citrus industry, being commonly applied three or more times a year (Durigan and Timossi, 2002; Gravena *et al.*, 2012). The widespread use of this herbicide is due to its efficiency as a desiccant of weeds, among other things, allowing their use as mulch after using the mower.

Moreover, glyphosate has the advantage of not being phytotoxic if applied on the trunk of adult citrus trees. In Valencia orange budded on Rangpur lime and Swingle citrumelo (*Poncirus trifoliata* L. Raf_× *Citrus paradise* Macf), Gravena *et al.* (2012) found that glyphosate doses of 0, 90, 180, 260, 540, 1080 and 2160 g a.i./ha, directly sprayed on the trunk at a height of 5 cm above the budding zone did not cause visual damage on the trees. However, when glyphosate was sprayed directly over the canopy, at dosages of 0, 0.036, 0.36, 3.6, 36, 360 and 720 g a.i./ha, toxicity symptoms were found on new shoots at a dose above 360 g a.i./ha. Little or no effect was observed in mature leaves. All trees affected by glyphosate recovered within 6–12 months after application. Victoria Filho *et al.* (1991) found no influence on tree development, production or the fruit quality of Pera orange budded on Rangpur lime with continuous use for 6 years of following herbicides and respective doses in kg a.i./ha: terbacil at 3.2; simazine at 4.0; ametryne + secbumetone at 4.5; dichlobenil at 5.0; diuron at 3.2; bromacil at 3.2; bromacil + diuron at 3.2; paraquat at 0.6; glyphosate at 1.61; and methanoarseniate monosodium acid (MSMA) at 1.77.

However, the application of the same active ingredient for several consecutive years can lead to the selection of weed species (Ramos and Durigan, 1996). Some weed species have their importance enhanced due to tolerance to glyphosate. These species occupy the space left by other species that are efficiently eliminated by the herbicide (Theisen and Theisen, 2011). Species such as *Conyza* spp., *Cynodon dactylon* L., *Commelina virginica* L., *Ipomoea* sp., *Cyperus* sp., *Tridax protumbens*, *Spermacoce latifolia* and *Synedrellopsis grisebachii* (Ramos and Durigan, 1996; Theisen and Theisen, 2011) require constant rotation of herbicides for effective control in citrus orchards.

Intercropping

The cost of planting an orchard of Tahiti lime is high and the return on invested capital may take about 5–6 years. Therefore, in the early years after planting the orchard, when there is still enough space in the inter-rows, it is possible to use it to exploit secondary crops in order to reduce planting and formation cost of the orchard and improve soil conditions in the inter-rows (Carvalho *et al.*, 2005b). The choice of secondary crop will depend on the commercial demand in the region where it is to be produced, the age of the Tahiti lime trees and the management method that will be used in the orchard. Crops such as soybean, maize, bean, watermelon, cassava, pumpkin, okra, sunflower, groundnut, leafy vegetables, sweet potato, pineapple, papaya and passion fruit can be intercropped with Tahiti lime (Carvalho *et al.*, 2005b) (Fig. 8.7).

A short cycle and not being the host of diseases and pests of Tahiti lime trees are the conditions that a species must have to be used in intercropping. If a tall tree is chosen, only one planting row is used in each inter-row of the orchard. Moreover, it is necessary to respect the nutritional requirements of Tahiti lime trees and of the intercropped cultures; therefore, it is mandatory to adopt a fertilizer management system specific to each species grown in the area.

The distribution of the roots of Tahiti lime in depth and in distance from the trunk varies according to the rootstock used, in which 50–90% of them are concentrated in the depth of 0–0.4 m (Machado and Coelho, 2000; Neves *et al.*, 2004; Abrêu and Salviano, 2007). Approximately 50–75% of the roots of citrus trees are present within the radius of canopy projection (Machado and Coelho, 2000; Neves *et al.*, 2004). Thus, the intercropped culture should be planted at a minimum distance of 1.5–2.0 m from the canopy of Tahiti lime trees. After a certain age, root growth, shading and lack of space caused by the growth of the canopy of Tahiti lime are the limiting factors for intercropping. In Tahiti lime tree at 11 years after planting, Neves *et al.* (2004) found that the effective distance of the roots on the line ranged from 1.69–2.54 m on the rows and 3.12–3.48 m in the inter-rows. The radius of the canopy on the row ranged from 2.53–3.26 m and 2.86–3.76 m in thr inter-rows, making intercropping unfeasible.

Special attention should be given to soil management for the intercrop, adopting techniques such as no-tillage or minimum tillage, when possible, to avoid revolving the soil and promoting erosion, and to prevent damage to the roots of Tahiti lime trees.

Fertilizer Management

For efficient management of soil fertility, it is necessary to know the chemical and physical characteristics of soils to ensure the efficient use of correctives and fertilizers as well, ensuring greater economic returns and reducing environmental impacts.

Parts of the Tahiti lime tree store different amounts of nutrients, and the leaf stores the largest amounts of N and Ca, while the fruit stores

Fig. 8.7. Intercropping of okra (A) and tomatoes (B) with young Tahiti lime trees.

contents of K slightly higher than the leaves (Table 8.2).

Soil analysis

The first step in evaluation of soil fertility is to perform an adequate sampling, because based on the chemical analysis of this sample available nutrient stores for root uptake will be evaluated (Cantarutti *et al.*, 1999; Quaggio *et al.*, 2005). The correct soil sampling results in the analysis, interpretation and development of appropriate doses of lime and fertilizers. Incorrect sampling with inadequate tools, or sampling that is unrepresentative of the area under cultivation, may cause economic losses and environmental damage (Cantarutti *et al.*, 1999).

In areas where new orchards will be planted, the first thing to be done is to divide the area into homogeneous plots of approximately 10 ha, taking into account the relief position, colour and texture of the soil, as well as the existing vegetation in the area or practices carried out before the orchard establishment. Sampling should be carried out several months before planting the nursery trees, to enable the reaction of the lime with the soil and, if necessary, the cultivation of green manures for improving soil quality. From each plot, 15–20 simple samples should be collected from 0–0.2 m depth, for the recommendation of appropriate fertilization and liming, and from 0.2–0.4 m, to identify possible chemical barriers, such as low contents of calcium and/or excess of aluminium. A mixture of simple samples from each depth will be the composite samples, which will be sent to the laboratory for analysis. In orchards that are already established, the division of plots should consider the same characteristics. Samples should be taken at 0.50 m inwards from the canopy projection and another 0.50 m outwards from the canopy projection. It is not recommended to perform sampling before 60 days after the last fertilization of the soil (Vitti and Cabrita, 1998; Mattos Junior *et al.*, 2003a; Quaggio *et al.*, 2005).

Table 8.2. Mean values of availability of soil nutrients, concentration in the leaf and fruit and fruit weight of 11 orchards of Tahiti lime in São Paulo, Brazil.

Soil		Leaf			Fruit			Fruit
K	Ca	N	K	Ca	N	K	Ca	weight
$mmol_c/dm^3$		----------------g/kg----------------						g
2.5	19.9	20.3	13.8	24.4	10.1	14.6	7.5	71.8

Source: Extracted from Mattos Junior *et al.*, 2010

Leaf analysis

The difficulty in preserving soil samples for N analysis and the lack of effective methods to quantify the availability in the soil make leaf analysis an important tool in the diagnosis and management of this nutrient in the field (Quaggio *et al.*, 1998; Quaggio *et al.*, 2006). The objective of leaf analysis is to determine the total content of nutrients in newly mature leaves (Vitti and Cabrita, 1998). The nutrient contents in the leaves do not depend only on the availability of the element in the soil, because they are influenced by other factors such as rate of tree growth, leaf age, canopy and rootstock combinations, and interactions between nutrients (Mattos Junior *et al.*, 2003a; Quaggio *et al.*, 2005).

To ensure representative sampling, leaves of the same age should be collected in the four quadrants of trees grown under similar conditions. In 10 ha plots, it is recommended to sample at least 25 trees and to collect four leaves per tree at 1.5 m above ground level, or at the medium height of the canopy, 30 days after the last spraying, at least.

Sampling is carried out on branches with fruit, collecting the third or fourth leaf, at 6 months of age, counted from the fruit towards the base of the branches. The samples should be placed in paper bags and stored at approximately 5°C until shipment to the laboratory (Quaggio *et al.*, 2005). For the interpretation of leaf analysis, laboratorial results should be compared with the respective levels shown in Table 8.3.

Correction of soil acidity – liming

Soil acidity is one of the major factors responsible for low crop yield. It is often caused by the toxicity of exchangeable Al^{3+}, by the low contents

Table 8.3. Appropriate levels of nutrient elements in the dry matter of leaves of Tahiti lime at 6 months of age, collected from branches with fruit.

Macronutrients	Appropriate levels (g/kg)	Micronutrients	Appropriate levels (mg/kg)
Nitrogen	18–22	Boron	36–100
Phosphorus	1.8–2.2	Copper	4.1–10.0
Potassium	15–20	Iron	50–120
Calcium	35–45	Manganese	35–50
Magnesium	2.5–4.0	Zinc	35–50
Sulfur	2.0–3.0	molybdenum	0.10–1.00

Source: Extracted from Mattos Junior *et al.*, 2003a

of Ca and Mg, and by reduction in the availability of other nutrients such as P, causing acidification in the superficial layers of the soil, where most tree roots are concentrated (Machado and Coelho, 2000; Neves *et al.*, 2004; Quaggio *et al.*, 2005; Abrêu and Salviano, 2007).

Thus, liming is the first practice to be adopted in order to increase the productivity of orchards. Besides correcting acidity, liming reduces the leaching of nutrients, increases the availability of most nutrients, decreases Al^{3+} toxicity, and provides Ca and Mg, in addition to stimulating microbial activity in the soil (Vitti and Cabrita, 1998; Alvarez and Ribeiro, 1999).

Thus, the need for liming may be defined as the amount of correction required to achieve maximum economic efficiency of the crop, which means to have a sufficient amount of available Ca and Mg in the soil and appropriate conditions of pH for a good availability of nutrients in general (Alvarez and Ribeiro, 1999). When soil acidity needs to be corrected prior to establishing Tahiti lime orchards, this should be done over the total area, with the deepest possible incorporation of limestone.

In already established orchards, this incorporation may present some issues due to the use of the harrow, such as the injury of the roots and subsequent risk of infections, and the spread of diseases and perennial weeds in the orchard (Silva *et al.*, 2007). The application of limestone in strips may be an interesting alternative for drive input in the most acidic areas of orchards that are already established (Quaggio *et al.*, 2005).

Silva *et al.* (2007) evaluated the superficial application of five doses of limestone (0, 0.552, 1.104, 1.656 and 2.208 t/ha) in an orchard of Pera orange (*C. sinensis* L. Osbeck) budded on Cleopatra mandarin (*C. reshni* hort. ex Tanaka) in 7 × 5 m spacing, and the dose of 1.104 t/ha was recommended to raise the base saturation by 70% in a medium texture soil. It was found that the effects of application of lime at the soil surface could reach the 0.2–0.4 m layer. The maximum reaction of lime occurred between 12 and 18 months after application.

It seems that the benefits of liming in citrus orchards depend on the rootstock used, and Rangpur lime is well adapted to acid soils. Auler *et al.* (2011) evaluated the effect of limestone incorporation at 0.20 m before planting and superficial reapplication of limestone 4 years after planting in a Valencia orange (*C. sinensis* L. Osbeck) orchard budded on Rangpur lime (*C. limonia* Osbeck), Cleopatra mandarin (*C. reshni* hort. ex Tanaka) and *Poncirus trifoliate* L. Raf. Fourteen years after planting, the authors found that the Rangpur lime was better adapted to acidity, to Al^{3+} and to lower levels of Ca and Mg in the soil, and it did not respond to treatment with lime. However, because of the liming, *P. trifoliata* increased by 126% in terms of quantity of roots and 26.4% in fruit production in comparison with the other rootstocks.

Besides the dose and choice of suitable limestone, attention should be paid to the application factors, uniformity, incorporation, time and place of application (Vitti and Cabrita, 1998). The lime requirement can be calculated based on the calibration curve established for citrus, to raise the base saturation by 70% in the 0–0.2 m depth (Quaggio *et al.*, 2005).

Application of calcium promotes the quality of Tahiti lime, which in turn produces fruit with better texture and firmness (Martinez-Romero *et al.*, 1999), as 60% of Ca is located in the cell wall (middle lamella), responsible for the structural function of the cell (Mattos Junior *et al.*, 2010).

Planting and young tree fertilization

The required amount of nutrients for the Tahiti lime depends on the developmental stage of the tree, scion–rootstock combination, the expected production, the orchard nutritional status and soil fertility.

Upon orchard planting, along with liming, the application of phosphorus in the furrows is also recommended at doses ranging from 20–80 g of P_2O_5 per linear metre, according to the soil analysis. The demand for P by young trees of Tahiti lime is proportionately larger than that of the mature trees because in the first years after citrus planting, the rate of tree growth is higher than that of a mature tree. In addition, the root system of the young tree exploits a small volume of soil; consequently, the absorption of P, which occurs by diffusion, is hindered (Quaggio *et al.*, 1998).

After planting, the recommended doses of N, P_2O_5 and K_2O are aimed at tree formation, and take into consideration the age of the orchard and the results of soil analyses for phosphorus and potassium. Potassium is required in small amounts for the growth of young trees of citrus, but is required in larger quantities by mature trees (Mattos Junior *et al.*, 2003b). However, Quaggio *et al.* (2004) concluded that the lack of potassium fertilization reduces the growth of young trees (up to 5 years of age). Furthermore, during orchard formation and the onset of citrus production, the response to NPK fertilization may vary according to the rootstock used. In trees budded on Swingle citrumelo, fruit production was greater as doses of N and K were increased, compared with that of trees budded on Rangpur lime and Cleopatra mandarin.

Fertilization of mature trees

The objective of the fertilization in the production phase of the tree is to get maximum productivity and fruit quality, aiming at meeting the needs of trees with no occurrence of excess or shortage. Analysis of soil, leaf analysis and the expected productivity must be taken into account to define the amounts of fertilizer to be added to the orchard.

Nutrient importance

Nitrogen, phosphorus and potassium

Nitrogen is regarded as the most important nutrient for citrus trees, with effects on the leaf area, directly affecting tree growth, yield and good quality fruit (Mattos Junior *et al.*, 2005). In general, citrus trees store large amounts of N in biomass, which can be redistributed, mainly to developing organs such as leaves and fruit (Mattos Junior *et al.*, 2003c). Nitrogen fertilization should be adjusted based on leaf analysis. Excess or lack of N affects fruit size and quality (Mattos Junior *et al.*, 2003a).

Nitrogen fertilization in non-irrigated orchards should be carried out in the rainy season, in order to avoid losses by volatilization or leaching, especially in sandy soils. To ensure the efficiency of nitrogen fertilizer, it must be split over the year. In non-irrigated orchards in the state of São Paulo, Brazil, it is recommended to apply 40% of the dose after flowering, with the remainder split over the next 5 months (Mattos Junior *et al.*, 2003a). Drip fertigation systems ensure better results in N absorption and lower leaching losses (Quiñones *et al.*, 2003).

In most tropical soils, P availability is low, being a frequent limitation to the growth and productivity of trees as well as affecting directly the magnitude of responses to fertilization with other nutrients (Wang *et al.*, 2010). In citrus orchards with P deficiency, trees grow slowly, the leaves lose their natural shine, presenting a tan colour on the distal portion and on the margin of the leaf blade and drop prematurely as well. The branches become leafless on the basal part due to the redistribution of nutrients from the older leaves to the younger ones (Quaggio *et al.*, 2006).

The sources of P applied to the soil should preferably be soluble in water, for maximum efficiency. In orchards of Tahiti planted in soils deficient in P, the correction of its deficiency is more efficient at a single dose and with the incorporation of fertilizer after application to ensure absorption by the roots, as this mineral has low mobility in the soil profile, especially in clay soils (Quaggio *et al.*, 2005).

Citrus are usually very demanding of K, and the amount required by these trees varies according to the rootstock used. Its deficiency is easily seen in the fruit, which become small, with thin and smooth peel and early ripening (Quaggio *et al.*, 2005). Fertilization with N and K exerts a great influence on the size of the fruit of Sicilian lemon, with negative and positive correlations with the size of the fruit for doses of N and K, respectively. However, for soluble solids, no effect of applied doses of nutrients was observed (Quaggio *et al.*, 2002).

Micronutrients

Citrus are demanding of B, Zn, Mn and Fe, and the deficiency of these micronutrients is common. In tropical conditions, deficiencies of B and Zn are the most frequent. However, iron deficiency is restricted to citrus grown in soils originating from calcareous substrate, such as those found in countries such as Spain, Italy and Morocco (Quaggio *et al.*, 2003; Quaggio *et al.*, 2005). The conditions for the occurrence of deficiency of these elements are as follows: sandy soils; soils poor in organic matter; excessive use of lime and phosphorus fertilization for Zn and Mn; waterlogging and prolonged drought period (Vitti and Cabrita, 1998).

Citrus trees with Zn deficiency have few sprouts, remain with old and few vigorous leaves and have the growth of the canopy and yield reduced. Rootstocks such as Cleopatra (*C. reshni* hort. ex Tanaka) and Sunki (*Citrus sunki* (Hayata) hort. ex Tanaka) mandarin trees are more demanding of Zn and therefore when budded with Tahiti lime tree they might need complementary applications of this nutrient (Quaggio *et al.*, 2005).

When B deficiency occurs, trees do not develop satisfactorily and due to the breaking of apical dominance, excessive sprouting of axillary buds occurs, resulting in short branches with small leaves. Manganese deficiency is also common in citrus orchards, but reduction in tree yield only occurs if this deficiency is severe (Mattos Junior *et al.*, 2003a).

The most commonly used method of micronutrient application in citrus is through the leaf due to the small amount required by trees and also to prevent the adsorption of the elements in organic and inorganic soil colloids, reducing the availability for trees (Quaggio *et al.*, 2003). The low mobility in the phloem limits redistribution of most micronutrients in the tree, so applications should be performed during the development of the main vegetation flows when the leaves are still young and the cuticle is little developed, which enhances the absorption of nutrients (Boaretto *et al.*, 2004). Therefore, the best time for leaf fertilization is from spring to late summer, the period of more intense vegetation flow of the trees.

The most recommended sources of micronutrients are metal salts formed with chloride, sulfate and nitrate ions. The success of leaf application also depends on other factors such as compatibility with the products and nutrients when mixed, occurrence of rainfall after application and factors associated with a good distribution over the tree (Vitti and Cabrita, 1998).

As for boron, there are indications that its application into the soil in the form of boric acid is also effective. Boric acid is compatible with pesticides and it can be dissolved in the spray volume of herbicides, such as glyphosate (Quaggio *et al.*, 2005), which is a practical and efficient way to apply the element in the soil since this herbicide is applied twice or more per year (Gravena *et al.*, 2012).

References

Abrêu, F.L.G. and Salviano, A.A.C. (2007) Sistema radicular de três porta-enxertos para lima ácida 'Tahiti' no estado do Piauí. *Semina: Ciências Agrárias* 28, 187–194.

Alvarez, V.V.H., Ribeiro, A.C. (1999) Calagem. In: Ribeiro, A.C., Guimarães, P.R.G. and Alvarez V.V.H. (eds) *Recomendações para o uso de corretivos e fertilizantes em Minas Gerais– 5ª Aproximação*. UFV, Viçosa, Minas Gerais, Brazil, pp. 43–60.

Andrade, D.J., Pattaro, F.C., Morais, M.R., Barbosa, C.L. and Oliveira, C.A.L. (2013) Aspectos técnicos e econômicos da poda e do controle químico de Brevipalpus phoenicis no manejo da leprose dos citros. *Revista Brasileira de Fruticultura* 35, 409–424.

Auler, P.A.M., Fidalski, J., Pavan, M.A. and Neves, C.S.V. (2008) Produção de laranja 'Pera' em sistemas de preparo de solo e manejo nas entrelinhas. *Revista Brasileira de Ciência do Solo* 32, 363–374.
Auler, P.A.M., Neves, C.S.V., Fidalski, J. and Pavan, M.A. (2011) Calagem e desenvolvimento radicular, nutrição e produção de laranja 'Valência' sobre porta-enxertos e sistemas de preparo do solo. *Pesquisa Agropecuária Brasileira* 46, 254–261.
Azevedo, F.A., Lanza, N.B., Sales, C.R.G., Silva, K.I., Barros, A.L. and De Negri, J.D. (2013) Poda na citricultura. *Citrus Research and Technology* 34, 17–30.
Boaretto, A.E., Boaretto, R.M., Contin, T.L.M. and Muraoka, T. (2004) É móvel ou imóvel o boro em laranjeiras? *Laranja* 25, 195–208.
Cantarutti, R.B., Alvarez, V. and Ribeiro, A.C. (1999) Amostragem do solo. In: Ribeiro, A.C., Guimarães, P.R.G. and Alvarez, V.V.H. (eds) *Recomendações para o uso de corretivos e fertilizantes em Minas Gerais – 5ª Aproximação*. UFV, Viçosa, Minas Gerais, Brazil, pp. 13–38.
Carvalho, J.E.B, Caldas, R.C., Cardoso, S.S. and Costa Neto, A. (1993) Influência das épocas de controle das plantas daninhas na produção de laranja 'Pera'. *Planta Daninha* 11, 49–54.
Carvalho, J.E.B., Souza, L.S., Jorge, L.A.C., Ramos, W.F., Neto, A.O.C., Araújo, A.M.A., Lopes, L.C. and Jesus, M.S. (1999) Manejo de coberturas do solo e sua interferência no desenvolvimento do sistema radicular da laranja 'Pêra'. *Revista Brasileira de Fruticultura* 21, 140–145.
Carvalho, J.E.B., Jorge, L.A.C., Ramos, W.F. and Araújo, A.M.A. (2001) Manejo de cobertura do solo e desenvolvimento do sistema radicular da combinação laranja 'Pera'/ limão 'Cravo' na Bahia e em Sergipe. *Laranja* 22, 259–269.
Carvalho, J.E.B., Souza, L.S., Caldas, R.C., Antas, P.E.U.T., Araújo, A.M.A., Lopes, L.C., Santos, R.C., Lopes, N.C.M. and Souza, A.L.V. (2002) Leguminosa no controle integrado de plantas daninhas para aumentar a produtividade da laranja 'Pêra'. *Revista Brasileira de Fruticultura* 24, 82–85.
Carvalho, J.E.B., Paes, J.M.V., Neves, C.S.V.J., Menegucci, J.L.P. and Silva, J.A.A. (2005b) Práticas culturais. In: Mattos Júnior, D., De Negri, J.D., Pio, R.M. and Pompeu Junior, J. (eds) *Citros*. Instituto Agronômico e Fundag, Campinas, São Paulo, Brazil, pp. 449–482.
Donadio, L.C. and Stuchi, E.S. (2001) *Adensamento de plantio e ananicamento de citros*. Funep, Jaboticabal, São Paulo, Brazil (Boletim Citrícola, 16), p. 70.
Durigan, J.C. and Timossi, P.C. (2002) *Manejo de plantas daninhas em pomares cítricos*. EECB, Bebedouro, São Paulo, Brazil (Boletim Citrícola, 22), p. 53.
FAOSTAT (2014) FAO Database. Food and Agriculture Organization of the United Nations. *FAOSTAT*. Available at: http://faostat.fao.org (accessed 20 October 2016).
Fidalski, J. and Stenzel, N.M.C. (2006) Nutrição e produção da laranjeira 'Folha Murcha' em porta-enxertos e plantas de cobertura permanente na entrelinha. *Ciência Rural* 36, 807–813.
Fidalski, J., Marur, C.J., Auler, P.A.M. and Tormena, C.A. (2006) Produção de laranja com plantas de cobertura permanente na entrelinha. *Pesquisa Agropecuária Brasileira* 41, 927–935.
Figueiredo, J.O., Stuchi, E.S., Donadio, L.C., Teófilo Sobrinho, J., Laranjeira, F.F., Pio, R.M. and Sempionato, O.R. (2002) Porta-enxertos para a lima-ácida-'Tahiti' na região de Bebedouro, SP. *Revista Brasileira de Fruticultura* 24, 155–159.
Gravena, R., Victoria Filho, R., Alves, P.L.C.A., Massafera, P. and Gravena, A.R. (2012) Glyphosate has low toxicity to citrus plants growing in the field. *Canadian Journal of Plant Science* 92, 119–127.
Grizotto, R.K., Silva, J.A.A., Miguel, F.B., Modesto, R.T. and Vieira Júnior, J.B. (2012) Qualidade de frutos de laranjeira Valência cultivada sob sistema tecnificado. *Revista Brasileira de Engenharia Agrícola e Ambiental* 16, 784–789.
Intrigliolo, F. and Roccuzzo, G. (2011) Modern trends of *Citrus* pruning in Italy. *Advances in Horticultural Science* 25, 187–192.
Loussert, R. (1992) *Los agrios*. Ediciones Mundi-Prensa, Madrid.
Machado, D.L.M. (2014) *Porta-enxertos e densidade de plantio para a limeira ácida 'Tahiti' cultivada na região norte de Minas Gerais*. Tese de Mestrado, Universidade Federal de Viçosa, Brazil.
Machado, C.C. and Coelho, R.D. (2000) Estudo da distribuição espacial do sistema radicular do limão 'Cravo' enxertado com lima ácida 'Tahiti'. *Laranja* 21, 359–380.
Martínez-Romero, D., Valero, D., Serrano, M., Martínez-Sánchez, F. and Riquelme, F. (1999) Effects of post-harvest putrescine and calcium treatments on reducing mechanical damage and polyamines and abscisic acid levels during lemon storage. *Journal of the Science of Food and Agriculture* 79, 1589–1595.
Matheis, H.A.S.M., Azevedo, F.A. and Victoria Filho, R. (2006) Adubaçãoverde no manejo de plantasdaninhasnacultura de citros. *Laranja* 27, 101–110.

Mattos Junior, D., Quaggio, J.A. and Cantarella, H. (2003a) Manejo da fertilidade do solo. In: Mattos Junior, D., De Negri, J.D. and Figueiredo, J.O. (eds) *Lima ácida Tahiti*. Instituto Agronômico, Campinas, São Paulo, Brazil, pp. 67–80.
Mattos Junior, D., Quaggio, J.A., Cantarella, H. and Alva, A.K. (2003b) Nutrient content of biomass components of Hamlin sweet orange trees. *Scientia Agricola* 60, 155–160.
Mattos Junior, D., Graetz, D.A. and Alva, A.K. (2003c) Biomass distribution and ^{15}N-nitrogen partitioning in citrus trees on a sandy Entisol. *Soil Science Society of America Journal* 67, 555–563.
Mattos Junior, D., Bataglia, O.C. and Quaggio, J.A. (2005) Nutrição dos citros. In: Mattos Júnior, D., De Negri, J.D., Pio, R.M. and Pompeu Junior, J. (eds) *Citros*. Instituto Agronômico e Fundag, Campinas, São Paulo, Brazil, pp. 197–219.
Mattos Junior, D., Milaneze, T.F., Azevedo, F.A. and Quaggio, J.A. (2010) Soil nutrient availability and its impact on fruit quality of Tahiti acid lime. *Revista Brasileira de Fruticultura* 32, 335–342.
Matuo, T. (1990) *Técnica de aplicação de defensivos agrícolas*. FUNEP/UNESP, Jaboticabal, São Paulo, Brazil, p. 139.
Medina, C.L. and Silva, J.A.A. (2003) Implantação do pomar e tratos culturais. In: Mattos Junior, D., De Negri, J.D. and Figueiredo, J.O. (eds) *Lima ácida Tahiti*. Instituto Agronômico, Campinas, São Paulo, Brazil, pp. 47–65.
Minatel, A.L.G., Andrioli, I., Centurion, J.F. and Natale, W. (2006) Efeitos da subsolagem e da adubação verde nas propriedades físicas do solo em pomar de citros. *Engenharia Agrícola* 26, 86–95.
Neves, C.S.V.J., Dechen, A.R., Feller, C. and Gonzalez, M.G.N. (1998) Influência de sistemas de manejo de solo em pomar de tangerina 'Poncã' sobre limão 'Cravo' em Latossolo roxo. *Revista Brasileira de Fruticultura* 20, 367–374.
Neves, C.S.V.J., Murata, I.M., Stenzel, N.M.C., Medina, C.C., Borges, A.V., Okumoto, S.H., Lee, R.H.C. and Kanai, H.T. (2004) Root distribution of rootstocks for 'Tahiti' lime. *Scientia Agricola* 61, 94–99.
Normative Instruction 48 (2013) Brasil. Instrução Normativa n° 48, de 24 de setembro de Estabelece as normas de produção e comercialização de material de propagação de citros – *Citrus* spp, *Fortunella* spp, *Poncirus* spp, e seus híbridos, bem como seus padrões de identidade e de qualidade, com validade em todo o Território Nacional. *Diário Oficial da República Federativa do Brasil,* Brasília; Brazil, no. 186, Seção 1, p. 38.
Oliveira, R.P., Scivittaro, W.B. and Petry, H.B. (2011) Poda de citros. In: Oliveira, R.P. and Scivittaro, W.B. (eds) *Cultivo de citros sem sementes*. Embrapa Clima Temperado (Sistema de produção 21), Pelotas, Rio Grande do Sul, Brazil, pp. 156–169.
Paes, A.R. and Esperancini, M.S.T. (2006) Análise de rentabilidade da citricultura da região sul paulista, sob condições de risco, em três densidades de plantio. *Engenharia Agrícola* 21, 18–33.
Quaggio, J.A., Cantarella, H. and Raij, B.V. (1998) Phosphorus and potassium soil test and nitrogen leaf analysis as a base for citrus fertilization. *Nutrient Cycling in Agroecosystems* 52, 67–74.
Quaggio, J.A., Mattos Junior, D., Cantarella, H., Almeida, E.L.E. and Cardoso, S.A.B. (2002) Lemon yield and fruit quality affected by NPK fertilization. *Scientia Horticulturae* 96, 151–162.
Quaggio, J.A., Mattos Junior, D., Cantarella, H. and Tank Junior, A. (2003) Fertilização com boro e zinco no solo em complementação à aplicação via foliar em laranjeira Pêra. *Pesquisa Agropecuária Brasileira* 38, 627–634.
Quaggio, J.A., Mattos Junior, D., Cantarella, H. and Stuchi, E.S. (2004) Sweet orange trees grafted on selected rootstocks fertilized with nitrogen, phosphorus and potassium. *Pesquisa Agropecuária Brasileira* 39, 55–60.
Quaggio, J.A., Mattos Junior, D. and Cantarella, H. (2005) Manejo da fertilidade do solo na citricultura. In: Mattos Júnior, D., De Negri, J.D., Pio, R.M. and Pompeu Junior, J. (eds) *Citros*. Instituto Agronômico e Fundag, Campinas, São Paulo, Brazil, pp. 483–507.
Quaggio, J.A., Mattos Junior, D. and Cantarella, H. (2006) Fruit yield and quality of sweet oranges affected by nitrogen, phosphorus and potassium fertilization in tropical soils. *Fruits* 61, 293–302.
Quiñones, A., Bañuls, J., Millo, E.P. and Legaz, F. (2003) Effects of ^{15}N application frequency on nitrogen uptake efficiency in *Citrus* trees. *Journal of Plant Physiology* 160, 1429–1434.
Ramos, H.H. and Durigan, J.C. (1996) Avaliação da eficiência da mistura pronta de glyphosate + 2,4-D no controle da *Commelina virginica* L. em citros. *Planta Daninha* 14, 33–41.
Santarosa, E., Koller, O.C., Petry, H.B. and Casamali, B. (2010) Frequência e intensidade de poda em pomar jovem de laranjeiras 'Valência' sob manejo orgânico. *Ciência Rural* 40, 2081–2085.
Santarosa, E., Koller, O.C., Casamali, B. and Petry, H.B. (2013) Produção e qualidade físico-química de frutos de laranjeiras 'Valência' em diferentes intensidades e frequências de poda. *Revista Brasileira de Fruticultura* 35, 790–798.

Sediyama, M.A.N., Santos, M.R., Vidigal, S.M., Santos, I.C. and Salgado, L.T. (2010) Ocorrência de plantas daninhas no cultivo de beterraba com cobertura morta e adubação orgânica. *Planta Daninha* 28, 717–725.

Severino, F.J. and Christoffoleti, P.J. (2001) Efeitos de quantidades de fitomassa de adubos verdes na supressão de plantas daninhas. *Planta Daninha* 19, 223–228.

Shiratsuchi, L.S. and Fontes, J.R.A. (2002) *Tecnologia de aplicação de herbicidas*. Embrapa Cerrados, Planaltina, Distrito Federal, Brazil (Documentos 78).

Silva, M.A.C., Natale, W., Prado, R.M., Corrêa, M.C.M., Stuchi, E.S. and Andrioli, I. (2007) Aplicação superficial de calcário em pomar de laranjeira 'Pêra' em produção. *Revista Brasileira de Fruticultura* 29, 606–612.

Silva, E.B., Franco, J.C., Vasconcelos, T. and Branco, M. (2010) Effect of ground cover vegetation on the abundance and diversity of beneficial arthropods in citrus orchards. *Bulletin of Entomological Research* 100, 489–499.

Stenzel, N.M.C. and Neves, C.S.V.J. (2004) Rootstocks for 'Tahiti' lime. *Scientia Agricola* 61, 151–155.

Stuchi, E.S., Donadio, L.C. and Sempionato, O.R. (2002) Produtividade e tamanho das plantas do clone CNPMF-01, premunizado contra a tristeza dos citros, da limeira-ácida 'Tahiti' em Bebedouro, SP. *Revista Laranja* 23, 221–230.

Teófilo Sobrinho, J., Salibre, A.A., Fiqueiredo, J.O. and Schinor, E.H. (2002) Adensamento de plantio para laranjeira 'Hamlin' sobre limoeiro 'Cravo' em Cordeirópolis (SP). *Laranja* 23, 439–452.

Teófilo Sobrinho, J., Pompeu Júnior, J. and Figueiredo, J.O. (2012) Adensamento de plantio da laranjeira 'Valência' sobre trifoliata – resultados de 18 anos de colheita. *Citrus Research and Technology* 33, 49–58.

Theisen, G. and Theisen, S. (2011) Manejo de plantasdaninhas. In: Oliveira, R.P. and Scivittaro, W.B. (eds) *Cultivo de citros sem sementes*. Embrapa Clima Temperado(Sistema de produção 21), Pelotas, Rio Grande do Sul, Brazil, pp. 192–227.

Tucker, D.P.H., Wheaton, T.A. and Muraro, R.P. (1994) *Citrus tree pruning principles and practices*. Fact Sheet HS-144, Horticultural Sciences Dept, Florida Cooperative Extension Service, Institute of Food and Agricultural Sciences, University of Florida, Gainesville, Florida.

Van Zyl, J.G., Sieverding, E.G., Viljoen, D.J. and Fourie, P.H. (2014) Evaluation of two organosilicone adjuvants at reduced foliar spray volumes in South African citrus orchards of different canopy densities. *Crop Protection* 64, 198–206.

Victoria Filho, R., Moreira, C.S., Shimoama, N.Y. and Shinohar, R.K. (1991) Uso contínuo de herbicidas em citrus (*Citrus sinensis* L.) Osbeck). II. efeitos no desenvolvimento, produção e qualidade dos frutos. *Planta Daninha* 9, 102–113.

Vitti, G.C. and Cabrita, J.R.M. (1998) *Nutrição e adubação de citros*. Funep, Jaboticabal, São Paulo, Brazil (Boletim Citrícola, 4), p. 31.

Wang, Y., Wang, E., Wang, D., Huang, S., Ma, Y., Smith, C.J. and Wang, L. (2010) Crop productivity and nutrient use efficiency as affected by long-term fertilisation in North China Plain. *Nutrient Cycling in Agroecosystems* 86, 105–119.

Whitney, J.D. and Wheaton, T.A. (1984) Tree spacing affects citrus fruit distribution and yield. *Proceedings of the Florida State Horticultural Society* 97, 44–47.

Yildirim, B., Yesiloglu, T., Incesu, M., Kamiloglu, M., Özgüven, F., Tuzcu, Ö. and Kaçar, Y.A. (2010) The effects of mechanical pruning on fruit yield and quality in 'Star Ruby' grapefruit. *Journal of Food, Agriculture and Environment* 8, 834–838.

9 Precision Agriculture in Lime: Potential for Application of Precision Agriculture Technologies in Lime Cropping Systems

Aitazaz A. Farooque[1]*, Qamar U. Zaman[2], Arnold W. Schumann[3] and Travis J. Esau[2]

[1]*School of Sustainable Design Engineering, University of Prince Edward Island, Charlottetown, Prince Edward Island, Canada;* [2]*Engineering Department, Faculty of Agriculture, Dalhousie University, Halifax, Nova Scotia, Canada;* [3]*Citrus Research and Education Center, University of Florida, Lake Alfred, Florida, USA*

Introduction

Lime belongs to the *Rutaceae* family and has a similar genotype to citrus. Limes primarily originated from India and China, followed by expansion to the Mediterranean regions and to the Americas (Reuther *et al.*, 1967). Total production (both lemon and lime) around the globe has been reported to be 15.14 billion kg which is three times higher than the production levels in the 1980s (US International Trade Commission, 2013). Brazil and México are the world's largest lime-producing countries (United Nations Food and Agriculture Organization, 2003). The harvesting of limes occurs multiple times throughout the year (Reuther *et al.*, 1967). The elevating production trends in lemon and lime crops are partially caused by improved management practices (irrigation, fertilization, fungicides/pesticides, pruning and improved harvesting methods). However, the majority of inputs in lime production are applied uniformly, ignoring the substantial variations that occur in tree size, soil properties, leaf nutrients, insect and disease pressure, weather conditions and nutrient requirements. Uniform management of crop inputs can increase production costs, reduce profitability, and enhance environmental risks such as via surface and sub-surface leaching, runoff and volatilization (Schumann *et al.*, 2003; Farooque *et al.*, 2013). In order to apply crop inputs on an as need basis, map and sensor-based precision agriculture (PA) technologies could be implemented in lime production for effective management of inputs to improve crop productivity.

Precision Agriculture Technologies

Soil and crop conditions vary spatially and temporally within fields and orchards (Zaman *et al.*, 2005). PA offers innovative and sophisticated tools to identify, quantify, locate and record substantial spatial variability in soil and crop conditions. Once spatial and temporal variability are properly explained, they can aid in site-specific application of agricultural inputs. Morgan and Ess (2003) defined PA as management of agricultural inputs on as need basis to increase farm profitability and maintain environmental sustainability. Khosla (2001) defined PA as the implementation

*E-mail: afarooque@upei.ca

 The Lime: Botany, Production and Uses (eds M.M. Khan, R. Al-Yahyai and F. Al-Said)

of advanced systems to improve crop productivity, while lowering the environmental risks. Thrikawala *et al.* (1999) reported PA as a technology to manage existing spatial variability in most crop production systems.

PA has several components, i.e. a differential global positioning system (DGPS), a geographical information system (GIS), analysis of tree characteristics and field conditions, real-time measurement of soil properties, crop stress, pest pressure and disease using advanced sensors, control systems for automation of field operations, variable rate (VR) technology and yield monitoring systems. PA technologies have been successfully implemented in citrus production to optimize agricultural inputs (fertilizers, spraying, irrigation, fertigation, yield monitoring, etc.) (Zaman *et al.*, 2003; Zaman *et al.*, 2005; Schumann *et al.*, 2012). As lime trees are similar to citrus and belong to the same genotype, these technologies can be modified and adopted in lime production to enhance productivity, reduce production cost and ensure environmental sustainability.

The ever-increasing population and food scarcity trends around the globe demand efficient use of farm resources, which can be achieved by using PA technologies. These innovative technologies have been fostered in developed countries to achieve optimal input use efficiency since the 1980s. Uniform management of agricultural resources lacks the ability to achieve optimal nutrient efficiency in terms of maximum yield or minimum cost of production (Khanna *et al.*, 1999). Uniform management decisions are not informationally driven (ignoring field variability) or need-based, but are made to overcome nutritional deficiency in crops, which usually results in over- or under-application of inputs. Excessive use of fertilizers can increase production cost and surface/sub-surface leaching of nutrients, which results in environmental degradation. Under-application of fertilizers can adversely affect the nutritional requirements of plants leading to a reduction in yield (Schumann *et al.*, 2003). Overall, uniform management of inputs has consequential impacts on plants, environment, economy, sustainability and food security.

PA systems rely on crop management according to spatial variability and site-specific conditions (Seelan *et al.*, 2003). PA technologies are categorized into the following groups: map-based, sensor-based or a combination of maps and sensors. Data collected using PA systems can aid in decision making on a spot-specific basis for effective management of resources. Spot-specific decision making based on soil and crop variability includes site-specific fertilization, pesticides and herbicides, seeding density, irrigation and tillage (Daberkow and McBride 1998). The ultimate goal of spot-specific management is to increase profitability and mitigate environmental risks (Fountas *et al.*, 2005).

Industrial evolution and technology advancements have made it easy for the agriculture sector to adopt these innovative systems, primarily developed for industries, and incorporate them into their machines for efficient resource allocation. Advancements in information technology have led to engineered products and automation of the equipment to measure and map spatial and temporal variability in the field and the tree canopy to achieve sustainable management of agricultural inputs. PA is recognized as a powerful tool to reorganize modern agriculture to achieve an efficient production system with low input use and improved productivity (Shibusawa, 1998). The management of soil and crop variability within the field to ensure sustainable agrochemical use is an ultimate objective of PA technology (Whelan *et al.*, 1997). Lime-producing countries can implement these innovative technologies (specially developed for citrus) to manage soil and crop variability within lime orchards, both in the spatial and temporal domains. Site-specific management of crop inputs in lime orchards can save a significant amount of agrochemicals and increase crop productivity.

Soil and Crop Variability in Spatial and Temporal Dimensions

Change in soil properties and crop characteristics over a certain distance and time is characterized as spatio-temporal variability (Morgan and Ess, 1997). Soil and crop properties are highly variable within a field and among fields on a farm. The driving forces behind this spatio-temporal variability include climatic conditions, soil properties, management practices, topographic features and site characteristics (Patzold *et al.*, 2008).

Spatial variability in fruit yield is primarily controlled by the heterogeneity in physical and chemical properties of the soil and weather fluctuations (Wong and Asseng, 2006). PA is concerned with management of soil variability to enhance crop productivity and reduce losses to the environment (Haefele and Wopereis, 2005). Spatial variability on an agricultural farm can be divided into six pillars, i.e. yield, field, soil, topography, crop and anomalous factors. Variability in fruit yield is a dependent variable in cropping systems, whereas the other factors are treated as independent variables. Like other cropping systems, lime groves exhibit spatio-temporal variability, which requires management using innovative PA technologies. Mapping of spatial variability using geo-referenced soil and crop sampling would be the first step to introduce PA technologies in the lime crop.

PA manages this spatio-temporal variability to meet the nutrient and water requirements of crops on an as need basis. The applications of PA technology to address spatial variability are mainly focused on cereals, soybean and cotton crops (Blackmore *et al*., 2003; Dobermann and Ping, 2004; Gemtos *et al*., 2004). The opportunities to extend the scope of PA technology to high value fruit crops has been tremendous over the past decade (Schumann *et al*., 2003; Zaman *et al*., 2011; Farooque *et al*., 2013; Esau *et al*., 2014). The chance of payback of the initial investment is very high in high value fruit crops.

Proper characterization and quantification of spatial variability for any crop requires geo-referenced soil and crop data, which need to be mapped in GIS to delineate management zones (MZs) for VR applications (fertilizer, irrigation, pesticides, etc.). Geo-referenced prescription maps equipped with VR technology (sprayers and spreaders) can facilitate spot applications of agricultural inputs in different MZs on an as need basis. VR technology has been developed and used for targeted application of agrochemicals (herbicides for weed control and fertilizer and fungicide application on plants only) for various cropping systems, including citrus, to foster crop growth and productivity. The VR technology developed for citrus crops could easily be modified to carry out spot applications in lime groves after proper characterization and quantification of spatio-temporal variability using either map- or sensor-based PA systems.

Applications of PA technology to map spatial variability have expanded to various crops including citrus, palm trees, grapes, olives, wild blueberries, pears, apples and peaches (Tagarakis *et al*., 2006; Zaman *et al*., 2008; Ampatzidis *et al*., 2009; Mazloumzadeh *et al*., 2009; Mann *et al*., 2010; Aggelopoulou *et al*., 2011; Fountas *et al*., 2011; Esau *et al*., 2014). However, the involvement of PA technology in addressing variability in lime crops has been very limited. Characterization and quantification of spatial and temporal variability is a starting point to implement PA technology for lime crops. The lime orchards need to be divided into productivity zones (low, medium and high) to receive individual management attention in terms of the nutrient requirements of the lime trees within the grove. Each productivity zone receives an adequate quantity of crop inputs including water, fertilizers, seed rate and pesticides, on an as need basis for an individual tree within the orchard. The MZs are delineated on the basis of stable data for soil, crop and yield attributes over several years (Ping and Dobermann, 2005). Other methods to delineate MZs include remotely sensed aerial images, vegetation indices, topographic features and combinations of various sources of data.

Boydell and McBratney (2002) developed MZs using remotely sensed aerial images for site-specific applications in cotton crops. Whitney *et al*. (1999) examined the potential for site-specific management of inputs due to high yield and canopy variations in citrus. Zaman and Schumann (2006) developed MZs using spatial variability in soil properties in combination with tree performance for VR fertilization. Mann *et al*. (2010) characterized and quantified variability in fruit yield, electrical conductivity and elevation, normalized difference vegetation index (NDVI) and tree volume to develop MZs for VR fertilization in Florida citrus. Schumann (2010) used remotely sensed colour-infrared aerial photography and the NDVI to delineate five MZs for VR applications in citrus (Fig. 9.1). VR fertilization in citrus crop using spatially variable canopy size-based prescription maps resulted in a 40% fertilizer saving according to Zaman *et al*. (2005). The results of this study confirmed the existence of large spatial variability in citrus orchards emphasizing the need to manage this variability using innovative PA technologies.

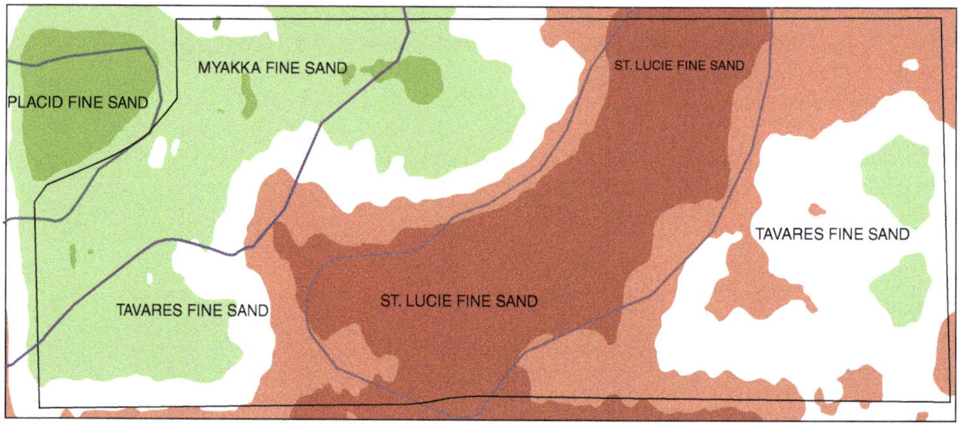

Fig. 9.1. Map of five productivity zones developed from aerial photography and NDVI for Florida citrus crop. Dark red is the least productive and dark green is the most productive. (Adapted from Schumann, 2010, with permission.)

The results of the above-cited citrus studies suggest that these methods can be adopted in lime production systems to delineate MZs for spot-specific management of crop inputs to improve input use efficiency and crop productivity.

Granados *et al.* (2004) developed prescription maps based on spatial variability in leaf nutrients for VR application of fertilizers to achieve optimum productivity. Farooque *et al.* (2012) reported fluctuations in wild blueberry fruit yield in relation to soil and crop variability, and created MZs for VR fertilization on an as need basis. Fountas *et al.* (2011) incorporated spatial variability in soil chemical properties to generate a prescription map for spot applications of fertilizers. Aggelopoulou *et al.* (2011) combined soil nutrients and yield data to generate prescription maps for VR fertilization in apple orchards. Molin *et al.* (2010) compared the uniform and VR applications, and studied the relationship between VR fertilization and crop productivity for a coffee crop. Results suggested that VR fertilization saved significant amounts of fertilizers and increased yield by 34%.

Like other tree crops, the lime production system also contains spatial variability in soil properties, tree sizes, disease infestations, distance between trees, nutrient requirements of each tree and topography. Fig. 9.2 shows the variability in tree size for a citrus orchard. Uniform management of lime trees with inadequate attention to these spatial variations not only increases production costs but also poses a serious threat to the environment. PA technology offers an innovative solution to overcome the problems caused by these spatial variations to manage crop inputs on an as need basis and to improve crop productivity. Application of PA technologies to map spatial variability in lime crops for site-specific management has been very limited. Geo-referenced soil sampling and analysis and yield mapping could be a starting point to ameliorate the existence of productive and unproductive areas within lime orchards for VR applications.

Management of Spatial Variability to Optimize Crop Productivity

Site-specific management of soil/crop variations can be addressed in two ways, i.e. map-based management, sensor-based management and/or their combination. Map-based management can be implemented using remote sensing, differential global positioning system, GIS, yield monitoring, crop and soil sensing, and soil sampling and analysis. A map-based approach for VR application is a logical starting point for sustainable lime production. The operational mechanism of site-specific crop management is illustrated in Fig.9.3. A sensor-based approach requires real-time estimations of soil/crop characteristics using sensing and control systems for on-the-go VR applications. In some cases, map

Fig. 9.2. A typical citrus orchard showing the variability in tree size.

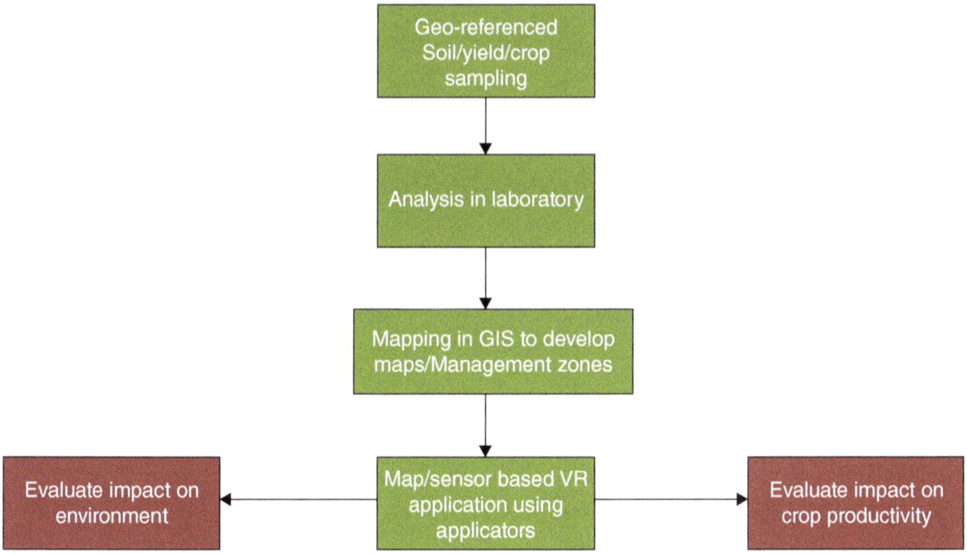

Fig. 9.3. Mechanism of map/sensor-based site-specific applications of agricultural inputs.

and sensor technologies are coupled to perform VR applications. The success of map/sensor-based VR technologies in citrus encourages their application in lime crops due to their similar genotype and tree structure. Yield maps can be developed to perform VR fertilization. These maps can be used as a tool to estimate the productivity potential, disease infestations and crop growth within a lime orchard. Applications of PA technologies in lime could provide a way for risk assessment and rational farm work scheduling to improve crop productivity.

The key component in PA is exact delivery of nutrients in a spot-specific fashion, while accounting for spatial and temporal variations within orchards. In order to achieve the best management practices, Florida citrus growers are in the process of adopting new PA systems for effective growth of individual trees to reach the maximum productivity potential. The beneficial aspects of PA technologies are very promising, which paves its path to lime production. The involvement of PA technologies in lime production would start with the proof of the concept, followed by education as to its key benefits for farmers, which should lead to successful implementation of PA systems in lime production. The adoption rates of PA technologies are very slow around the globe as producers have a particular mindset, which requires an exceptional effort from researchers, academia and extension agents to overcome to ensure adoption of these technologies in a new crop.

The site-specific crop management (SSCM) technologies for citrus production, with slight modifications, would perfectly fit lime production systems. Using site-specific technologies, growers would be able to manage their input applications, which would help them to maintain production costs at an acceptable level. Adoption trends for PA systems might be higher for crops that require intensive applications of inputs (Daberkow, 1997) – lime may be a likely candidate. GPS coupled with VR applicators offers an opportunity to perform site-specific applications of agrochemicals using prescription map or real-time sensing technologies or both (Morgan, 1995; Buick, 1997) for efficient use of agricultural resources.

Map based management for variable rate applications

Problems associated with soil, crop and yield variability can be addressed by implementing the concept of MZs for effective and efficient management of crop inputs. The MZ is a division of a field into subregions with relatively similar attributes (soil properties, landscape, tree sizes, etc.) in order to perform VR applications (Ferguson *et al.*, 2003). Fruit yield, soil fertility, remotely sensed data, soil survey maps and tree size can be used to develop MZs for SSCM (Boydell and McBratney, 2002; Farooque *et al.*, 2012). Currently, producers manage their fields uniformly by ignoring soil and crop variability (Schueller *et al.*, 1999).

Uniform application of agricultural inputs could result in under/over-application, which may lead to increased production costs, lower crop productivity and elevated risks of environmental contamination (Schumann *et al.*, 2003). Spatially variable fertilizer applications can achieve the specific demands of individual trees in an economical way compared with uniform management (Mann *et al.*, 2010). Site-specific fertilization requires proper understanding of the attributes (soil, crop, topography, weather, etc.) causing fluctuations in crop yield. Soil nutrient variability is considered to be a dominating factor influencing crop growth and yield in spatio-temporal dimensions (Ping *et al.*, 2005).

Many researchers have characterized and quantified spatio-temporal soil, crop and yield variability to develop MZs for different crops (McBratney and Pringle 1999; Zaman *et al.*, 2005; Wong and Asseng, 2006; Farooque *et al.*, 2012; Mann *et al.*, 2010). However, very little attention has been paid to the lime production system to date. This situation emphasizes the need to develop MZs for the lime production system to achieve targeted application of agrochemicals. Delineation of MZs for SSCM in lime production has the potential to ensure economic and environmental sustainability, which is extremely important in light of rapid urbanization around the globe. However, increased profitability, reduced production cost and minimal environmental risks cannot be achieved if the lime orchards are managed uniformly, ignoring the substantial soil and crop variability that exist within fields. VR technology is an essential component of PA to achieve SSCM. VR technology incorporates the application of inputs as per soil and crop demands in a variable fashion within a field to facilitate effective use of inputs. VR technology for site-specific fertilization is well known but not well adopted by the growers. It has been developed and tested for citrus production in Florida and Brazil as both of the regions are leading producers.

VR technology provides better management practices and is recognized to provide many benefits compared with uniform applications in citrus crops. Whitney *et al.* (1999) illustrated the potential of implementing a site-specific nutrient management strategy in citrus production, which led to high yield and canopy variations. The high spatial variability in the landscape and soil properties of citrus crops (Oliveira *et al.*, 2009) and their impact on yield (Farias *et al.*, 2003) also demonstrate a requirement for site-specific fertilization in citrus crops.

The effect of VR fertilization on yield, soil and plant characteristics is essential information for growers who intend to adopt PA technology as a management strategy for their crop production. Miller *et al.* (2003) performed VR fertilization in a citrus orchard using GPS guided productivity maps. Schumann and Zaman (2005) developed and tested an automated ultrasonic sensor-based system for real-time estimation of citrus tree volume. This system was capable of developing canopy volume-based prescription maps/MZs to accomplish VR fertilization using a fertilizer spreader on an individual tree basis.

Zaman *et al.* (2005) developed a GPS-guided prescription map using ultrasonically sensed tree volume for VR fertilization in Florida citrus (Fig. 9.4). The tree canopy size varied from <25 m^3 to >225 m^3 (Fig. 9.4) within the selected grove. The fertilizer rates were allocated based on the variation in tree size to optimize crop productivity and to ensure environmental sustainability. The citrus trees with the largest canopy size received a uniform rate (270 kg/ha/year), and the remaining trees received diminishing amounts of fertilizer based on tree volume variability (236, 202, 168 and 135 kg/ha/year). Zaman *et al.* (2005) saved 40% nitrogen through VR fertilization using canopy size-based prescription maps. Molin *et al.* (2010) showed 23% savings on phosphate fertilizer and a 13% increase in potassium fertilizer when using VR technology, with a 34% increase in yield. The ultrasonic volume and prescription maps generated by Zaman *et al.* (2005) are presented in Fig. 9.5.

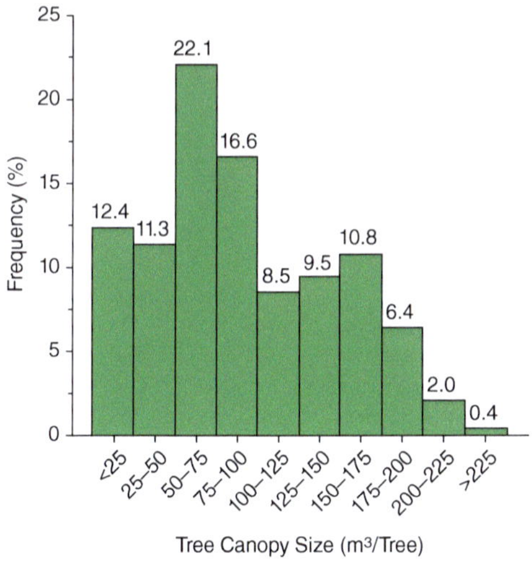

Fig. 9.4. Distribution of citrus canopy estimated from an ultrasonic sensor within a citrus grove. (From: Zaman *et al.* (2005). Copyright 2005 American Society of Agricultural and Biological Engineers.)

Geo-referenced crop, yield and soil sampling coupled with GIS can be used to develop MZs (prescription maps) for site-specific management of agricultural inputs within lime crops. Manual and ultrasonically sensed tree volume measurements can aid in VR fertilization to meet the nutrient requirements of variable sizes in the lime cropping system. Disease infestations can also be mapped with positioning devices for SSCM of agricultural resources. Moreover, PA technologies developed for citrus can be easily refined for lime trees to improve crop productivity and mitigate environmental risks. Mapping of trees and allocation of fertilization rates for citrus groves are presented in Fig. 9.5. Similar strategies can be developed and implemented in lime production for efficient use of agricultural resources.

Sensor-based management

Young trees usually require smaller amounts of input than larger trees based on foliage density/tree size within a row, which creates the need for an accurate application rate. VR applicators consist of sensors and control systems to differentiate target and non-target areas for spot application of agrochemicals on an as need basis. Multiple sensors are widely adopted in PA systems to measure and map soil and plant characteristics in real time for VR applications. PA systems involve a wide spectrum of sensors, hardware

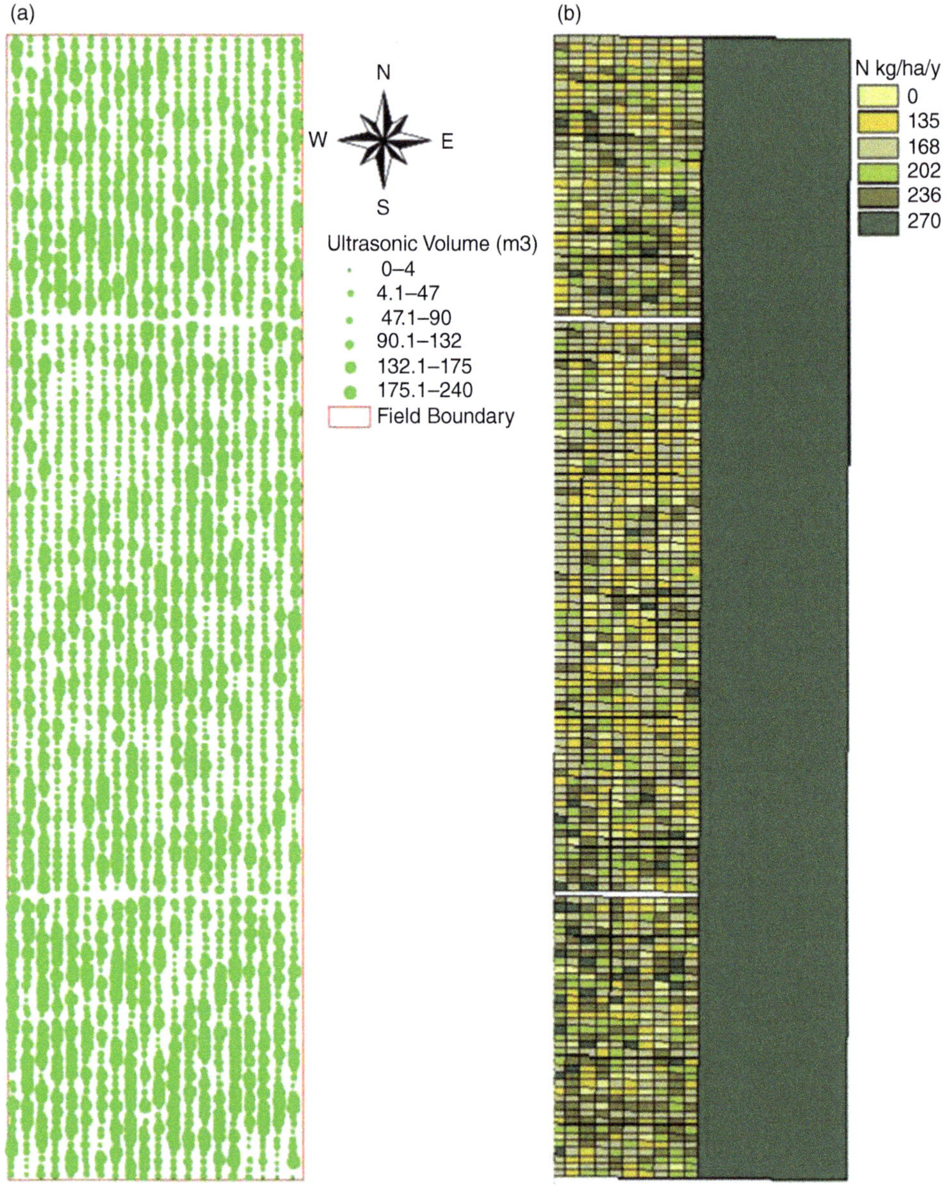

Fig. 9.5. (a) Ultrasonic volume of each tree and (b) prescription map of the grove showing variable N rates for citrus orchard. (From: Zaman *et al.* (2005). Copyright 2005 American Society of Agricultural and Biological Engineers.)

and software for data acquisition, automated recording, analogue and digital processing and analysis of data all within a framework to assess the spatial variations within the field. This information is processed to accomplish real-time spot applications.

The intensive geo-referenced data collected by the ground-based sensors, mounted on the farm equipment, can overcome the problems linked with satellite-based sensing technologies (Bausch and Delgado, 2003). The spectral data retrieved from the satellite platforms are complex to process, the quality deteriorates as it is affected by the weather conditions, and the data are not readily available to implement SSCM. Obtaining up-to-date aerial images is very expensive and the data processing to perform VR applications is complicated. Sensing and control systems are essential pillars for real-time application of agricultural inputs. The acquisition of up-to-date information for on-the-go application of agrochemicals is a major advantage of sensor-based management over map-based technology.

Map-based management of agricultural inputs can be influenced by the rapid variability in tree size and site conditions (removal of diseased trees, hurricane, freeze damage, etc.). Sensors and controllers have been used in citrus to detect tree characteristics and sizes for VR applications (Zaman *et al.*, 2005; Schumann, 2010). The technologies available to measure soil, crop and yield attributes include ground sensors, spectroradiometers, digital imaging, satellite imagery and remote sensing. These technologies have been tested and evaluated in citrus production to perform spot-specific applications. Slight modifications in the software and hardware components of these PA systems could facilitate implementation of PA technologies in lime cropping systems for the wise use of crop inputs.

Ultrasonic sensors, digital colour cameras, photoelectric sensors and lasers have been used in citrus to implement VR applications. With small modifications these sensors have strong potential to be incorporated into agricultural machinery for lime crops for on-the-go application of agrochemicals. Ultrasonic sensors can detect the tree by calculating the reflection of ultrasonic voltage (Zaman *et al.*, 2005; Farooque *et al.*, 2013), which enables VR technology to apply agrochemicals based on tree size. Photoelectric and laser sensors use a near-infrared light source and narrow scanning laser beam to detect trees for spot applications. The canopy information gathered can be used to estimate the canopy volume for lime trees. This canopy sensing information can be linked with the applicators through electro-mechanical design and VR controllers for real-time application of agrochemicals in lime crops. The lag time created by the sensors can be compensated for by look ahead features of the application software.

Most commonly used applicators in citrus/lime groves are air-blast pesticide sprayers or herbicide boom sprayers. The synchronization of sensed information with the applicator is very important to implement VR applications. VR application of agrochemicals (spray and fertilizers) is becoming increasingly popular in speciality crops due to advances in sensing and control systems and higher computing capabilities. Moshou *et al.* (2006) sensed disease infestations for wheat crop using optical sensors and discriminated nutritional and pathological stresses. Gil *et al.* (2007) connected the ultrasonic sensors with sprayer valves and adjusted the flow rate through nozzles to perform real-time VR applications based on the variability in vine canopy width.

Zaman *et al.* (2011) invented a cost-effective ultrasonic sensor-based VR sprayer for real-time tall weed detection and site-specific application of herbicides in wild blueberry fields. Esau *et al.* (2013) incorporated a sensing and control system into a commercial sprayer to achieve VR application of fungicide only on plants, in wild blueberry fields. Pai *et al.* (2009) used a laser scanner coupled with an air blast sprayer and mechanical valves to perform site-specific application of agrochemicals in citrus. Fox *et al.* (2008) tested the spray patterns and jet velocities of sprayers and identified the need for optimization of spray volume based on tree size for accurate and targeted applications.

Schumann *et al.* (2010) developed a VR sprayer using ultrasonically sensed tree volume information for site-specific application of agrochemicals in Florida citrus. The dispensing nozzle rates were decided based on the tree volumes estimated from ultrasonic sensors. The working principle of Schumann's automated system is described in Fig. 9.6. Zaman *et al.* (2003) developed a VR spreader based on the tree volume measurement using ultrasonic sensors, to implement VR nitrogen fertilizer applications (Zaman *et al.*, 2005). Zaman *et al.* (2006) estimated citrus fruit yield using sensing technologies. Miller *et al.* (2003) tested the performance of a map and a photoelectric sensor-based VR spreader for spot application of fertilizer in citrus crop.

Advances in sensing and control systems and their incorporation into agricultural machines/applicators have paved the path towards sustainable agriculture via targeted applications on an as need basis. The incorporation of sensing and control systems into traditional applicators, combined with computing and GPS technology can measure the lime trees along the row and make

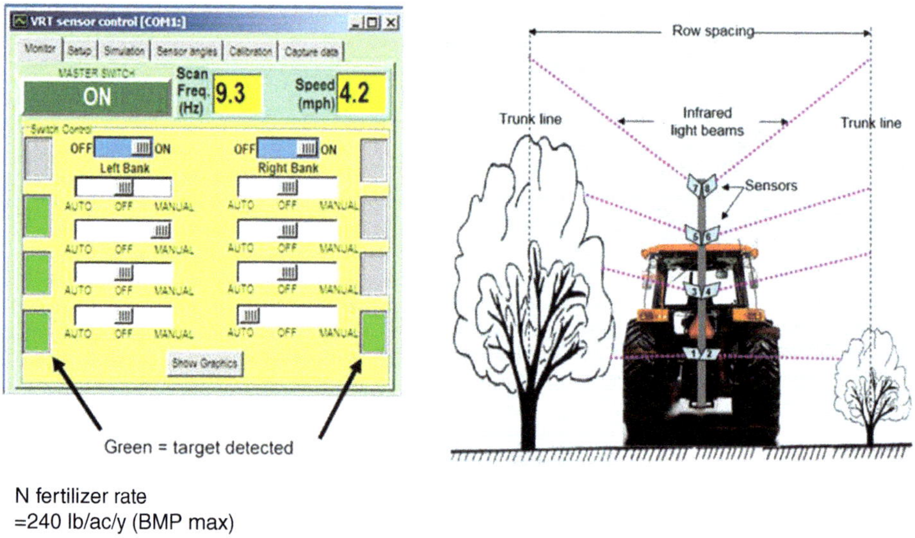

Fig. 9.6. Tree volume measurement using ultrasonic sensors for variable rate applications. (Adapted from Schuman, 2010 with permission.)

adjustments to the amount of agrochemicals delivered to individual trees on an as need basis. Also, with the implementation of PA systems, agrochemicals will not be applied in the spaces in between the trees or to missing trees, which will improve farm profitability and mitigate environmental pollution. Targeted application of agrochemicals in tree crops, based on tree volume and site characteristics has a great potential to save agrochemicals, ranging from 15 to 50% in mature groves and >50% in young groves (Zaman *et al.*, 2006; Schumann, 2010). Implementation of VR technology in lime cropping can offer significant savings in agrochemicals (spray and fertilizers) by increasing input use efficiency.

Many innovative systems comprising sensing and control systems have been developed for speciality crops to perform spot applications; however, the application of these technologies to the lime production systems has been very limited. Because of their similarities, the technologies developed for citrus could easily be modified for lime crops. The development of an automated system for lime crops to implement VR application (spraying and fertilization) based on site and tree size variability has great potential to save significant amount of agrochemicals, with the added advantage of lower environmental risks.

Benefits of Precision Agriculture in the Lime Group

The expected benefits of implementing PA technology in lime production systems are profitability and environmental sustainability. PA technologies have great potential to save agrochemicals, fuel, labour and machinery use. They provide farmers with an opportunity to change the distribution and timing of agricultural resources based on the spatio-temporal variability of the crops to meet nutrient demands. Intelligent decision making regarding agrochemical application based on variability in soil properties and yield, tree size and crop requirements could help to achieve improved crop productivity in lime crops. The concept of profitability in any crop assumes that the savings from VR technologies is more than the additional costs associated with acquiring specialized equipment and labour.

VR technologies have proved to be profitable when compared with uniform rate applications (Griffin *et al.*, 2000; Esau *et al.*, 2014). The uncertainty of profitability is compounded by feasibility considerations and high investment costs. PA technologies also offer the opportunity to increase crop quality and yield (Bramley and Hamilton 2007). Uniform management of agricultural resources in lime production not only increases the cost of production but also poses a serious threat to the environment. The adoption of PA technologies may prove to be a very profitable and environmentally sustainable option for lime crop production.

Strict environmental legislations in developed countries demand precise application to ensure environmental sustainability. Uniform application of agricultural resources results in nutrient leaching and runoff, which leads to deterioration in surface and sub-surface water quality. Uniform applications also have an impact on air quality. The implementation of PA technologies can provide a means of precise and targeted applications in lime crop on an as need basis. Spot applications of agricultural resources can minimize environmental risks.

Summary

Spatial and temporal variations in soil properties, leaf nutrients and tree size in lime groves demand VR management of crop inputs to improve growth and productivity. Site-specific management of agricultural inputs using innovative PA technologies, based on a proper understanding of the soil and crop variability, can improve farm profitability and environmental efficiency. Characterization and quantification of soil-related factors influencing lime tree performance for optimal productivity would be a first step to implementing PA technologies in lime cropping systems. Large lime groves with acreage varying from 0.5 ha to several hundred hectares can be divided into MZs based on productivity (low, medium and high), and prescription maps can be generated for VR fertilization on an as need basis.

Substantial variation in lime tree size, soil properties and crop characteristics emphasizes the need for VR nutrient management to achieve optimum growth and productivity. Tree size can be measured using sensors and MZs can be delineated to implement spot-specific fertilization to meet each individual tree's nutrient requirements. GPS-guided soil-based prescription maps can also be developed to perform spot-specific fertilization within lime groves. Incorporation of sensors and control systems to sense targets on-the-go and spot spraying can be an adequate way to perform VR applications. The VR application of agrochemicals can serve as a powerful tool to implement best management practices for lime cropping systems. Once implemented in lime crops, PA technologies have great potential to provide optimal nutrients on an individual tree basis for effective tree growth and productivity. These technologies can save significant amounts of agrochemicals when compared with traditional uniform applications. Additionally, environmental sustainability can be ensured by implementing VR applications of agrochemicals in lime crops.

References

Aggelopoulou, A., Bochtis, D., Fountas, S., Swain, K.C., Gemtos, T. and Nanos, G. (2011) Yield prediction in apples based on image processing. *Precision Agriculture* 12(3), 448–456.

Ampatzidis, Y.G., Vougioukas, S.G., Bochtis, D.D. and Tsatsarelis, C.A. (2009) A yield mapping system for hand-harvested fruits based on RFID and GPS location technologies: field testing. *Precision Agriculture* 10(1), 63–72.

Bausch, W.C. and Delgado, J.A. (2003) Ground-based sensing of plant nitrogen status in irrigated corn to improve nitrogen management. In: VanToai, T. *et al.*, (eds) *Digital Imaging and Spectral Techniques: Application to Precision Agriculture and Crop Physiology*. ASA Special Publication 66. Madison, Wisconsin, pp. 145–157.

Blackmore, S., Godwin, R.J. and Fountas, S. (2003) The analysis of spatial and temporal trends in yield map data over six years. *Biosystems Engineering* 84(4), 455–466.

Boydell, B. and McBratney, A.B. (2002) Identifying potential within field management zones from cotton yield estimates. *Precision Agriculture* 3, 9–23.

Bramley, R.G.V. and Hamilton, R.P. (2007) Terroir and precision viticulture: are they compatible? *Journal international des Sciences de la Vigne et du Vin* 41(1), 1–8.

Buick, R.D. (1997) Precision agriculture: an integration of information technologies with farming. *New Zealand Plant Protection Society Paper from 50th Conference Proceedings, Palmerston North, New Zealand*. New Zealand Plant Protection Society Publisher, Palmerston North, New Zealand, pp. 176–184.

Chattha, H.S., Zaman, Q.U., Chang, Y.K., Read, S., Schumann, A.W., Brewster, G.R. and Farooque, A.A. (2014) Variable rate spreader for real-time spot-application of granular fertilizer in wild blueberry. *Computers and Electronics in Agriculture* 100, 70–78.
Daberkow, S. (1997) *Adoption Rates for Selected Crop Management Practices: Implications for Precision Farming*. Choices (3rd Quarter)/ University of Minnesota Press, St. Paul, Minnesota, pp. 26–30.
Daberkow, S.G. and McBride, W.D. (1998) Socio-economic profiles of early adopters of precision agriculture technologies. *Agribusiness* 16(2), 151–168.
Dobermann, A. and Ping, J.L. (2004) Geostatistical integration of yield monitor data remote sensing improves yield maps. *Agronomy Journal* 96, 285–297.
Esau, T.J., Zaman, Q.U., Chang, Y.K., Schumann, A.W., Percival, D.C. and Farooque, A.A. (2013) Spot-application of fungicide for wild blueberry using an automated prototype variable rate sprayer. *Precision Agriculture* 15(2), 147–161.
Esau, T.J., Zaman, Q.U. Chang, Y.K., Groulx, D., Schuman, A.W. and Farooque, A.A. (2014) Prototype variable rate sprayer for spot-application of agrochemicals in wild blueberry. *Applied Engineering in Agriculture* 30(5), 717–725.
Farias, P.R.S., Nociti, L.A.S., Barbosa, J.C. and Perecin, D. (2003) Agricultura de Precisão: precision agriculture: yield mapping in citrus groves using geostatistics. *Revista Brasileira de Fruticultura* 25(2), 235–241.
Farooque, A.A., Zaman, Q.U., Schumann, A.W., Madani, A. and Percival, D.C. (2012) Response of wild blueberry yield to spatial variability of soil properties. *Soil Science* 177(1), 56–68.
Farooque, A.A., Chang, Y.K., Zaman, Q.U., Groulx, D., Schumann, A.W. and Esau, T.J. (2013) Performance evaluation of multiple ground based sensors mounted on a commercial wild blueberry harvester to sense plant height, fruit yield, and topographic features in real-time. *Computer and Electronics in Agriculture* (91), 135–144.
Ferguson, R.B., Lark, R.M. and Slater, G.P. (2003) Approaches to management zone definition for use of nitrification inhibitors. *Soil Science Society of America Journal* 67(3), 937–947.
Fountas, S., Blackmore, S., Ess, D., Hawkins, S., Blumhoff, G. and Lowenberg-Deboer, J. (2005) Farmer experience with precision agriculture in Denmark and the US eastern corn belt. *Precision Agriculture* 6, 121–141.
Fountas, S., Aggelopoulou, A., Bouloulis, C., Nanos, G., Wulfsohn, D., Gemtos, T., Paraskevopoulos, A. and Galanis, M. (2011) Site-specific management in olive tree plantation. *Precision Agriculture* 12(2), 179–195.
Fox, R.D., Derksen, R.C., Zhu, H., Brazee, R.D. and Svensson, S.A. (2008) A history of air-blast sprayer development and future prospects. *Transactions of the ASABE* 51(2), 405–410.
Gemtos, T.A., Markinos, A., Toulios, L., Pateras, D. and Zerva, G. (2004) Precision farming applications in cotton fields of Greece. Presented in the CIGR Conference Beijing, China 11–14 October 2004 in Precision Agriculture Session No. 30. Proceedings in CD 30_ 096A.pdf.
Gil, E., Escolá, A., Rosell, J.R., Planas, S. and Val, L. (2007) Variable rate application of plant protection products in vineyard using ultrasonic sensors. *Crop Protection* 26(8), 1287–1297.
Granados, F.L., Exposito, M.J., Alamo, S. and Garcia-Torres, L. (2004) Leaf nutrient spatial variability and site-specific fertilization maps within olive (*Olea europaea* L.) orchards. *European Journal of Agronomy* 21, 209–222.
Griffin, T.W., Popp, J.S. and Buland, D.V. (2000) Economics of variable rate applications of phosphorus on a rice and soybean rotation in Arkansas. *Proceedings of the Fifth International Conference on Precision Agriculture (CD)*, Bloomington, Minnesota, 16–19 July.
Haefele, S.M. and Wopereis, M.C.S. (2005) Spatial variability of indigenous supplies for N, P and K and its impact on fertilizer strategies for irrigated rice in West Africa. *Plant Soil* 270, 57–72.
Khanna, M., Epouhe, O.E. and Hornbaker, R. (1999) Site-specific crop management: adoption patterns and incentives. *Review of Agricultural Economics* 21(2), 455–472.
Khosla, R. (2001) Zoning in on Precision Ag. *Colorado State University Agronomy Newsletter* 21(1), 2–4.
Mann, K.K., Schumann, A.W., Obreza, T.A., Harris, W.G. and Shukla, S. (2010) Spatial variability of soil physical properties affecting Florida citrus production. *Soil Science* 175 (10), 487–499.
Mazloumzadeh, S.M., Shamsi, Z.M. and Nezamabadi-pour, Z.H. (2009) Fuzzy logic to classify date palm trees based on some physical properties related to precision agriculture. *Journal of Precision Agriculture* 11(3), 258–273.
McBratney, A.B. and Pringle, M.J. (1999) Estimating average and proportional variograms of soil properties and their potential use in precision agriculture. *Precision Agriculture* 1(2), 125–152.

Miller, W.M., Schumann, A.W., Whitney, J.D. and Buchanon, S. (2003) *VRT Citrus Test Plot Applications of Granular Fertilizer*. ASAE, St. Joseph, Michigan.
Molin, J.P., Motomiya, A.V. de A., Frasson, F.R., Faulin, G.D.C. and Tosta, W. (2010) Test procedure for variable rate fertilizer on coffee. *Acta Scientiarum Agronomy* 32(4), 569–575.
Morgan, M.T. (1995) Precision farming: sensors vs. map-based. Available at: http://www.lars.purdue.edu/SSMC/Frames/sensors.html (accessed 24 October 2016).
Morgan, M. and Ess, D. (1997) *The Precision-Farming Guide for Agriculturalists*. John Deere Publishing, Moline, Illinois, p. 117.
Morgan, M. and Ess, D. (2003) *The Precision-Farming Guide for Agriculturists*, 2nd edn. John Deere Publishing, Moline, Illinois.
Moshou, D., Bravo, C., Wahlen, S., West, J., McCartney, A. and De Baerdemaeker, J. (2006) Simultaneous identification of plant stresses and diseases in arable crops using proximal optical sensing and self-organising maps. *Precision Agriculture* 7, 149–164.
Oliveira, P.C.G., Farias, P.R.S., Lima, H.V., Fernandes, A.R., Oliveira, F.A. and Pita, J.D. (2009) Spatial variability of soil chemical properties and citrus yield on the eastern Amazon. *Revista Brasileira de Engenharia Agrícola e Ambiental* 13(91), 708–715.
Pai, N., Salyani, M. and Sweeb, R. (2009) Regulating airflow of orchard airblast sprayer based on tree foliage density. *Transactions of the ASABE* 52(5), 1423–1428.
Patzold, S., Mertens, F.M., Bornemann, L., Koleczek, B., Franke, J., Feilhauer, H. and Welp, G. (2008) Soil heterogeneity at the field scale: a challenge for precision crop protection. *Precision Agriculture* 9, 367–390.
Ping, J.L. and Dobermann, A. (2005) Processing of yield map data. *Precision Agriculture* 6, 193–212.
Ping, J., Green, C., Bronson, K., Zartman, R. and Dobermann, A. (2005) Delineating potential management zones for cotton based on yields and soil properties. *Soil Science* 170(5), 371–377.
Reuther, W., Webber, H.J. and Batchelor, L.D. (1967) *The Citrus Industry Volume 1: History, World Distribution, Botany and Varieties*. University of California, Board of Regents, Division of Agricultural Sciences, Davis, California.
Seelan, S.K., Laguette, S., Casady, G.M. and Seielstad, G.A. (2003) Remote sensing applications for precision agriculture: a learning community approach. *Remote Sensing of Environment* 88, 157–169.
Schueller, J.K., Whitney, J.D., Wheaton, T.A., Miller, W.M. and Turner, A.E. (1999) Low-cost automatic yield mapping in hand-harvested citrus. *Computers and Electronics in Agriculture* 23, 145–153.
Schumann, A.W. (2010) Precise placement and variable rate fertilizer application technologies for horticultural crops. *HortTechnology* 20, 34–40.
Schumann, A.W. and Zaman, Q.U. (2005) Software for real-time mapping of ultrasonic tree canopy size. *Computers and Electronics in Agriculture* 47(1), 25–40.
Schumann, A.W., Fares, A., Alva, A.K. and Paramasivam, S. (2003) Response of 'Hamlin' orange to fertilizer source, rate and irrigated area. *Proceedings of the Florida State Horticulture Society*. 116(21), 372–381.
Schumann, A.W., Muraro, R., Hostler, H.K., Waldo, L. and Mann, K.K. (2012) *Designing Advanced Citrus Production Systems to Maximize Early Production*. Florida Citrus Show, Ft. Pierce, Florida.
Shibusawa, S. (1998) Precision Farming and Terra-mechanics. *Fifth ISTVS Asia-Pacific Regional Conference in Korea*, 20–22 October.
Tagarakis, A., Chatzinikos, T., Fountas, S. and Gemtos, T. (2006) Delineation of Management Zones in Precision Viticulture. In: *3rd HAICTA Conference, September 2006, Volos, Greece*.
Thrikawala, S., Weersink, A., Kachanoski, G. and Fox, G. (1999) Economic feasibility of variable-rate technology for nitrogen on corn. *American Journal of Economics* 81, 914–927.
United Nations, Food and Agricultural Organization, FAOSTAT database (2003) Available at: http://faostat3.fao.org/download/Q/QC/E. (accessed 25 October 2016).
Whelan, B.M., McBratney, A.B. and Boydell, B.C. (1997) The Impact of Precision Agriculture. *Proceedings of the ABARE Outlook Conference, 'The Future of Cropping in NW NSW'*, Moree, UK, p. 5. CSIRO Publishing.
Whitney, J.D., Miller, W.M., Wheaton, T.A., Salyani, M. and Schueller, J.K. (1999) Precision farming applications in Florida citrus. *Applied Engineering in Agriculture* 15(5), 399–403.
Wong, M.T.F. and Asseng, S. (2006) Determining the causes of spatial and temporal variability of wheat yields at sub-field scale using a new method of upscaling a crop model. *Plant Soil* 283(1–2), 209–221.
Zaman, Q.U. and Schumann, A.W. (2006) Nutrient management zones for citrus based on variation in soil properties and tree performance. *Precision Agriculture* 7(1), 45–63.
Zaman, Q.U., Shumann, A.W. and Shiel, R.S. (2003) Possibilities of precision fertilization with P and K based on varying nutrient content and yield potential. *Pakistan Journal of Agricultural Science* 40(1–2), 7–10.

Zaman, Q.U., Schumann, A.W. and Miller, W.M. (2005) Variable-rate nitrogen application in Florida citrus based on ultrasonically sensed tree size. *Applied Engineering in Agriculture* 21(3), 331–335.
Zaman, Q.U., Schumann, A.W. and Shibusawa, S. (2006) Impact of variable rate fertilization on nitrate leaching in citrus orchards. *8th International Precision Agriculture Conference Minnesota,* 24–26 July 2006.
Zaman, Q.U., Schumann, A.W., Percival, D.C. and Gordon, R.J. (2008) Estimation of wild blueberry fruit yield using digital color photography. *Transactions of the ASABE* 51(5), 1539–1544.
Zaman, Q.U., Esau, T., Schumann, A.W., Percival, D.C., Chang, Y.K., Read, S. and Farooque, A.A. (2011) Development of cost-effective prototype variable rate sprayer for spot-specific application of agrochemicals in wild blueberry cropping systems. *Computer and Electronics in Agriculture* 76, 175–182.

10 Plant Protection: Lime Diseases and Insect Pests

A.M. Al-Sadi[1]*, Renan B. Queiroz[2], Philip Donkersley[3], A. Nasehi[1] and Simon L. Elliot[3]
[1]*Department of Crop Sciences, College of Agricultural and Marine Sciences, Sultan Qaboos University, Al Khod, Oman;* [2]*Instituto Capixaba de Pesquisa, Assistência Técnica e Extensão Rural, Linhares, Espírito Santo, Brazil;* [3]*Department of Entomology, Universidade Federal de Viçosa, Viçosa, Minas Gerais, Brazil*

Acid lime is affected by several diseases and pests, resulting in variable losses, which range from negligible effects on leaves, roots, stems or fruits to reduction in growth and yield or complete decline. The distribution of acid lime pests and diseases and their severity are affected by several factors, including conduciveness of the environmental conditions, aggressiveness of pathogen strains, type of insect pests, and resistance of limes and rootstocks. Phytoplasma-induced witches' broom disease of lime is a serious disease, especially in the Middle East. Citrus canker, Huanglongbing disease and several fungal and viral diseases are common in different parts of the world. Leafhoppers, psyllids and leaf miners are among the most damaging pests of limes.

This chapter will be divided into two parts. The first part will address some of the most common fungal, viral and prokaryotic diseases of acid lime. Information will be given about symptoms of each disease, causal agent(s) and management. The second part will focus on the main acid lime pests with their occurrence, biology and description, damage and some management strategies.

Acid Lime Diseases

Citrus tristeza virus

Citrus tristeza virus (CTV) is a serious disease of citrus. It resulted in the decline of millions of citrus trees especially those that are grafted on sour orange rootstocks in Brazil, the USA, Spain, Argentina, Israel, Peru and Venezuela (Moreno *et al.*, 2008). CTV is widespread in most citrus-producing countries (Timmer *et al.*, 2000; Al-Sadi *et al.*, 2012).

Symptoms

CTV can cause three distinct syndromes according to different strains. The first is tristeza or rapid decline, which is characterized by the decline of citrus trees grafted on sour orange rootstocks. The second type of symptom is characterized by stem pitting (pits or groves) on trunks and twigs (Brlansky *et al.*, 2002). The same strain can also reduce fruit production and quality (Garnsey and Lee, 1988) and result in stunting (Fig. 10.1). Stem pitting symptoms are common in acid lime, which is very susceptible

*E-mail: alsadi@squ.edu.om

Fig. 10.1. Stunting in sweet orange infected by CTV. (Photo taken by A.M. Al-Sadi.)

to the disease. They are also common in grapefruit and sweet orange (Lee and Bar-Joseph, 2000). The third symptom, seedling yellows, results in leaf chlorosis and stunting (Fraser, 1952; McClean, 1960). Temperatures above 30°C are known to suppress field symptoms of CTV to some extent (Mathews *et al.*, 1997).

Causal agent

The causal agent, CTV, is phloem limited and it is one of the longest viruses. Its RNA is single stranded (Bar-Joseph *et al.*, 1989; Karasev *et al.*, 1995).

Transmission

CTV dispersal can be either by propagation of virus-infected buds that can cause an introduction of CTV into a new area (Bar-Joseph and Lee, 1989; Al-Sadi *et al.*, 2012), or by vector transmission. The standard vectors of CTV are members of the *Aphididae*, *Toxoptera citricida*, *Aphis gossypii* and *Aphis spiraecola* (Roistacher and Bar-Joseph, 1987).

Management

All citrus-growing countries should have certification programmes to ensure that CTV is not spread with bud wood or seedlings that are used for commercial propagation (Navarro *et al.*, 1988; Al-Harthi *et al.*, 2013). EPPO recommends forbidding importation of citrus seedlings from countries where CTV occurs (EPPO, 1990). Cross-protection by inoculating healthy seedlings with certain mild strains of CTV is also recommended (Agrios, 2005). It is necessary to spray citrus seedlings in the nursery with insecticides to control vectors of the disease.

Witches' broom disease of lime

Witches' broom disease of lime (WBDL) is one of the most destructive diseases of acid lime, especially in the Middle East. WBDL was first observed in Oman in the 1970s (Waller and Bridge, 1978). Later, the disease was reported in the United Arab Emirates (UAE), Iran, India and Saudi Arabia (Bove *et al.*, 1988; Garnier *et al.*, 1991; Chung *et al.*, 2006). Over 50% of the area cultivated with lime was lost in Oman, mainly due to WBDL (Al-Yahyai *et al.*, 2015). The disease also destroyed hundred thousands of lime trees in Iran.

Symptoms

Symptoms of WBDL are characterized by dense production of branches and smaller leaves that are light green to yellow in colour (Fig. 10.2). Symptomatic branches usually do not produce any fruits, resulting in decline of tree productivity with time. The symptomatic branches usually accumulate high levels of sodium (Al-Ghaithi *et al.*, 2016). Infected trees usually die within few years (Chung *et al.*, 2006; Al-Yahyai *et al.*, 2015).

Causal agent

The causal agent of WBDL is a wall-less Gram-positive prokaryote called *Candidatus* Phytoplasma aurantifolia. Phytoplasmas have a spheroidal to ovoid shape and they are found mainly in the phloem sieve tubes of vascular plants (Sugio and Hogenhout, 2012). Phylogenetic analysis of *Ca.* P. aurantifolia placed it in the 16S rRNA Group II-B (Al-Ghaithi *et al.*, 2016).

Transmission

Phytoplasmas can be generally transmitted by leafhoppers or psyllids. *Hishimonus phycitis* (leafhopper) has been reported to transmit *Ca.* P. aurantifolia (Salehi *et al.*, 2007). In addition, phytoplasmas can be transmitted experimentally between citrus and periwinkle via dodder. Layering has been suggested to be one of the mechanisms that contributed to the widespread nature of *Ca.* P. aurantifolia in Oman (Al-Sadi *et al.*, 2012).

Fig. 10.2. A lime tree developing witches' broom disease symptoms. (Photo taken by A.M. Al-Sadi.)

Management

Quarantine measures should be implemented to prevent the movement of WBDL into new areas. Certification programmes should be implemented to produce disease-free seedlings. Controlling the insect vectors and removing symptomatic branches may help reduce disease levels and pathogen inoculum.

Citrus canker

Citrus canker affects all important citrus crops (Das, 2003). The disease is widespread in several Asian countries including Japan, India, Iran, Iraq, the UAE, Saudi Arabia and Yemen and Oman, and in the USA and many other countries (Gottwald *et al.*, 1997; Gottwald *et al.*, 2002; Das, 2003; Polek, 2007). Lime, grapefruit, orange, mandarin, pummelo, lemon, tangerine, tangelo, Rough lemon, sour orange, Calamondin, trifoliate orange and kumquat are hosts of this disease (Gottwald *et al.*, 2002).

Symptoms

Canker can affect young leaves, stems and fruits, and produces water-soaked lesions that are variable in size (Fig. 10.3). Developing lesions form a raised pustule surrounded by a chlorotic halo (Ploetz, 2003). The disease can cause defoliation, shoot dieback and fruit drop under highly favourable conditions (Gottwald *et al.*, 2002).

Fig. 10.3. Citrus canker symptoms developing on acid lime leaves. (Photo taken by A.M. Al-Sadi.)

Causal agent

The causal agent of citrus canker is *Xanthomonas citri* ssp. *citri* (Grygiel *et al.*, 2014; Redondo *et al.*, 2015; Caicedo *et al.*, 2016; Carvalho *et al.*, 2016). This bacterium is rod-shaped and Gram negative. Colonies of the bacterium are usually yellow on media.

Management

Implementing strict quarantine measures is important, especially in places where citrus canker is not present. Moreover, eradication efforts should be made when new localized infections are identified. Frequent applications of copper sprays may be used to reduce disease severity (Timmer et al., 2000; Gottwald *et al.*, 2002). In a study on acid lime, pruning of infected twigs followed by the application of copper oxychloride, Bordeaux mixture and neem seed kernel extract was found to be effective in reducing disease severity (Bulbule *et al.*, 2007).

Huanglongbing (greening)

Huanglongbing (HLB) is a devastating insect-vectored disease of citrus (da Graça *et al.*, 2016). It is also referred to as citrus greening. The disease probably originated in China. It has been reported in Asia, Africa and America (Halbert and Manjunath, 2004).

Symptoms

Early symptoms of HLB appear as yellowing of citrus shoots (Haapalainen, 2014). The bacterium moves slowly within the tree, causing progressive yellowing on the entire canopy. Symptoms also include lopsided, small and poorly coloured fruit, and premature fruit drop. The fruits contain aborted seeds (Wang and Trivedi, 2013; Haapalainen, 2014).

Causal agent

HLB is caused by a phloem-limited bacterium, which is in the alpha-proteobacteria division (Gram-negative bacteria). The disease is caused by three species (Haapalainen, 2014). *Candidatus* Liberibacter africanus is associated with the heat-sensitive African form of HLB (greening symptoms only develop when the temperature is in the range 20–25°C). *Ca.* L. asiaticus is associated with the heat-tolerant Asian form of HLB (greening symptoms can develop at temperatures of up to 35°C). The third form is the American form, *Ca.* L. americanus, which is also heat sensitive as is the case with the African form.

Transmission

HLB is vectored by two psyllid species, the Asian (*Diaphorina citri*) and the African (*Trioza erytreae*) citrus psyllids (Bové, 2006). HLB is also graft-transmissible, even when individual leaves are used from infected trees (Hilf and Lewis, 2016).

Management

Quarantine regulations should be directed towards eliminating the spread of HLB disease to the many areas that are disease-free. Starting with disease-free seedlings and removing infected trees from fields is recommended to eliminate the source of infection and reduce the chance for further spread of HLB (Haapalainen, 2014). Chemical control can reduce psyllid vectors (Tiwari *et al.*, 2012; Boina and Bloomquist, 2015; Yan *et al.*, 2015). Biological control of the psyllid vectors by specific hymenopterous ectoparasites is another solution to the problem. The parasitic wasps *Tamarixia* (ex *Tetrastichus*) *dryi* and *Tamarixia radiantus* have been effectively used to control HLB (Timmer *et al.*, 2000). Plant defence inducers have also been proved to reduce the severity of HLB (Li *et al.*, 2016).

Gummosis and foot/root rot

Gummosis and foot/root rot diseases are among the destructive fungal diseases of citrus (Davies and Albrigo, 1994; Spiegel-Roy and Goldschmidt, 1996; Brentu and Vicent, 2015). They have a worldwide distribution and are responsible for 10–30% losses in citrus around the world (Timmer *et al.*, 2000; Al-Sadi *et al.*, 2014). Various species of citrus including lemons, acid limes, sweet orange and grapefruit are susceptible to gummosis and foot/root rot.

Symptoms

Foot rot is caused by an injury to bark while gummosis is a rotting of the bark. The affected area is often surrounded by callus tissue (Timmer *et al.*, 2000; Savita and Nagpal, 2012). Girdling is often observed in the infected trees resulting in the death of young trees or defoliation, twig dieback and short growth flushes in older trees (Timmer *et al.*, 2000).

Causal agents

Phytophthora spp. (*Phytophthora nicotianae* and *Phytophthora citrophthora*) are the main cause of citrus gummosis and foot rot (Sadeghy *et al.*, 2014; Brentu and Vicent, 2015; Das *et al.*, 2016). Other common *Phytophthora* species on citrus include *Phytophthora citricola*, *Phytophthora dreschleri* and *Phytophthora palmivora* (Erwin and Ribeiro, 1996; Naqvi and Singh, 2002).

Management

The occurrence of gummosis and foot/root rot diseases on citrus is dependent on several factors such as the presence of the inoculum, suitable conditions, poor horticultural practices, the use of *Phytophthora*-sensitive rootstocks such as acid lime and excessive irrigation. In addition, the presence of wounds on the base of stem, which are induced by farmers during the removal of new shoots from the base of tree and during weeding, can contribute to the initiation of infection in the stem base for both causal agents.

Trees should be irrigated during the day for short periods. Removal of soil around the collar can prevent infection by these pathogens. It is recommended to use resistant rootstocks such as trifoliate orange (Cacciola and Sanlio, 2008). In the nursery, plants should be grown in separate containers, and plants with gummosis symptoms must be eliminated.

Chemical treatment is an effective method for managing gummosis. Mefenoxam, Al ethylphosphyte or fosetyl-Al and dimethomorph are commonly used for management of *Phytophthora* infection (Matheron and Porchas, 2002; Cacciola and Sanlio, 2008).

Lime anthracnose

Lime anthracnose has been reported in humid regions and some semi-arid regions of the world, especially during rainy seasons (Orozco-Santos, 1996; Orozco-Santos *et al.*, 2006). It has a devastating effect on lime production (Orozco-Santos, 1996; Timmer, 2000; Orozco-Santos *et al.*, 2006).

Symptoms

Disease severity ranges from tiny lesions to shoot and inflorescence blight and shoot tip dieback. In addition, anthracnose can affect leaves, twigs, fruits and flower petals, with late infections often producing larger lesions accompanied by fruit distortion (Orozco-Santos, 1996; Timmer, 2000; Orozco-Santos *et al.*, 2006). Localized necrotic lesions can develop on leaves and fruits under less favourable conditions.

Causal agent

Colletotrichum acutatum J.H. Simmonds causes lime anthracnose (Chung *et al.*, 2003; Peres *et al.*, 2008). Colonies of *C. acutatum* are usually white during the first days and later become pink to orange. Conidia are usually fusiform and ellipsoid (Adaskaveg, 2000).

Management

Anthracnose is difficult to control especially during favourable conditions. Removal of symptomatic tissues may help reduce pathogen inoculum. However, the use of fungicides such as trifloxystrobin + tebuconazole mixture is recommended, especially under favourable conditions for the disease (Silva-Junior *et al.*, 2014). Biological control using beneficial microorganisms (Klein *et al.*, 2013; Lopes *et al.*, 2015) and the use of growth regulators (Chen *et al.*, 2006) have also been shown to be effective in managing some citrus diseases caused by *C. acutatum*.

Melanose

Melanose is a common disease of citrus that is prevalent in different countries and on most citrus species, including acid lime in Australia, Barbados, Cook Islands, Florida, Hawaii, Niue, Samoa, and South Africa (Farr and Rossman, 2016).

Symptoms

Symptoms develop on immature leaves, young branches, stalks, stems and fruit. Small, dark and sunken spots develop on these organs under high moisture, and later develop into raised, corky, superficial, necrotic areas. Affected leaves may become distorted and fall prematurely. Symptoms on fruits are in the form of dark-brown specks that may induce the formation of wound scab tissue and cracking and result in premature falling of fruits (Ploetz, 2003).

Causal agent

Melanose is caused by *Diaporthe citri* F.A. Wolf (Huang *et al.*, 2013). Pycnidia are ostiolate, black and up to 600 μm in diameter (Timmer *et al.*, 2000; Ploetz, 2003).

Management

Pruning of dried branches, dead wood and centres of infection is usually recommended to reduce disease severity. Application of fungicides, such as pyraclostrobin, can help reduce severity of melanose on citrus (Mondal *et al.*, 2007).

Citrus scab

Citrus scab affects the external quality of fruits in susceptible cultivars. The disease has been reported in several citrus species, including acid limes in Australia, Barbados, Brazil, Brunei Darussalam, Cook Islands, El Salvador, Fiji, Florida, Hawaii, Honduras, Malaysia and Thailand (Farr and Rossman, 2016).

Symptoms

Pustules develop on both sides of young leaves and can also appear on green twigs or young stems. Pustules consist of the body of the fungus and host tissues. Scab symptoms can also develop on young fruits (Nelson, 2008).

Causal agent

Citrus scab is caused by *Elsinoe fawcettii* and *Elsinoe australis* (Chung, 2011; Hou *et al.*, 2014). Conidia are hyaline, one-celled and elliptical. Citrus scab is usually more severe at moderate temperatures (Agostini *et al.*, 2003).

Management

It is recommended to avoid overhead irrigation for the production of fresh fruits of susceptible cultivars to reduce the chance of infection. Young fruits are more susceptible to scab, therefore fungicides (e.g. copper-based) should be applied during this period to control the disease (Timmer and Zitko, 1997).

Acid Lime Pests

Leafhopper – *Hishimonus phycitis* (Distant 1908) (Hemiptera: Cicadellidae)

Occurrence

This leafhopper is classically discussed as a pest of aubergine, but has moved hosts to lime in Oman, UAE, Iran, India and Pakistan (Bové *et al.*, 2000). Based on analysis of genetic markers (cytochrome *c* oxidase I and microsatellites) *H. phycitis* populations have been found to be genetically distinct in Oman and in Iran (Shabani *et al.*, 2013). Likely this genetic differentiation may be attributable to climate and topographical barriers. Genetic variation such as this may lead to evolution of novel tritrophic interactions (host plant–pathogen–insect vector), and thus the management strategies may have to be adapted to local conditions.

Biology and description

Adults have a yellow body and transparent wings with a characteristic brown circle on the forewings, which overlap at rest. Nymphs are also yellow with brown spots on the abdomen (Fig. 10.4(A)); they are wingless and move quickly through the branches and leaves, hiding when disturbed. Adults lay eggs close to the midrib of leaves (Fig. 10.4(B)).

Damage

As with other leafhoppers, *Hishimonus phycitis* is a phloem-feeding insect. The main damage caused by this insect is as a vector, transmitting a phytoplasma (wall-less bacterial plant pathogens) (Salehi *et al.*, 2007.). One of these organisms is *Candidatus* Phytoplasma aurantifolia, the causative organism of WBDL. See WBDL disease aspects above.

Control methods

Insecticide applications are the most common control method, as no efficient biological control method has been developed (Hogenhout *et al.*, 2008b). Chemical control must follow the manufacturer's recommendations and be in accordance with the regulations of each country for its correct use. Monitoring using yellow sticky traps to evaluate the *H. phycitis* population density can help shape effective management strategies. A survey of *H. phycitis* population dynamics conducted in Oman on acid lime orchards shows that they remained at high populations throughout the year (Queiroz *et al.*, 2016).

Asian citrus psyllid – *Diaphorina citri* (Kuwayama) (Hemiptera: Liviidae)

Occurrence

Diaphorina citri originated in south-western Asia (Beattie *et al.*, 2009) and currently is present on almost all continents, except for Europe, which has been confirmed Asian citrus psyllid-free (EPPO, 2013).

Fig. 10.4. (A) *Hishimonus phycitis* nymphs collected on acid lime (*Citrus aurantifolia*) in Oman. (B) Eggs are laid near the midrib. (Photo provided by Renan B. Queiroz.)

Biology and description

The adults are small (2.7–3.3 mm in length) with mottled brown wings. They usually have a brown-grey or yellow abdomen. Adults primarily feed on citrus leaves or young shoots (Fig. 10.5(A)). Females are able to lay 500–800 eggs every 2 months, preferentially on young leaves and new shoots (Nava *et al.*, 2007; Fig. 10.5(B)). They are initially oval and yellow and develop two distinct red eyespots before hatching (Hall *et al.*, 2013). Nymphs have five instars and feed on young leaves and stems; during feeding they secrete waxy white material, which helps with their field identification (Fig. 10.5(C)). Depending on host plant and temperature, the complete life cycle can vary between 14 and 28 days, reaching up to ten generations per year (Tsai and Liu, 2000).

Damage

The Asian citrus psyllid feeds on phloem using piercing sucking mouthparts (Hall *et al.*, 2013), which causes a leaf distortion that impacts normal plant development (Fig. 10.6). The main problem related to this insect is its capacity as a vector of major plant pathogens of lime: HLB or citrus greening is the most destructive citrus disease in the world. The disease is associated with three species of *Candidatus* Liberibacter: *asiaticus*, *africanus*, *americanus*. To date, almost 100 million trees have been affected and destroyed globally (Gottwald *et al.*, 2007). The majority of HLB cases worldwide are caused by *Ca.* Liberibacter asiaticus transmitted by *D. citri* (Hall and Gottwald, 2011).

Control methods

Chemical control using insecticides (i.e. imidacloprid, fenpropathrin, chlorpyrifos and dimethoate) is the primary method for reducing the impact *D.* citri has on crops (Hall *et al.*, 2013). Generally, however, these applications have been found to be ineffective at preventing the spread of HLB in new citrus orchards (Bergamin-Filho *et al.*, 2009; Ichinose *et al.*, 2010), which has increased the necessity of finding alternative control methods. Biological control of *D. citri* has been a successful alternative to intensive use of insecticides. Several natural enemies have been studied to control *D. citri*, of which two parasitoids (*Tamarixia radiata* and *Diaphorencyrtus aligarhensis*) show promise (McFarland and Hoy 2001; Rohrig *et al.*, 2012). Furthermore, several entomopathogenic fungi may also work as efficient pathogens against *D. citri*, such as the ascomycete fungi *Isaria fumosorosea* and *Hirsutella citriformis* (Subandiyah *et al.*, 2000). Cultural control through the removal of alternative host plants, such as *Murraya paniculata* (Rutaceae) has been suggested. In São Paulo and Paraná states (Brazil), state and municipal laws prohibit the planting of *M. paniculata*. Moreover, if a HLB-infected plant is identified within an orchard, all other citrus plants within a 30-m radius must be removed (Paraná 2008).

Fig. 10.5. (A) *Diaphorina citri* adults with yellowish colour abdomen and their bodies making a 45° angle from the plant surface. (B) Female oviposition on young shoots. (C) Waxy white material produced from nymphs during feeding. (Photo provided by Renan B. Queiroz.)

Fig. 10.6. Leaf distortion caused by *Diaphorina citri* feeding. (Photo provided by Renan B. Queiroz.)

An integrated pest management approach using all the above described methods is essential for effective control of *D. citri*. However, a lowering of expectations is necessary as even using these integrated control methods it is still difficult to contain this insect vector and the HLB disease, since it is impossible to achieve 100% insect mortality and only a few insects are needed to be capable of pathogen transmission.

Whitefly – *Bemisia tabaci* (Gennadius 1889) (Hemiptera: Aleyrodidae)

Occurrence

Bemisia tabaci is a cosmopolitan invasive species that is believed to have originated from Asia (Mound and Halsey 1978), and is now present and widespread worldwide (Dinsdale *et al.*, 2010; De Barro *et al.*, 2011).

Biology and description

Bemisia tabaci is part of a complex of 11 groups containing at least 24 morpho-species (Dinsdale *et al.*, 2010; De Barro *et al.*, 2011). Adults are ~1 mm in length and males are smaller than females. The body and wings are covered with a waxy secretion, which is white to yellow in colour (Fig. 10.7(A)). A complete life cycle lasts approximately 20 days depending on the temperature and host plant. Females can lay up to 160 eggs each and they are typically laid in circular groups on the underside of leaves (Fig. 10.7(B)), with up to 11 to 15 generations in a year. The nymphs are light yellow in colour, translucent with an oval outline in the scale form. Usually, first instar nymphs move for a few hours searching for an appropriate place to fix. In the fourth and latter instar occurs a phase known as 'puparium', within which the metamorphosis to adult occurs (Gerling *et al.*, 1995).

Damage

Bemisia tabaci is an extremely polyphagous species, with a host range of approximately 600 different plant species, posing a threat to ornamental, vegetable, grain legume and cotton production (Jones 2003). It causes direct damage through feeding and is able to transmit approximately 128 plant viruses, mainly the begomoviruses (Hogenhout *et al.*, 2008a). Adults and nymphs are phloem-sucking, causing chlorotic spots on the leaves surface. In instances of high infestation, these spots may coalesce to total coverage of the leaf surface. This feeding also results in honeydew production that covers the leaves surface and can cause a reduction in photosynthetic potential when colonized by the *Capnodium elaeophilum* fungus (Gerling *et al.*, 1995).

Control methods

The use of insecticides (mainly with neonicotinoids and insect growth regulators) has been the main control method for whiteflies in the field and glasshouses. However, owing to the polyphagous characteristic of *B. tabaci* and its high reproductive rate, insecticide resistance has become a significant issue (Cahill *et al.*, 1996; Elbert and Nauen 2000; Gerling *et al.*, 2001; Horowitz *et al.*, 2005). Integrated management of whiteflies is necessary for future control strategies. Biological control using natural enemies and entomopathogenic fungi, such as the parasitoid wasp *Encarsia formosa* and the ascomycete fungus *Verticillium lecanii*, has been an alternative on cultivated crops, mainly in greenhouses.

Citrus leaf miner – *Phyllocnistis citrella* (Stainton 1856) (Lepidoptera: Gracillariidae)

Occurrence

The Citrus leaf miner originated from Asia and has subsequently been discovered in Florida (USA) in 1993, and currently is distributed

Fig. 10.7. (A) *Bemisia tabaci* adults, and (B) eggs laid in circle. (Photo provided by Renan B. Queiroz.)

worldwide on all continents (Heppner 1993; CABI/EPPO 2003).

Biology and description

Citrus leaf miner adults are small white moths, about 2 mm in length. Adults can be easily identified by their fringed wings, with black and brown lines and an apical black spot (Heppner, 1993). Adults lay eggs individually on the underside of leaves and are a translucent greenish-yellow colour. Eggs hatch within 2–10 days and, upon hatching, the larvae will burrow into the leaf surface to begin tunnelling and feeding. The result of this is serpentine mines that can be seen on the ventral surface of young leaves. Typically, one mine is present per leaf but heavy infestations may have two or three mines per leaf. Larvae have four instars and development takes 5–20 days. Pupation may take between 6 and 22 days, during which the insect remains inside a specialized pupal cell at the edge of the leaf (Heppner, 1993; CABI/EPPO, 2003).

Damage

The damage is caused by larvae through mining under the leaf surface; in high infestations they can attack both surfaces and occasionally fruit and stems. The serpentine mine has a silvery appearance and reaches a length of 50–100 mm. Typically, young leaves are attacked and mines can cause leaf curl. The serpentine mine form is characteristic of this species and can help with its identification. Citrus leaf miner causes direct and indirect damage to plants, reducing photosynthetic area and causing injuries that facilitate the infection by the *Xanthomonas citri* bacteria, the causal agent of citrus canker (Chagas *et al.*, 2001).

Control methods

There are some efficient control methods for *P. citrella*: pheromone traps have been used with some success, evidence suggests a single deployment may be capable of effective mating disruption for an entire growing season in Florida (USA) (Lapointe *et al.*, 2015). Furthermore, a parasitoid wasp, *Ageniaspis citricola* was introduced to Florida citrus groves in 1994–1995; it has had success as a natural enemy and efficiently suppresses leaf miner populations (Smith and Hoy, 1995). This particular parasitoid species showed effective parasitism rates of 60–80, 76, 89 and 98% in the USA, Brazil, Argentina and Peru, respectively (Chagas *et al.*, 2002). Another efficient parasitoid, *Cirrospilus quadristriatus*, has also been introduced to Florida (Smith and Hoy, 1995).

Citrus thrips – *Scirtothrips citri* (Moulton 1909) (Thysanoptera: Thripidae)

Occurrence

Scirtothrips citri is present in the USA (Arizona, California, Florida, Washington State) and in northern México, and also in limited areas of China (CABI, 2015b).

Biology and description

Adult *S. citri* thrips are small, orange-yellow in colour with fringed wings. The nymphs have four stages and are pale yellow to opaque white. The females and nymphs measure 0.6–0.88 and 0.5–0.9 mm, respectively. Each female produces about 250 eggs during her lifetime. Under favourable conditions, it may take only 15 days to complete a single generation and 10–12 generations typically occur per year (Dreistadt, 2012).

Damage

Damage by *Scirtothrips citri* feeding is primarily on young citrus fruit, causing a characteristic ring scar (Grafton-Cardwell *et al.*, 1998; Grafton-Cardwell *et al.*, 2003). They may also feed on new citrus buds, causing twisting and abnormal growth (Grafton-Cardwell *et al.*, 1998); for example, feeding by thrips on young leaves causes thick and grey streaks on both sides of the midrib and the leaves often become distorted when they expand (Dreistadt, 2012). Heavy infestations can cause buds to die and drop from the plant (Grafton-Cardwell *et al.*, 1998). Damage caused by second instar nymphs is the most problematic as they feed mainly under the sepals of young fruit.

Control methods

Insecticides (carbamates, organophosphates and pyrethroids) have been widely used to control

S. citri in California for many years (Khan and Morse, 1997). However, thrips have demonstrated an ability to rapidly develop resistance to chemicals that are used for their control (Khan and Morse, 1998), thus control on non-fruit-bearing young trees is not recommended (Grafton-Caldwell *et al.*, 1998). *S. citri* is capable of resurgence after non-species-specific treatments that destroy populations of natural enemies, especially predaceous mites, such as *Euseius tularensis* (Grafton-Cardwell and Ouyang, 1993), which has an important role in the control of this pest.

Brown citrus aphid – *Toxoptera citricida* (Kirkaldy 1907) (Hemiptera: Aphididae)

Occurrence

Toxoptera citricida is native to Asia where citrus originated and currently has a worldwide distribution (CABI, 2015a).

Biology and description

Toxoptera citricida adults are shiny black and nymphs may be grey or red-brown. Alate and apterous adult females are 1.1–2.6 and 1.5–2.8 mm in length, respectively (CABI, 2015a). In tropical regions, these aphid populations are almost entirely female, reproducing by parthenogenesis. The life cycle duration varies from 1–2 weeks, and there may be dozens of generations per year. Initially, colonies consist of the apterous forms, and after high population growth timed with citrus shoot maturation, alate adults are produced to colonize new shoots or host plants and establish new colonies. Normally, the *T. citricida* population peak coincides with the shoot production on citrus plants.

Damage

T. citricida is an oligophagous species, of which the primary hosts are citrus and citrus relatives (Rutaceae). The insects feed from phloem in shoots, and inject salivary phytotoxic substances, which cause leaf growth distortion. However, the major damage is related to its transmission of CTV; acid lime is known to be the most susceptible citrus species to CTV (Yokomi *et al.*, 1994; Moreno *et al.*, 2008).

Control methods

Natural enemies are important methods for controlling the citrus aphid population; focusing control on migrant vector populations before they spread through susceptible crops is particularly important (Mackauer, 1976). Parasitoids such as *Lysiphlebia japonica*, *Lysiphlebus mirzai* and *Lysiphlebus testaceipes* have been tested in Japan (Takanashi, 1990), China (Liu and Tsai, 2002) and Florida (Evans and Stange, 1997), respectively. Several predators including lacewings (*Chrysoperla plorabunda*), syrphid fly (*Pseudodorus clavatus*) and coccinellid beetles (*Coelophora inaequalis*, *Coccinella septempunctata*, *Cycloneda sanguinea*, *Harmonia axyridis*, *Hippodamia convergens*) have been evaluated in the field and laboratory (Michaud, 2000; Michaud, 2001; Michaud and Belliure, 2001; Wang and Tsai, 2001). Chemical control of *T. citricida* has not been tested in its efficacy to slow spread of CTV. Thus, the best strategy is to use CTV-resistant/tolerant rootstock such as mandarins (*Citrus reticulata*), pummelos (*Citrus maxima*), tangelos (*C. reticulata* × *C. maxima*) and tangor (*C. reticulata* × *C. sinensis*). When planting susceptible scion varieties, pre-infection with a cross-protecting CTV strain can afford some resistance to the disease (Garnsey *et al.*, 1998).

References

Adaskaveg, J. (2000) Occurrence and management of anthracnose epidemics caused by colletotrichum species on tree fruit crops in California. In: Prusky, D., Freeman, S. and Dickman, M.B. (eds) *Colletotrichum: Host Specificity, Pathology, and Host–pathogen Interaction*. APS Press, St. Paul, Minnesota, pp. 317–336.

Agostini, J.P., Bushong, P.M., Bhatia, A. and Timmer, L.W. (2003) Influence of environmental factors on severity of citrus scab and melanose. *Plant Disease* 87, 1102–1106.

Agrios, G. (2005) Plant diseases caused by fungi. *Plant Pathology*, fifth edition, pp. 385–614.

Al-Ghaithi, A.G., Hanif, M.A., Al-Busaidi, W.M. and Al-Sadi, A.M. (2016) Increased sodium and fluctuations in minerals in acid limes expressing witches' broom symptoms. *SpringerPlus* 5, 418. doi: 10.1186/s40064-016-2049-0.

Al-Harthi, S.A., Al-Sadi, A.M. and Al-Saady, A.A. (2013) Potential of citrus seedlings originating in the Middle East as sources of citrus viroids. *Crop Protection* 48, 13–15.

Al-Sadi, A., Al-Hilali, S., Al-Yahyai, R., Al-Said, F., Deadman, M., Al-Mahmooli, I. and Nolasco, G. (2012) Molecular characterization and potential sources of Citrus Tristeza Virus in Oman. *Plant Pathology* 61, 632–640.

Al-Sadi, A.M., Al-Moqbali, H., Al-Yahyai, R., Al-Said, F. and Al-Mahmooli, I. (2012) AFLP data suggest a potential role for the low genetic diversity of acid lime (*Citrus aurantifolia*) in Oman in the outbreak of witches' broom disease of lime. *Euphytica* 188, 285–297.

Al-Sadi, A.M., Al-Ghaithi, A.G., Al-Fahdi, N. and Al-Yahyai, R.A. (2014) Characterization and pathogenicity of fungal pathogens associated with root diseases of Citrus in Oman. *International Journal of Agriculture and Biology* 16, 371–376.

Al-Yahyai, R.A., Al-Sadi, A.M., Al-Said, F.A., Al-Kalbani, Z., Carvalho, C.M., Elliot, S.L. and Bertaccini, A. (2015) Development and morphological changes in leaves and branches of acid lime (*Citrus aurantifolia*) affected by witches' broom disease. *Phytopathologia Mediteranea* 54, 133–139.

Bar-Joseph, M. and Lee, R. (1989) Citrus tristeza virus. AAB descriptions of plant viruses. Commonwealth Mycological Institute/Association of Applied Biology, Kew, UK, p. 353.

Bar-Joseph, M., Marcus, R. and Lee, R.F. (1989) The continuous challenge of citrus tristeza virus control. *Annual Review of Phytopathology* 27, 291–316.

Beattie, G.A.C., Holford, P., Haigh, A.M. and Broadbent, P. (2009) On the origins of Citrus, Huanglongbing, *Diaphorina citri* and *Trioza erytreae*. In: Gottwald, T.R. and Graham, J.H. (eds). *Proceedings of the International Research Conference on Huanglongbing*. Plant Management Network, Orlando, Florida, pp. 23–56.

Bergamin-Filho, A., Gasparoto, M.C.G., Bassanezi, R.B. and Amorim, L. (2009) Relationship between insecticide sprays and Huanglongbing progress in a citrus orchard in São Paulo, Brazil. In: Gottwald, T.R. and Graham, J.H. (eds). *Proceedings of the International Research Conference on Huanglongbing*. Plant Management Network, Orlando, Florida, pp. 282–284.

Boina, D.R. and Bloomquist, J.R. (2015) Chemical control of the Asian citrus psyllid and of huanglongbing disease in citrus. *Pest Management Science* 71, 808–823. doi: 10.1002/ps.3957.

Bové, J.M. (2006) Huanglongbing: a destructive, newly-emerging, century-old disease of citrus. *Journal of Plant Pathology* 88, 7–37.

Bové, J., Garnier, M., Mjeni, A. and Khayrallah, A. (1988) Witches' broom disease of small-fruited acid lime trees in Oman: first MLO disease of citrus. *Proceedings of the International Organization of Citrus Virologists*, IOCV, Riverside, California, 10, pp. 307–309.

Bové, J.M., Danet, J.L., Bananej, K., Haasanzadeh, N., Taghizadeh, M., Salehi, M. and Garnier, M. (2000) Witches' broom disease of lime (WBDL) in Iran. In: da Graça, J.V., Lee, R.F. and Yokomi, R.K. (eds). *Proceedings of the 14th Conference of the International Organization of Citrus Virology*. IOCV, Riverside, California, pp. 207–212.

Brentu, F.C. and Vicent, A. (2015) Gummosis of citrus in Ghana caused by *Phytophthora citrophthora*. *Australasian Plant Disease Notes* 10, 1–3. doi: 10.1007/s13314-015-0184-z.

Brlansky, R., Howd, D., Broadbent, P. and Damsteegt, V. (2002) Histology of sweet orange stem pitting caused by an Australian isolate of Citrus tristeza virus. *Plant Disease* 86, 1169–1174.

Bulbule, S.V., Gohil, S.N., Thakare, C.S., Patil, R.V., Jadhav, V.B. and Jagtap, D.D. (2007) Studies on integrated management of canker in acid lime. *Ecology, Environment and Conservation* 13, 637–640.

CABI (2015a) Invasive Species Compendium. Available at: http://www.cabi.org/isc/datasheet/54271 (accessed 11 October 2015).

CABI (2015b) Plantwise. Available at: http://www.plantwise.org/KnowledgeBank/Datasheet.aspx?dsid=49063 (accessed 9 October 2015).

CABI/EPPO (2003) *Phyllocnistis Citrella. Distribution Maps of Plant Pests*. CAB International, Wallingford, UK.

Cacciola, S.O. and Sanlio, G.M. (2008) Management of citrus diseases caused by *Phytophthora* spp. In: Ciancio, A. and Mukerji, K.G. (eds) *Integrated Managment of Diseases Caused by Fungi, Phytoplasma and Bacteria*. Springer Science & Business Media, New Delhi, India, pp. 61–84.

Cahill, M., Denholm, I., Ross, G., Gorman, K. and Johnston, D. (1996) Relationship between bioassay data and the simulated field performance of insecticides against susceptible and resistant adult *Bemisia tabaci* (Homoptera: Aleyrodidae). *Bulletin of Entomological Research* 86, 109–116.

Caicedo, J.C., Villamizar, S., Ferro, M.I.T., Kupper, K.C. and Ferro, J.A. (2016) Bacteria from the citrus phylloplane can disrupt cell-cell signalling in *Xanthomonas citri* and reduce citrus canker disease severity. *Plant Pathology* 65, 782–791. doi: 10.1111/ppa.12466.
Carvalho, F.M.S., Oliveira, J.C.F., Laia, M.L., Jacob, T.R., Ferreira, R.M., Ferro, M.I.T., Tezza, R.I.D., Zingaretti, S.M., Silva, C.F. and Ferro, J.A. (2016) Mapping and validation of *Xanthomonas citri* subsp *citri* genes regulated by putative plant-inducible promoter box (PIP-box). *Genetics and Molecular Research* 15, gmr8428. doi: 10.4238/gmr.15028428.
Chagas, M.C.M., Parra, J.R.P., Namekata, T., Hartung, J.S. and Yamamoto, P.T. (2001) *Phyllocnistis citrella* Stainton (Lepidoptera: Gracillariidae) and its relationship with the citrus canker bacterium *Xanthomonas axonopodis* pv. *citri* in Brazil. *Neotropical Entomology*, 30, 55–59.
Chagas, M.C.M., Parra, J.R.P., Milano, P., Nascimento, A.M., Parra, A.L.G.C. and Yamamoto P.T. (2002) *Ageniaspis citricola*: criação e estabelecimento no Brasil. In: Controle biológico no Brasil: Parasitóides e predadores. Parra, J.R.P. Botelho, P.S.M., Corrêa-Ferreira, B.S. and Bento, J.M.S. (eds) São Paulo, Brazil. *Editora Manole*, pp. 377–391.
Chen, H.Q., Dekkers, K.L., Cao, L., Burns, J.K., Timmer, L.W. and Chung, K.R. (2006) Evaluation of growth regulator inhibitors for controlling postbloom fruit drop (PFD) of citrus induced by the fungus *Colletotrichum acutatum*. *HortScience* 41, 1317–1321.
Chung, K. (2011) Elsinoë fawcettii and Elsinoë australis: the fungal pathogens causing citrus scab. *Molecular Plant Pathology* 12, 123–135. doi: 10.1111/j.1364-3703.2010.00663.x.
Chung, K.R., Shilts, T., Ertürk, U., Timmer, L.W. and Ueng, P.P. (2003) Indole derivatives produced by the fungus colletotrichum acutatum causing lime anthracnose and postbloom fruit drop of citrus. *FEMS Microbiology Letters* 226, 23–30. doi: 10.1016/s0378-1097(03)00605-0.
Chung, K., Khan, I. and Brlansky, R. (2006) *Citrus Diseases Exotic to Florida: Witches' Broom Disease of Lime (WBDL)*. IFAS Extentention, Gainesville, Florida.
da Graça, J.V., Douhan, G.W., Halbert, S.E., Keremane, M.L., Lee, R.F., Vidalakis, G. and Zhao, H. (2016) Huanglongbing: an overview of a complex pathosystem ravaging the world's citrus. *Journal of Integrative Plant Biology* 58, 373–387. doi: 10.1111/jipb.12437.
Das, A. (2003) Citrus canker-a review. *Journal of Applied Horticulture* 5, 52–60.
Das, A.K., Nerkar, S., Kumar, A. and Bawage, S. (2016) Detection, identification and characterization of *Phytophthora* spp. Infecting citrus in India. *Journal of Plant Pathology* 98, 55–69. doi: 10.4454/jpp.v98i1.038.
Davies, F.S. and Albrigo, L.G. (1994) *Citrus*. CAB International, Wallingford, UK.
De Barro, P.J., Liu, S.-S., Boykin, L.M. and Dinsdale, A.B. (2011) *Bemisia tabaci*: a statement of species status. *Annual Review of Entomology* 56, 1–19.
Dinsdale, A, Cook, L., Riginos, C., Buckley, Y.M. and De Barro, P. (2010) Refined global analysis of *Bemisia tabaci* (Hemiptera: Sternorrhyncha: Aleyrodoidea: Aleyrodidae) mitochondrial cytochrome oxidase 1 to identify species level genetic boundaries. *Annals of the Entomological Society of America* 103, 196–208.
Dreistadt, S. (2012) *Integrated Pest Management for Citrus*, 3rd edn. University of California, Division of Agriculture and Natural Resources, Oakland, California. p. 275.
Elbert, A. and Nauen, R. (2000) Resistance of *Bemisia tabaci* (Homoptera: Aleyrodidae) to insecticides in southern Spain with special reference to neonicotinoids. *Pest Management Science* 56, 60–64.
EPPO (1990) Specific quarantine requirements. European and Mediterranean Plant Protection Organization (EPPO) Technical Documents No. 1008.
EPPO (2013) PQR database. European and Mediterranean Plant Protection Organization, Paris. Available at: http://www.eppo.int/DATABASES/pqr/pqr.htm (accessed 30 October 2016).
Erwin, D.C. and Ribeiro, O.K. (1996) *Phytophthora Diseases Worldwide*. APS Press, St. Paul, Minnesota.
Evans, G.A. and Stange, L.A. (1997) *Parasitoids Associated with the Brown Citrus Aphid, Toxoptera Citricida, in Florida (Insecta: Hymenoptera)*. Florida Department of Agriculture and Consumer Services, Gainesville. Florida, *Entomology Circular* 384.
Farr, D. and Rossman, A. (2016) Fungal Databases, U.S. National Fungus Collections. ARS, USDA. Available at: http://nt.ars-grin.gov/fungaldatabases/ (accessed 23 April 2016).
Fraser, L.R. (1952) Seedling yellows, an unreported virus disease of citrus. *Agricultural Gazette New South Wales* 63, 125–131.
Garnier, M., Zreik, L. and Bové, J. (1991) Witches' broom, a lethal mycoplasmal disease of lime trees in the sultanate of Oman and the United Arab Emirates. *Plant Disease* 75, 546–551.
Garnsey S. and Lee, R. (1988) Tristeza. In: Whiteside, J.O., Garnsey, S.M. and Timmer, L.W. (eds) *Compendium of Citrus Diseases*. APA Press, St. Paul, Minnesota, pp. 48–50.

Garnsey, S.M., Gottwald, T.R. and Yokomi, R.K. (1998) Control strategies for *Citrus Tristeza Virus*. In: Hadidi, A., Khetarpal, R.K. and Koganezawa, H. (eds) *Plant Viral Disease Control: Principles and Practices*. APS Press, St. Paul, Minnesota, pp. 639–658.
Gerling D and Mayer, R.T. (1995) *Bemisia: Taxonomy, Biology, Damage, Control and Management*. Intercept Ltd, Andover, UK.
Gerling, D., Alomarb, Ò. and Arnó, J. (2001) Biological control of *Bemisia tabaci* using predators and parasitoids. *Crop Protection* 20, 779–799.
Gottwald, T., Graham, J. and Schubert, T. (1997) An epidemiological analysis of the spread of citrus canker in urban Miami, Florida, and synergistic interaction with the Asian citrus leafminer. *Fruits* 52, 383–390.
Gottwald, T.R., Graham, J.H. and Schubert, T.S. (2002) Citrus canker: the pathogen and its impact. *Plant Health Progress* 10. doi:10.1094/PHP-2002-0812-01-RV.
Gottwald, T.R., da Graça, J.V. and Bassanezi, R.B. (2007) Citrus huanglongbing: the pathogen and its impact. *Plant Health Progress* 6. doi:10.1094/PHP-2007-0906-01-RV.
Grafton-Cardwell, E.E. and Ouyang, Y. (1993) Toxicity of four insecticides to various populations of the predacious mite *Euseius tularensis* from San Joaquin Valley California citrus. *Journal of Applied Entomology* 10, 21–29.
Grafton-Cardwell, E.E., Morse, J.G. and Gjerde, A. (1998) Effect of insecticide treatments to reduce infestation by citrus thrips (Thysanoptera: Thripidae) on growth of nonbearing citrus. *Journal of Economy Entomology* 91, 235–242.
Grafton-Cardwell, E.E., O'Connell, N.V., Kallsen, C.E. and Morse, J.G. (2003) *Photographic Guide to Citrus Fruit Scarring*. University of California, Oakland, California, Publ. 8090, p. 8.
Grygiel, P., Seny-Couty, A., Abdou Hassani, F., Boyer, C., Boyer, K., Vernière, C., Pruvost, O. and Hamza, A.A. (2014) First report of *Xanthomonas citri* pv. *citri* pathotype a causing Asiatic citrus canker in Grande Comore and Anjouan. *Plant Disease* 98, 1739. doi: 10.1094/pdis-06-14-0624-pdn.
Haapalainen, M. (2014) Biology and epidemics of candidatus liberibacter species, psyllid-transmitted plant-pathogenic bacteria. *Annals of Applied Biology* 165, 172–198. doi: 10.1111/aab.12149.
Halbert, S.E. and Manjunath, K.L. (2004) Asian citrus psyllids (Sternorrhyncha: Psyllidae) and greening disease of citrus: a literature review and assessment of risk in Florida. *Florida Entomologist* 87, 330–353.
Hall, D.G. and Gottwald, T.R. (2011) Pest management practices aimed at curtailing citrus huanglongbing disease. *Outlooks on Pest Management* 22, 189–192.
Hall, D.G., Richardson, M.L., Ammar, E.D. and Halbert, S.E. (2013) Asian citrus psyllid, *Diaphorina citri*, vector of citrus huanglongbing disease. *Entomologia Experimentalis et Applicata* 146, 207–223.
Heppner, J.B. (1993) Citrus leafminer, *Phyllocnistis citrella*, in Florida (Lepidoptera: Gracillariidae: Phyllocnistinae). *Tropical Lepidoptera* 1, 49–64.
Hilf, M.E. and Lewis, R.S. (2016) Transmission and propagation of 'candidatus liberibacter asiaticus' by grafting with individual citrus leaves. *Phytopathology* 106, 452–458. doi: 10.1094/phyto-09-15-0221-r.
Hogenhout, S.A., Ammar, E., Whitfield, A.E. and Redinbaugh, M.G. (2008a) Insect vector interactions with persistently transmitted viruses. *Annual Review of Phytopathology* 46, 327–359.
Hogenhout, S.A., Oshima, K., Ammar, E.-D., Kakizawa, S., Kingdom, H.N. and Namba, S. (2008b) Phytoplasmas: bacteria that manipulate plants and insects. *Molecular Plant Pathology* 9, 403–423.
Horowitz, A.R., Kontsedalov, S., Khasdan, V. and Ishaaya, I. (2005) Biotypes B and Q of *Bemisia tabaci* and their relevance to neonicotinoid and pyriproxyfen resistance. *Archives of Insect Biochemistry and Physiology* 58, 216–225.
Hou, X., Huang, F., Zhang, T.Y., Xu, J.G., Kevin, H.D. and Li, H.Y. (2014) Pathotypes and genetic diversity of Chinese collections of elsinoë fawcettii causing citrus scab. *Journal of Integrative Agriculture* 13, 1293–1302. doi: 10.1016/s2095-3119(13)60522-5.
Huang, F., Hou, X., Dewdney, M.M., Fu, Y., Chen, G., Hyde, K.D. and Li, H. (2013) Diaporthe species occurring on citrus in China. *Fungal Diversity* 61, 237–250. doi: 10.1007/s13225-013-0245-6.
Ichinose, K., Miyazi, K., Matsuhira, K., Yasuda, K., Sadoyama, Y., Do, H.T. and Doan, V.B. (2010) Unreliable pesticide control of the vector psyllid *Diaphorina citri* (Hemiptera: Psyllidae) for the reduction of microorganism disease transmission. *Journal of Environmental Science and Health B* 45, 466–472.
Jones, D.R. (2003) Plant viruses transmitted by whiteflies. *European Journal of Plant Pathology* 109, 195–219.
Karasev, A., Boyko, V., Gowda, S., Nikolaeva, O., Hilf, M., Koonin, E., Niblett, C., Cline, K., Gumpf, D. and Lee, R. (1995) Complete sequence of the citrus tristeza virus RNA genome. *Virology* 208, 511–520.
Khan, I. and Morse, J.G. (1997) Toxicity of pesticide residues to citrus thrips (Thysanoptera: Thripidae). *Journal of Agricultural Entomology* 14, 409–420.

Khan, I. and Morse, J.G. (1998) Citrus thrips (Thysanoptera: Thripidae) resistance monitoring in California. *Journal of Economic Entomology* 91, 361–366.
Klein, M.N., Silva, A.C., Lopes, M.R. and Kupper, K.C. (2013) Application of microorganisms, alone or in combination, to control postbloom fruit drop in citrus. *Tropical Plant Pathology* 38, 505–512. doi: 10.1590/s1982-56762013000600006.
Lapointe, S.L., Keathley, C.P., Stelinski, L.L., Urrutia, W.H. and Mafra-Neto, A. (2015) Disruption of the leafminer *Phyllocnistis citrella* (Lepidoptera: Gracillariidae) in citrus: effect of blend and placement height, longevity of disruption and emission profile of a new dispenser. *Florida Entomologist* 98, 742–748.
Lee, R. and Bar-Joseph, M. (2000) Tristeza. In: Timmer, L.W., Garnsey, S.M and Graham. J.H. (eds) *Compendium of Citrus Diseases*, 2nd edition. APS Press, St. Paul, Minnesota, pp. 61–63.
Li, J., Trivedi, P. and Wang, N. (2016) Field evaluation of plant defense inducers for the control of citrus huanglongbing. *Phytopathology* 106, 37–46. doi: 10.1094/phyto-08-15-0196-r.
Liu, Y.H. and Tsai, J.H. (2002) Effect of temperature on development, survivorship and fecundity *Lysiphlebia mirzaii* (Hymenoptera: Aphidiidae). *Environmental Entomology* 31, 418–424.
Lopes, M.R., Klein, M.N., Ferraz, L.P., da Silva, A.C. and Kupper, K.C. (2015) Saccharomyces cerevisiae: a novel and efficient biological control agent for *colletotrichum acutatum* during pre-harvest. *Microbiological Research* 175, 93–99. doi: 10.1016/j.micres.2015.04.003.
Mackauer, M. (1976) *Myzus persicae* Sulz., an aphid of world importance. In: Delucchi, V.L. (ed.) *Studies in Biological Control. International Biological Programme* 9. Cambridge University Press, Cambridge, UK, pp. 51–119.
Matheron, M.E. and Porchas, M. (2002) Comparative ability of six fungicides to inhibit development of *Phytophthora gummosis* on citrus. *Plant Disease* 86, 687–690.
Mathews, D., Riley, K. and Dodds, J. (1997) Comparison of detection methods for citrus tristeza virus in field trees during months of nonoptimal titer. *Plant Disease* 81, 525–529.
McClean, A.P.D. (1960) Seedling-yellows in South African citrus trees. *South African Journal of Agriculture Science* 3, 259–279.
McFarland, C.D. and Hoy, M.A. (2001) Survival of *Diaphorina citri* (Homoptera: Psyllidae) and its two parasitoids, *Tamarixia radiata* (Hymenoptera: Eulophidae) and *Diaphorencyrtus aligarhensis* (Hymenoptera: Encyrtidae), under different relative humidities and temperature regimes. *Florida Entomologist* 84, 227–233.
Michaud, J.P. (2000) Development and reproduction of ladybeetles (Coleoptera: Coccinellidae) on the citrus aphids *Aphis spiraecola* Patch and *Toxoptera citricida* (Kirkaldy) (Homoptera: Aphididae). *Biological Control* 18, 287–297.
Michaud, J.P. (2001) Evaluation of green lacewings, *Chrysoperla plorabunda* (Fitch) (Neuroptera Chrysopidae), for augmentative release against *Toxoptera citricida* (Homoptera Aphididae) in citrus. *Journal of Applied Entomology* 125, 383–388.
Michaud, J.P. and Belliure, B. (2001) Impact of syrphid predation on production of migrants in colonies of the brown citrus aphid, *Toxoptera citricida* (Homoptera: Aphididae). *Biological Control* 21, 91–95.
Mondal, S.N., Vicent, A., Reis, R.F. and Timmer, L.W. (2007) Efficacy of pre- and postinoculation application of fungicides to expanding young citrus leaves for control of melanose, scab, and alternaria brown spot. *Plant Disease* 91, 1600–1606. doi: 10.1094/pdis-91-12-1600.
Moreno, P., Ambros, S., Albiach-Martí M.R., Guerri, J. and Pena, L. (2008) Citrus tristeza virus: a pathogen that changed the course of the citrus industry. *Molecular Plant Pathology* 9, 251–268.
Mound, L.A. and Halsey, S.H. (1978) *Whiteflies of the World, a Systematic Catalogue of the Aleyrodidae (Homoptera) with Host Plant and Natural Enemy Data.* British Museum (Natural History), London.
Naqvi, S. and Singh, S. (2002) Fungal diseases of citrus–diagnosis and management. *Technical Bulletin* 5, 61.
Nava, D.E., Torres, M.L.G., Rodrigues, M.D.L., Bento, J.M.S. and Parra, J.R.P. (2007) Biology of *Diaphorina citri* (Hemiptera: Psyllidae) on different hosts and at different temperatures. *Journal of Applied Entomology* 131, 709–715.
Navarro, L., Juarez, J., Pina, J., Ballester, J. and Arregui, J. (1988) The citrus variety improvement program in Spain after eleven years. *Proc 10th Conference International Organization Citrus Virol IOCV*. Balaband, Rehovot, Israel and Margrat Publishers, Weikersheim, Germany, p. 400.
Nelson, S. (2008) Citrus scab. *Plant Disease* 60, 1–6.
Orozco-Santos, M. (1996) *Enfermedades presentes y potenciales de los cítricos en México*. Universidad Autónoma Chapingo, México.

Orozco-Santos, M., Medina-Urrutia, V., Robles-González, M., Orozco-Romero, J., Pérez-Zamora, O., Veláquez-Monreal, J., Timmer, L. and Guzmán-Gonzáles, S. (2006) Biología y manejo integrado de antracnosis del limón mexicano en el trópico seco de México. INIFAP, Folleto Técnico Núm. 2. Tecomán, Colima, México.
Paraná (2008) Lei Estadual n° 15.953, de 24 de setembro de 2008. Proíbe o plantio, comércio, transporte e produção da planta Murta (*Murraya paniculata*), por ser vegetal hospedeiro da bactéria *Candidatus* Liberibacter ssp., disseminada pelo inseto vetor *Diaphorina citri*, transmissor da praga denominada Huanglongbing (HLB – Greening). Diário Oficial do Estado do Paraná, Curitiba, PR.
Peres, N.A., MacKenzie, S.J., Peever, T.L. and Timmer, L.W. (2008) Postbloom fruit drop of citrus and key lime anthracnose are caused by distinct phylogenetic lineages of Colletotrichum acutatum. *Phytopathology* 98, 345–352. doi: 10.1094/phyto-98-3-0345.
Ploetz, R.C. (2003) *Diseases of Tropical Fruit Crops*. CAB International, Wallingford, UK.
Polek, M. (2007) *Citrus Bacterial Canker Disease and Huanglongbing (Citrus Greening)* UCANR Publications, University of California, Oakland, California.
Queiroz, R.B., Donkersley, P., Silva, F.N., Al-Mahmmoli, I.H., Al-Sadi, A.M., Carvalho, C.M. and Elliot, S.L. (2016) Invasive mutualisms between a plant pathogen and insect vectors in the Middle East and Brazil. *Royal Society Open Science* 3(12), 160557. doi:10.1098/rsos.160557.
Redondo, C., Sena-Vélez, M., Gell, I., Ferragud, E., Sabuquillo, P., Graham, J.H. and Cubero, J. (2015) Influence of selected bactericides on biofilm formation and viability of Xanthomonas citri subsp. citri. *Crop Protection* 78, 204–213. doi: 10.1016/j.cropro.2015.09.010.
Rohrig, E.A., Hall, D.G. and Stansly, P.A. (2012) Field releases in Florida of *Diaphorencyrtus aligarhensis* (Hymenoptera: Encyrtidae) an endoparasitoid of *Diaphorina citri* (Homoptera: Psyllidae) from mainland China. *Florida Entomologist* 95, 476–478.
Roistacher, C. and Bar-Joseph, M. (1987) Aphid transmission of citrus tristeza virus: *a review. Phytophylactica* 19, 163–167.
Sadeghy, B., Rohani, M. and Aghasi, N. (2014) Distribution of Phytophthora citrophthora R.E. Sm. and E.H. Sm. and P. parasitica Dastur within citrus orchards in Kerman province, Iran. *Archives of Phytopathology and Plant Protection* 47, 254–258. doi: 10.1080/03235408.2013.808850.
Salehi, M., Izadpanah, K., Siampour, M., Bagheri, A. and Faghihi, S.M. (2007) Transmission of 'Candidatus Phytoplasma aurantifolia' to Bakraee (Citrus reticulate hybrid) by feral Hishimonus phycitis leafhoppers in Iran. *Plant Disease* 91, 466.
Savita, V.G. and Nagpal, A. (2012) Citrus diseases caused by Phytophthora species. *GERF Bulletin Biosciences* 3, 18–27.
Shabani, M., Bertheau, C., Zeinalabedini, M., Sarafrazi, A., Mardi, M., Naraghi, S.M., Heshmatollah Rahimian, H. and Shojaee, M. (2013) Population genetic structure and ecological niche modeling of the leafhopper *Hishimonus phycitis*. *Journal of Pest Science* 86, 173–183.
Silva-Junior, G.J., Spósito, M.B., Marin, D.R. and Amorim, L. (2014) Efficacy and timing of application of fungicides for control of citrus postbloom fruit drop. *Crop Protection* 59, 51–56. doi: 10.1016/j.cropro.2014.01.012.
Smith, J.M. and Hoy, M.A. (1995) Rearing methods for *Ageniaspis citricola* (Hymenoptera: Encyrtidae) and *Cirrospilus quadristriatus* (Hymenoptera: Eulophidae) released in a classical biological control program for the citrus leafminer *Phyllocnistis citrella* (Lepidoptera: Gracillariidae). *Florida Entomologist* 78, 600–608.
Spiegel-Roy, P. and Goldschmidt, E.E. (1996) *Biology of Citrus*, 1st edn. Cambridge University Press, Cambridge, UK.
Subandiyah S., NIkoh, N., Sato, H., Wagiman, F., Tsyuymyu, S. and Fukatsu, T. (2000) Isolation and characterization of two entomopathogenic fungi attacking *Diaphorina citri* (Homoptera, Psylloidea) in Indonesia. *Mycoscience* 41, 509–513.
Sugio, A. and Hogenhout, S.A. (2012) The genome biology of phytoplasma: modulators of plants and insects. *Current Opinion in Microbiology* 15, 247–254. doi: 10.1016/j.mib.2012.04.002.
Takanashi, M. (1990) Development and reproductive ability of *Lysiphlebus japonicus* Ashmead (Hymenoptera: Aphidiidae) parasitizing the citrus brown aphid, *Toxoptera citricidus* (Kirkaldy) (Homoptera: Aphidiidae). *Japanese Journal of Applied Entomology and Zoology* 34, 237–243.
Timmer, L. (2000) Anthracnose diseases. In: Timmer, L.W., Garnsey, S.M. and Graham, J.H. (eds) *Compendium of Citrus Diseases,* 2nd edition. APS Press, St. Paul, Minnesota, pp. 21–23.
Timmer, L. and Zitko, S. (1997) Evaluation of fungicides for control of Alternaria brown spot and citrus scab. *Proceedings – Florida State Horticultural Society, Florida State Horticultural Society*. 110, 71–76.

Timmer, L.W., Garnsey, S.M. and Graham, J.H. (2000) *Compendium of Citrus Diseases*. APS Press, St. Paul, Minnesota.

Tiwari, S., Clayson, P.J., Kuhns, E.H. and Stelinski, L.L. (2012) Effects of buprofezin and diflubenzuron on various developmental stages of Asian citrus psyllid, Diaphorina citri. *Pest Management Science* 68, 1405–1412. doi: 10.1002/ps.3323.

Tsai, J.H. and Liu, Y.H. (2000) Biology of *Diaphorina citri* (Homoptera: Psyllidae) on four host plants. *Journal of Economic Entomology* 93, 1721–1725.

Waller, J.M. and Bridge, J. (1978) Plant diseases and nematodes in the sultanate of Oman. *PANS* 24, 313–326.

Wang, J.J. and Tsai, J.H. (2001) Development and functional response of *Coelophora inaequalis* (Coleoptera: Coccinellidae) feeding on brown citrus aphid, *Toxoptera citricida* (Homoptera: Aphididae). *Agricultural and Forest Entomology* 3, 65–69.

Wang, N. and Trivedi, P. (2013) Citrus huanglongbing: a newly relevant disease presents unprecedented challenges. *Phytopathology* 103, 652–665. doi: 10.1094/phyto-12-12-0331-rvw.

Yan, H., Zeng, J. and Zhong, G. (2015) The push–pull strategy for citrus psyllid control. *Pest Management Science* 71, 893–896. doi: 10.1002/ps.3915.

Yokomi, R.K., Lastra, R., Stoetzel, M.B., Damsteegt, V.D., Lee, R.F., Garnsey, S.M., Gottwald, T.R., Rocha-Pena, M.A. and Niblett, C.L. (1994) Establishment of the brown citrus aphid (Homoptera: Aphididae) in Central America and the Caribbean Basin and transmission of *Citrus Tristeza Virus*. *Journal of Economic Entomology* 87, 1078–1085.

11 Innovative Production Technologies

Muhammad Fakhar-ud-Din Razi[1]*, Rhonda Janke[2], Muhammad Mumtaz Khan[2] and Um-e-Ammara[2]

[1]Department of Horticulture, University of Agriculture, Faisalabad Sub-Campus Burewala, Vehari, Punjab, Pakistan; [2]Department of Crop Sciences, College of Agricultural and Marine Sciences, Sultan Qaboos University, Al Khod, Oman

Agriculture has been reshaped due to technological advancements in the past 100 years (Schultz, 1964) and this is likely to continue because of the ever-growing human population, decreasing cultivable lands and the cost effectiveness of modern and innovative production technologies. The total cultivated land in the USA shrunk from 141.7 to 129.55 million ha between 1920 and 1995; similarly, the farm labour force also decreased from 9.5 to 3.3 million people during the same period of time. However, the gross agricultural crop production in the USA during 1995 was recorded to be 3.3 times higher than in 1920 (US Bureau of the Census, 1975; US Bureau of the Census, 1980; US Bureau of the Census, 1998). Despite the fact that the world population has more than doubled from 1950 to 1998, incredible changes have been observed in the patterns of crop production in different regions. Interestingly, the harvested acreage per person dropped by half while the grain production increased by nearly 12% per person (Brown *et al.*, 1999). It is evident from the facts that agricultural productivity has increased; however, it is a cumulative effect of better and improved seeds, farm mechanization, enhanced insect pest and disease control and improved production methods. Innovation is a broad term and holds its own unique vocabulary. Innovation may be as simple as a method of trapping a specific insect or as advanced and complex as the latest site-specific crop management (SSCM), also known as precision farming. In relevance to agriculture, the term 'innovation' may be defined as 'a new method, technique or technology which reduces human effort, minimizes cost of implementation, protects soil, atmosphere, water resources, human health, wild life and flora while increasing crop yield and quality under a given socio-economic scenario'. It involves two main research streams: (i) research to generate an innovation, and (ii) analysis of adoptability of the research-oriented invention and innovation. The second is exclusively connected with the agricultural extension service and will not be discussed in this chapter. Innovations are differentiated into policies and models and are categorized as *embodied* and *disembodied*. The *embodied* innovations include capital, goods and/or products, for example, machines, chemicals such as fertilizers and pesticides, new varieties of seeds and other farm inputs, whereas *disembodied* innovations include comprehensive pest management techniques, efficient weed control etc., and are designed to develop public interest, motivation and investment in innovation generation (Sunding and Zilberman, 2001). Non-government organizations mostly avoid financing disembodied

*E-mail: fakharuddin_razi@hotmail.com

innovations because of marketing problems in disposing of the products. The agricultural innovations are classified as, but not confined to, mechanical, biological, chemical, agronomic, biotechnological and informational innovations that depend on computer technologies. None of these groups is constant and any may undergo a change or improvement either due to the introduction of new research or because of socio-economic or environmental considerations.

Digital Agriculture

Living in the digital era, agricultural technology is rapidly changing and reshaping field operations. The latest and digitalized equipment is connecting farm fields and satellites for real-time crop monitoring, leading to the highest level of precision in agricultural practices. Precision or 'site-specific agriculture' is an outcome of scientific revolution that has speedily developed a new approach in agricultural systems, which monitors, controls and performs diverse agricultural management practices in the field in accordance with the real-time field conditions. This is based on multidisciplinary concepts of systems approaches to identify and solve problems with the least human effort in a cropping zone. Robert *et al.* (1995) defined SSCM as a system completely based on information processed by a monitoring system and technology with the ability to detect, analyse and react to any spatial or temporal changes on the farm field, resulting in the best control measures for a wide range of problems. The system is designed not only to generate optimum and sustainable profits, but also to protect the environment from the deleterious effects of various technologies and field inputs (Robert *et al.*, 1995). Briefly, the system consists of several components including GPS to locate a specific point in the field through a connection with three (in actuality a fourth one is used for error correction) of 24 orbiting satellites at any time. The satellite data are received on the Earth and the information is used by the farm machinery equipped with onboard sensors to monitor crop load and give commands for crop inputs. Precision agriculture helps to reduce production costs by using the inputs (fertilizers, pesticides, etc.) as per actual plant needs. The technology may also have a positive impact on the environment through the reduction of fertilizer or pesticide application, which contaminates streams and underground water resources. GPS is the basis for technological developments for field mapping in small grids. It provides data and yield responses to different input levels, which are processed by geographic information system (GIS) computers installed with advanced software programs, which process the data and provide field operation guidelines. This information is used to apply variable rate technology (VRT), which automates the equipment to change application rates as per the exact field conditions and location of the equipment (Skotnikov and Robert, 1996). This crop management system (SSCM) is in the development phase; however, it covers different components of precision agriculture including yield monitoring, yield mapping, variable rate fertilizers, weed mapping, variable spray mapping, topography and boundaries, salinity mapping, guidance system and records and systems analysis.

Growing limes involves as many factors as one can imagine. The final crop yield depends upon plant genetic resources, rootstocks, weather and climate, soil–water relationships, soil physical and chemical properties, slope and related aspects of a given site, crop inputs, field history and cultural practices (Tom, 1997). When the permanent features of a site (weather, climate, soil, etc.) are suitable for a fruit crop, the yield will depend upon efficient plant water use or evapotranspiration (de Wit, 1958; Ritchie, 1983). Water availability to plants can explain up to 69% of yield fluctuations in agronomic crops (Batchelor and Paz, 1998) but the same cannot be assumed to be true for tree fruit crops of a perennial nature. Being extremely critical, crop water requirements for limes and other commercially grown citrus fruits should be well programmed according to local weather and climate, soil type and stages of plant growth.

Remote Sensing Technology in Citrus Groves

Remote sensing is generalised as a process of collecting data from distant place/s by the analysis of reflection and emission of energy waves from objects such as plants (Aggarwal, 2008). This technology consists of several components and is applicable to a wide range of real-time field assessments.

Fluorescence spectroscopy is one of the remote sensing techniques and is used to record disease incidence and plant nutrient insufficiencies (Belasque *et al.*, 2008). Fluorescence imaging is a modern technique in which real-time photography is subjected to analysis of fruit quality, level of photosynthetic activity, tissue anatomy and disease symptomatology in plants (Moshou, *et al.*, 2005; Chaerle *et al.*, 2007). Remote sensing is also applied to estimate crop load in citrus (Shrivastava and Gebelein, 2006), canopy size measurement (Schumann and Zaman, 2008) and crop maturity indices (Nageswara Rao *et al.*, 2004). Citrus groves were mapped for crop size estimation by Whitney *et al.* (2002) with the help of a mechanical ultrasonic system, whereas Zaman *et al.* (2006) applied a sensor-based crop load monitoring technique. Yield mapping based on remote sensing was also conducted by Schumann *et al.* (2004) and Schumann and Hostler (2008) using colour shooting and ultrasonic sensors, respectively. Trees in citrus orchards infested with soft bodied scales were identified by Hart and Myers (1968) with the help of colour-infrared (CIR) photography associated with spectral reflectance on the basis of development of sooty mould on the secretions of the insects. Citrus whitefly and black fly were also detectable using the same techniques (Everitt *et al.*, 1994). Remote sensing techniques have been tested to detect various diseases on citrus (Huang *et al.*, 2007) and are applicable to a varying extent in different citrus varieties under agro-climatic conditions.

Scope of Robotics in Production of Limes

The availability of farm labour has been continuously decreasing due to urbanization and greater attraction to industrial jobs across the world. This necessitates the automation of farm operations with optimum accuracy and efficiency. Several farm operations such as application of herbicides require involvement of chemicals that pollute the soil, water and air. Pruning is the removal of unnecessary wood and involves huge expenditures. Robots look to be an attractive alternative for these and many other field operations in lime plantations. Robotics in agriculture systems probably started with a robot meant for harvesting of tomatoes (Kawamura *et al.*, 1984). Several robotic harvesters for a wide range of horticultural crops, especially fruits, have been examined in research trials. These include cucumber (Van Henten *et al.*, 2002), cherry (Tanigaki *et al.*, 2008) and citrus (Hannan and Burks, 2004) harvesting robots. There has been little success in commercialization of harvesting robots except a machine meant for the harvesting of cherries as reported by Kondo *et al.* (2005). Several weed species are known to have developed resistance to herbicides (Grift *et al.*, 2008). Monitoring and removal of weeds is probably the easiest task for the robots and can be performed on the basis of weed morphological characteristics by using 'charge coupled device' (CCD) cameras (Tian, 2002). A prototype robot for weed control was developed in 1998, while Bak and Jakobsen (2004) developed a smart field robot that could be applied between the plants or crop rows and identify the level of the weed population. Other robots were developed by Hofstee *et al.* (2004) and Grift *et al.* (2005) for similar purposes. Although so far there are no robots commercially available to perform operations in lime groves specifically, the prospects of robotics application in limes are equal to many other fruit crops.

Use of Nanoparticles in Citrus Production

Several diseases are reported to infect citrus plantations throughout the world. Among these, some of the fungal diseases such as *Alternaria citri* cause post-harvest fruit rot. It has been reported that silver nanomaterials help control the fruit decay caused by *Penicillium* and *Alternaria* species (Sanzani *et al.*, 2012; Youssef *et al.*, 2012). Gram-negative bacteria cause three major diseases in citrus including Huanglongbing, citrus canker and variegated chlorosis. These diseases were thought to have no treatment options before the advent of nanoparticles. Nanomaterials are less than 100 nm in any of their three dimensions and are thought to be helpful in strengthening plant defence mechanisms. Several experiments are in progress to control bacterial diseases in citrus by using nanomaterials of zinc, copper, silver and other metals (Abdelmalek and Salaheldin, 2016). Novel engineering techniques of innate immune defence will be applied against the invasive pathogens with a three-step disease control strategy, i.e. identification of plant signals of pathogenic infection, signal interpretation and application of nanoparticles for patholysis (Gupta, 2014).

Crop Water Requirements and Irrigation Plan

Crop water requirements vary greatly as the soil, climatic conditions and plant growth phases change, and additional water may be needed even in regions with 900–1400 mm of annual rainfall (Jackson, 2011) because even 24 h of water stress may affect the yield very seriously. The topic is fully covered in Chapter 8; a few basic concepts of innovative irrigation techniques are discussed here. The irrigation schedules should be kept flexible in accordance with the soil water holding capacity, expected amount of rainfall, its distribution and reliability and evapotranspiration rates. The available water in different soils varies greatly and is estimated to be 0–66 mm per 1 m depth of sand and 160–177 mm for a clay loam soil. Irrigation plans should be based upon meteorological data and daily water loss from the soil. Evaporimeters – Type A – are used for estimating daily evaporation. Orchard floor management also helps to reduce water losses from the soil surface; evapotranspiration losses are decreased if the soil between the tree or vine rows is free from weeds and grass (Jackson, 2011).

Irrigation Decisions and Integrated Nutrient Management

Irrigation schedules are designed depending upon primary information about soil type, the amount of water available in the soil, seasonal rainfall and evapotranspiration. A poorly structured clay loam soil with a hardpan at 1 m depth holds about 165 mm moisture above permanent wilting point. Once such soils are at field capacity, a fully grown orchard will reach permanent wilting point after 21 days (Jackson, 2011) and daily evaporation is averaged to 8 mm. However, in a well structured silt loam soil with 2 m depth, about 370 mm of water would be available between field capacity and permanent wilting point, which would be able to meet plant water requirements for about 46 days when mean daily evapotranspiration is estimated to be 8 mm per day. While growing limes, a suitable irrigation system should be adopted as per local climatic conditions and soil characteristics. Flood irrigation wastes a lot of irrigation water and also washes away nutrients. Other systems like sprinklers (overhead and beneath tree), trickle or drip irrigation demand capital as well as engineering support with many other merits and demerits.

In citrus production, about 20–30% of the production cost every year is spent on fertilizers alone (Srivastava and Singh, 2003; Srivastava and Singh, 2005; Srivastava and Singh, 2008). Citrus production has been experiencing repeated crop failure and huge yield fluctuations during on and off years in developing countries (Smith, 1976; Rojas, 1998). Current studies have exclusively favoured application of mineral fertilizers to maintain plant nutrient levels, which is continuously falling due to demographic status, changing living standards and business opportunities (Reddy *et al.*, 2003). Crop nutrition must be planned according to integrated nutrient management (INM) prescriptions.

INM is devised for the best economic returns from an orchard while protecting soil physical, biochemical and microbial health (He *et al.*, 1997). In the absence of an appropriate INM, agricultural mining of the nutrients ultimately disturbs the balance between soil and plant nutrient demands as well as yearly nutrient recommendations and applicable doses. Citrus orchard soils differ from the soils of general crops, which are not cultivated for a few to several months each year. Organic matter in such soils undergoes gradual depletion; however, in soils of permanent plantations, the rate of biological oxidation of existing organic matter is slower (Srivastava *et al.*, 2002). As stated by Joa *et al.* (2006) the deficiency of multiple plant nutrients has an increasing effect on atmospheric carbon dioxide (CO_2). Therefore, soil organic matter content is thought to be a vital parameter to explain the long-term effects of INM as explained by Singh *et al.* (1999).

Soil fertilization

Fertilization in citrus orchards with commercial productivity is essential; however, one must understand that it is the trees, not the soil that must be fertilized. Roots are the main organs to absorb water and nutrients from the soil. The roots may pick up the nutrients differentially as dictated by sink strength such as when fruiting or when there is newly emerging vegetative growth (Boaretto *et al.*, 1999; Srivastava *et al.*, 2007).

Different techniques can be used for fertilizer application in orchard soils. Circle banding, strip band application and hole placement are commonly used for fertilizer application in citrus groves. In circular banding, furrows up to 30 cm deep and about 20 cm wide are made in a circular pattern along the edges of the canopy; while in strip banding furrows of the same dimensions are made in parallel between the lines. In hole application of fertilizers, holes are made in sets of four or five per tree with uniform sizes (30 cm deep with 8–10 cm radius). They are made along the outer edge of the tree canopies, filled with the fertilizers and irrigated (Srivastava, 2012).

Citrus nutrition requires regular and routine analysis of plant and soil nutrient status, without which nutrient doses cannot be optimized. Most of the lab and field instruments show some level of complexity and sophistication. Such equipment can both directly and indirectly measure the nutrients as well as trace elements with further ability to improve the results through statistical approaches and mathematical relationships with direct and secondary analytical techniques. These instruments are equally applicable to research and precision-based nutrition of citrus crops. Some of these include; high-temperature slow oxidation, microwave digestion (Soon and Kalra, 1994), the sequential multiple analyzer also sold as the AutoAnalyzer (Handson and Shelley, 1993), segmented flow analysis (SFA) (Ceballos *et al.*, 2006), flow injection technology (FIT) (Carrasco *et al.*, 2007), atomic absorption and inductively coupled plasma spectroscopy (Evans *et al.*, 2003; Husted *et al.*, 2011), coupled plasma mass spectrometry (CP-MS) (Laursen *et al.*, 2009), sensors (Min *et al.*, 2008) and near-infrared spectroscopy (Menesatti *et al.*, 2010). As well as these instruments, visual deficiency symptoms, plant growth patterns and production data are of significant importance. However, analysis of these demands a good level of competence as many of the nutrient deficiency symptoms may be misleading because of several biotic and abiotic stresses.

Innovative Fertilizer Application Techniques in Citrus

Direct or soil application of fertilizers has several drawbacks including leaching contamination of water resources, volatilization (Sherlock and Goh, 1985) oriented pollution in the atmosphere and direct or indirect financial losses from procuring huge quantities of chemical and organic fertilizers. The idea of feeding trees through their stem xylem tissues may be acceptable for citrus growers. The method was used for small-scale studies to resolve uptake or translocation issues of nutrients like iron or potassium. It has been proven that only a fraction of the applied fertilizer dose is picked up and consumed by the root system, whereas a major part, i.e. 62–85% N (Gaines and Gaines, 1994) and 80–95% phosphorus and potash, is lost through leaching from the root zone, physical and biological fixation and volatilization (Morgan *et al.*, 2006). The idea of administering fertilizers in injectable form into the tree trunk was tested and proved that only 5–10% of the typical fertilizer levels were sufficient for normal growth and yield. Growth sizes in mango (*Mangifera indica*) and grapevines (*Vitis vinifera*) were noted to be 20–25% and 32–49%, respectively, higher in trunk-injected plants as compared with soil-fertilized plants. It was illustrated that reducing sugars and ethanol in fresh grape juice increased by 7.5–11.9% and 41.4–50%, respectively, with a 6.2–19.7% decrease in pulp/juice acidity (Shaaban, 2009). Guava and kaki trees of 7 years of age produced 84% and 89% higher yield, respectively, with stem nutrition as compared to soil application. Trunk nutrition was reported to generate more growth in young citrus plants when nutrients were injected into the tree trunks suggesting the method to be simple, economically feasible and environment friendly.

Foliar Nutrition

Foliar nutrition of plants is based on the idea of feeding the plants rather than the soil and is the fastest technique of applying inorganic nutrients to the above ground plant parts. However, its effectiveness depends on many endogenous (leaf anatomy and tissue structure) and exogenous meteorological and ecological aspects. The concept of foliar feeding dates back to 1844 (Srivastava and Singh, 2003) and became popular by the mid-20th century. Now it is considered an important method of feeding plants, and has replaced the natural method of nutrient application through the soil (Marschner, 1995; Starck, 1997; Michałojć and Szewczuk, 2003). Although restriction factors such as plant species, mineral nutrient/s dose/s and inadequate translocation from leaves restrain plant nutrition through the foliage (Szewczuk and Michałojć, 2003a), it is still of crucial importance in removing nutrient deficiency, improving tree nutritional status, maximizing both yield and quality, and mitigating the shocks of drought and freeze injuries (Smoleń, 2012).

Leaf age, turgor and foliar nutrition

With greater leaf area, higher nutrient uptake is generally seen. However, young developing leaves accept nutrients and photosynthates more readily (Fageria *et al.*, 2009). Leaves in the developing phase have thinner cuticle surfaces and allow rapid infiltration of minerals, while mature leaves with greater photosynthetic rates nourish other plant parts. Most of the aerially applied nutrients taken up by mature leaves are therefore translocated to new growth, flowers, fruit and roots along with photosynthetic products. During leaf senescence, endogenous mineral nutrients are remobilized for storage in shoots; that is, foliar nutrition should be avoided as the uptake efficiency is decreased at this stage. Similarly, nutrients are leached from the leaf surface during rain and fog as cell membrane permeability is increased under such conditions (Marschner, 1995).

Leaf turgor is characterized by changes of a diurnal nature influenced by external factors including light, temperature and wind. These changes influence plant water relations. At turgidity just after sunrise and late afternoon, leaves exhibit the highest absorption and metabolic consumption of applied mineral nutrients. During midday when leaf turgor is lowest, the nutrient uptake and use efficiency is decreased or even stopped during a hot summer with ample sunshine. Foliar feeding remains effective when the leaves maintain their turgor for longer periods, particularly in the morning, during cloudy periods with high humidity and at moderate temperatures (Smoleń, 2012).

Nutrient deficiency and efficiency of foliar nutrition

Excessive foliar application and doses of nutrients are likely to show negative results. Mineral toxicity may be observed even if low concentrations are applied when the chemical form of a nutrient element is not suited to a specific plant species. Most of the micronutrients are limited in the soil because of low levels, unsuitable pH, nutrient antagonistic relationships, and soil moisture and carbon content.

A deficiency can be more quickly ameliorated by foliar application than soil application (Marschner, 1995). Under ideal conditions, mineral nutrients are absorbed 8–20 more times efficiently through foliage than soil application (Kuepper, 2003). Commonly, in the aerial plant parts nutrient deficiency (as in case of phosphorus and calcium) occurs due to impaired translocation rather than unavailability in the soil. Calcium deficiency causing citrus blossom-end rot is caused by disturbances in water relations and problems in its transport to the exterior of the leaves and berry rind. Citrus trees show Ca deficiency symptoms in highly acidic soils and under long-term cloudy or rainy regimes.

External factors affecting foliar nutrition

Srivastava and Singh (2003) divided external factors into two categories: environmental and spray solution factors. The most crucial exogenous features affecting foliar fertilization include light duration, intensity and quality, air temperature, wind speed, time of day, relative humidity, rainfall, drought, soil water, crop load and plant nutritional status. Factors associated with solution effectiveness include water (solvent) quality, solute (nutrient) concentration and pH of the final solution. Environmental factors directly influence plant physiological and biochemical activities in relation to foliar feeding efficiency. In bright sunlight the mist of working liquids acts as tiny lenses focusing sun radiation onto leaves and causing leaf burn and necrosis. High temperature causes turgor loss, which hampers biological reactions and impairs the uptake of nutrients. At higher air temperatures and low humidity spray droplets rapidly evaporate, reducing nutrient diffusion from the cuticle and increasing viscosity of the remaining working solution (Marschner, 1995). Prolonged high temperature, solar intensity and day length changes the structure of waxes in the cuticle. It is emphasized that temporary modifications in cuticle structures facilitate nutrient absorption (Komosa, 1990). Marschner (1995) reported that the rate of absorption by leaves increases during daytime as compared with the nocturnal period because of more rhythmic metabolism. However, the working solution droplets quickly dry at noon. Foliar nutrient application in the late afternoon was recommended by Szewczuk and Michałojć (2003a, b), and also Fageria *et al.* (2009) suggested late afternoon (2–3:00 pm) to be good. However, late afternoon applications keep the leaf surface wet and increase bacterial and fungal infections, while additions of fungicides may reduce nutrient absorption (Schönherr, 2002). The uptake efficiency of inorganic nutrients never reaches 100% by foliar application even in the case of urea (Bondada *et al.*, 2006). Significant quantities of the nutrients persist on the epidermis as 'dry deposit' after evaporation of the working solution droplets, and are generally absorbed due to redissolving by air moisture. The ionic strength of the working solution varies considerably depending upon the salts and concentrations used, particularly when multinutrients (mixtures) are applied. The ability to rehydrate by air moisture varies greatly for different compounds and air temperature might have very little effect on it (Kolthoff *et al.*, 1969). The nature and concentration of the compound may be important in this regard.

The permeability of the leaf cuticle layer mostly increases when relative humidity is high as it reduces evapotranspiration and increased leaf turgor (Schönherr, 2001; Schönherr and Luber, 2001; Schönherr, 2002). The efficiency of foliar nutrition also depends on solution pH (Marschner, 1995) and leaf damage is low if the pH of the working solution is low. Komosa (1990) stated that better nutrient absorption occurs at pH values from 3 and 4, while with alkaline pH values, ammonium ion toxicity is increased. Zekri and England (2010) stated pH values from 5.0 to 7.5 are optimum for foliar nutrition in citrus plants. Above or below the critical pH value nutrient uptake is decreased and it also causes damage or burns to leaves and fruit.

Synthetic Plant Growth Stimulants and Their Use in Citrus

In horticultural terms 'optimum' yield means production sustained year after year while utilizing land, energy, tree nutritional sources and inputs at the full potential. The 'optimum yield' is a function of plant density, tree canopy size, flowering, fruit set and finally the harvested fruit mass. This level of optimum production is achieved after 5–12 years of planting and depends on plant density and cultivars (Boswell *et al.*, 1970; Boswell *et al.*, 1975); the orchard shows equilibrium of growth and production unless disturbed by environmental conditions or other factors. Citrus growth and performance is a product of several internal and external factors. Among these, plant growth regulators (PGRs) have been used to maximize tree output in terms of fruit production and quality. Application of 2,4-dichlorophenoxy acetic acid (2,4-D) in plant multiplication, fruit growth, thinning and prevention of abscission layer in mature fruit is widely documented across regions and cultivars (El-Otmani *et al.*, 2000). Several plant growth regulating chemicals are used in citrus production worldwide from nursery production to regulation of plant growth phases and crop load management in the orchard.

Application of plant growth regulators in citrus nurseries

In citrus nurseries, PGRs are used to reduce seed germination time (hastened germination), and enhance subsequent growth of seedlings and budding/grafting success. The reports show that soaking of seeds in 40 ppm gibberellic acid (GA_3) or naphthalene acetic acid (NAA) solution for 12 h increase both germination rate and seedling growth. Lemons, limes and citron cuttings root easily when treated with indole butyric acid (IBA) as reported by Moss (1975), whereas, bud union success was increased by using IBA and IAA (Coggins and Hield, 1968), while 6-benzyladenine (BA) and 6-(benzylamine)-9-(2-tetrahydropyranyl)-9H-purine promoted bud swelling (Nauer *et al.*, 1979). Oslund and Davenport (1987) suggested that in Tahiti lime marcotts endogenously produced cytokinins might have some role in promoting root growth in marcotted branches. Diwaker and Katiyar (2013) reported that acid lime stem cuttings showed maximum sprouting (24.33%) and number of roots per cutting (7.67) in response to combined application of IBA and p-hydroxy benzoic acid (PHB) at 2000 and 1000 ppm, respectively. They also found primary root length and diameter (9.33 cm and 2.80 mm) to be the highest with the same treatment. However, sprout diameter (3.53 mm) and mean leaf size (length × width in cm) was the maximum (5.67 cm and 3.83 cm, respectively) with IBA 1500 ppm + PHB 1000 ppm. The use of growth regulators is very common in plant tissue culture, with a wide range of plant tissues used for callus induction. Jaskani *et al.* (1996) reported successful callus formation from endosperm triploid tissues using 2,4-D and kinetin. Successful *in vitro* regeneration of disease-free citrus plants through shoot tip grafting was reported by Navarro and Juarez (1977). Use of BA (0.1 mg/l) and higher concentrations are reported to inhibit root growth and promote root bud formation; however, IAA (1–10 mg/l) stimulated lateral root formation (Navarro *et al.*, 1975).

Citrus nursery plants gained better height with reduced stem girth in response to GA_3 application (Abdalla *et al.*, 1979), and in sour orange seedlings GA_3 (200 ppm) increased plant size and produced ready to bud plants within 8 months after germination (Khattab and Guindy, 1996).

PGRs and tree growth

In most citrus rootstock seedlings the apical meristem synthesizes auxins (IAA), which suppresses lateral growth and scion bud; hence the central shoot of the rootstock is either bent or removed to promote scion bud growth. This may also favour sprouting of basal buds on the rootstock, and the removal of these outgrowths is laborious and expensive. Rouse (1994) reported that the problem is addressed by a commercial product 'Trehold' containing NAA, which inhibits sprouting from the stems of citrus rootstocks in the nursery beds. NAA is effective in nurseries and non-bearing field plants as well as in controlling bud dormancy (Knapp, 1996). Trehold is recommended for tree trunk application as a spray or using light brushing, with the precaution that heavy application may damage bearing trees. It should not be applied at the onset of winter season so as to avoid post-freeze delay in tree sprouting, which is necessary for the trees to recover from freezing stress (El-Otmani *et al.*, 2000). Citrus trees may require pruning every year, which is very costly and time consuming. Growth retardants have been tested on a wide range of citrus cultivars. Dhillon and Sidhu (1992) and Deng and Zhen (1996) reported that gibberellins and auxins enhance tree vegetative growth at low concentrations (El-Otmani *et al.*, 1996), whereas higher concentrations of IAA and abscisic acid (ABA) suppressed vegetative growth (Bertling and Lovatt, 1996). Growth retardants including maleic hydrazide, morphactins and ammonium ethyl carbamoylphosphate suppressed tree growth and induced dieback and meristematic injuries rather than reducing internode length according to El-Otmani *et al.* (2000).

Effect of PGRs on flowering

Flowering is a product of genetic characteristics, environment and cultural practices. Under optimum conditions citrus trees exhibit heavy bloom reaching more than 100,000 flowers per tree (Agusti *et al.*, 1982) with an abscission of up to

90–99% of the blooms (El-Otmani *et al.*, 1992). GA_3 has an inhibitory effect on flowering subject to time of application (Davenport, 1990), whereas GA biosynthesis inhibitory triazole compounds are reported to increase flowering (Harty and van Staden, 1988). Different doses of GA_3 in most of the citrus cultivars including Tahiti lime (35 ppm) reduced flowering percentage by nearly 50% in comparison to the control when treated in mid-December in New Zealand conditions (Davenport, 1983). It should be kept in mind that different buds mature at different times so a single application of any PGR may not be sufficient to control bloom intensity; hence repetition may be needed according to climatic conditions. Persian limes are best sold during winter in México, so the spring flowering needs to be shifted to later months to enhance production during and after winter. For this purpose monthly application of GA_3 (40 ppm) from June through to August has been shown to delay flowering in Persian limes, and also foliar application of ethephon spray (0.1 gm/l) in October resulted in the desired blooming and fruit setting in December leading to a competitive winter harvest, as reported by Vargas (1993). In tropical climates Persian limes have a tendency to bloom, set and produce fruit throughout the year. The above mentioned treatments have been found to be beneficial to prevent summer through autumn blossoming and to stimulate late autumn bloom to produce a competitive winter crop. However, ethephon (500 mg/l) is reported to be necessary for enhanced flowering, while the lower doses remain ineffective (Vargas and Espinoza, 1993). In other citrus cultivars auxins are also reported to suppress flowering, e.g. 2.4-D (12 ppm) generated the same response as GA_3 at a concentration of 100 mg/l (Guardiola *et al.*, 1977). Foliar sprays of 2,4-D are widely applied to control abscission in ripening fruit and may coincide with early bud development causing serious damage to expected spring flush (Coggins, 2000).

Fruit retention, fruit size and tree yield

Fruit yield (kg/tree) depends on the number of fruit and size, where fruit size is a function of mitosis and cell enlargement. High numbers of fruit may not always result in good returns because marketable fruit size may not be achieved due to excessive fruit setting and retention. Excessive fruit drop due to impaired hormonal production during the early stages of fruit setting and development may also result in lower production, particularly when initial bloom is not sufficient. GA_3 was reported to improve early fruit growth with decreased button stage fruit drop. A foliar application of 10 ppm was effective during full bloom and petal fall (El-Otmani *et al.*, 1992). In some years blooms appear for a short period (about 10 days) and a single 15 ppm GA_3 is enough. However, during extended bloom periods it should be ensured that maximum coverage is achieved by splitting the above dose into two equal halves to be applied at early flowering and petal fall. A concentration above 15 mg/l is reported to cause excessive leaf shedding and twig dieback (Krezdorn and Jernberg, 1977). The pH of the working solution is not often considered except in South Africa and Israel where maximum efficiency is achieved at pH values between 3 and 5 (El-Otmani *et al.*, 2000). Foliar application of plant growth regulators in acid lime cv. Kaghzi (*Citrus aurantifolia*) resulted in high per tree yield (46.38 kg) in response to application of GA_3 (50 mg/l) where fruit volume (47.90 cm^3), diameter (4.54 cm) and weight (47.40 g) were also significantly higher. Per tree number of fruit was as high as 1020.33 in response to NAA 200 mg/l with significantly higher soluble solids (9.58°Brix.) and Vitamin C content (30.41 mg per 100 g pulp) for the same treatment, whereas number of seeds per fruit (6.13) and acidity (7.05%) were significantly decreased in response to GA_3 (10 mg/l) application (Jagtap *et al.*, 2013). Devi *et al.* (2011) reported that growth regulators modified the annual pattern in acid lime. They observed maximum (1.35) new flush in response to 2,4-D at the rate of 40 ppm, and maximum new growth in response to paclobutrazol (5 ml per m^2 of canopy) during December–January in Indian conditions. Flowering was regulated by applying paclobutrazol at 5 ml/m^2 of canopy and was noted to be the highest (13.68 per shoot) resulting in the highest number of fruit (195.67) per tree and yield (6.185 kg/tree) (Devi *et al.*, 2011). Acid lime (*C. aurantifolia* Swingle) produces three important flushes in India in January–February (Ambia bahar), June–July

(Mrig bahar) and September–October (Hasta bahar), where all the three flushes set fruit. The third (Hasta bahar) crop comes during summer and commands the best prices, however, it is difficult to produce fruit in September–October due to monsoon rains. Mukunda *et al.* (2014) treated acid lime trees with foliar sprays of GA_3 (50 and 100 ppm), cycocel (1000 ppm), KNO_3 (1% and 2%) and salicyclic acid (100 and 200 ppm) 4 months prior to June flowering and in September–October. The results showed that GA_3 (50 ppm in June) followed by CCC at 1000 ppm in September and KNO_3 1% in October produced the highest number of fruit per tree (529.34), fruit weight (41.12 g) and yield (24.08 kg/tree).

Both excessively small fruit and fruit abscission result in severe losses every year. Massive blooms and favourable climatic conditions promote excessive fruit setting, which results in extremely small sized fruit, breakage of overloaded branches, tree decline and even death of the trees, suggesting fruit thinning as a workable solution. An early fruit thinning significantly increases fruit size. Several fruit thinning agents including ethephon have been tested with different concentrations, combinations and stages of fruit development, suggesting a wide range of variations in agro-climatic conditions, citrus cultivars and extent of thinning (El-Otmani *et al.*, 1992). Use of PGRs is becoming as common as nutrient elements. However, care should be taken as most of the PGRs are used in much smaller amounts than those of even the plant micronutrients. Time, dose and method of application must be well thought out and maximum precision should be maintained. Labels should never be ignored and hazard management and recovery backup must be adopted as per local conditions. The weather forecast must be followed before taking the application decision with the emphasis that in a climatic change regime, every year is expected to be a different year with reference to global life.

Good Agricultural Practices in Lime Plantations

Good agricultural practices (GAP) not only promote food safety but also lead to higher and more sustainable fruit production with the least environmental threat. In lime production, the parameters of GAPs are considered uniform with a few modifications in different producing areas. These practices include planting density, pruning, irrigation, fertilizer application, harvesting and many more with the single objective of producing a safe and sufficient yield. The average yield of Persian limes is estimated to be 13.15 t/ha in México in comparison with 34.69 t/ha in the USA (FAO, 2010), which is related to poor cultural practices (Becerra *et al.*, 2009). In relation to the poor performance of Persian limes, six problem areas were identified including poor food safety standards, poorly performing rootstocks, suboptimal cultural practices, few diagnostic facilities, irrigation mismanagement, little institutional support and poor diffusion of the research-based technologies (REVIDEC, 2003). Studies conducted in Martinez de la Torre, Veracruz, emphasized that most of the Persian lime growers (82%) never practised GAP, and about 30.8% of the Persian lime orchards could barely meet the food safety prerequisites (Herbert, 2009). Adaptation of GAPs in Mexican lime orchards has led to improvement in yield and quality of Persian and Tahiti limes. Fernández Lambert *et al.* (2014) reported that seasonal GAP activities in the production of Persian lime improved the production and quality of the fruit.

The results presented by Fernández Lambert *et al.* (2015) proposed that a combination of fertilization and pruning increased orchard productivity. However, less than 50 mm of rainfall per month is unlikely to produce similar results, even when plant nutrition and pest control have been efficiently performed in Persian limes grown in Mexican provinces. Tahiti limes are popular in the São Paulo state of Brazil. Tree dwarfing and high density plantations were evaluated by Stuchi *et al.* (2003). They reported that 2500 trees/ha on Flying Dragon (*Poncirus trifoliata* var. *monstrosa*) rootstock with 4 m row to row and 1 m plant to plant distance achieved an average canopy width of 2.75 m with no difference in tree height with any other plant distance. The highest fruit yield (21.6 t/ha) was also produced with the same planting distance with slightly lower but still exportable fruit physicochemical attributes in terms of TSS (°Brix), acidity (%) fruit weight, diameter and height (Stuchi *et al.*, 2003).

México is the leading exporter of Persian limes, where only 25% of the pack houses own both the orchards and GAP certification. Larger fruit with less than 5% blemishes and 80–90% deep green colour are exported to Europe and Japan; however, US markets accept the produce with a minimum of 70% intense green fruit surface. Several certification systems including GAP, Euro-Retailer Produce Working Group and Good Agricultural Practices (EUREPGAP), Codex Alimentarius, SAGARPA (abbreviated in Spanish) and México Selected Quality of Persian Limes are operative in México (Rivera-Cabrera *et al.*, 2010).

Organic Production of Limes

According to the definition of the Codex Alimentarius:

> Organic agriculture is a holistic production management system which promotes and enhances agroecosystem health, including biodiversity, biological cycles and soil biological activity. It emphasizes the use of management practices in preference to the use of off-farm inputs, taking into account that regional conditions require locally adapted systems. This is accomplished by using where possible, agronomic, biological and mechanical methods, as opposed to using synthetic materials, to fulfil any specific function within the system. (Liu, 2003)

Citrus products labelled as 'organic' are those that have met certification standards and been approved and inspected by an accredited organization or government body, usually from the importing country.

Extent of production and economic considerations

The most recent data for the production of organic limes worldwide are difficult to find. Nearly all organic production reporting is done on a voluntary basis through a combination of organic certification organizations supplemented by government sources when available. The most recent data (Willer and Lernoud, 2015) are for the production year 2013. Of the 170 countries reporting, there were 43.1 million hectares of organic agricultural land worldwide and nearly 2 million producers. This represents about 1% of the total agricultural land. For citrus crops, there were 82,000 ha grown organically, or 0.9% of the world's total citrus area. Unfortunately, the data for organic area in the three largest citrus-producing countries, Brazil, Nigeria and India, were not available, so the organic area is likely an underestimate. Also, limes specifically are not separated out from the rest of the citrus crops in this data set.

Leading organic producers of organic citrus from the reporting areas are Italy (28,816 ha), México (11,917 ha), China (11,531 ha), USA (7528 ha), Ghana (6783 ha) and Spain (6332 ha). The types of citrus certified organic by category are oranges (52%), lemons and limes (12%), pomelos/grapefruit (5%), tangerines (5%) and unknown/not listed (26%). Overall growth in the organic citrus area has tripled since 2004, when only 28,500 ha of organic citrus were grown/reported.

As with most crops, profitability is the driver of growth in production. Again, data specifically on limes are difficult to find, but if we use other citrus crops as a proxy, a study for the Food and Agriculture Organization of the United Nations (FAO) (Igual and Izquierdo, 2001) sheds some light on this issue using production in Spain as a case study. Using data from a growers' cooperative of oranges and mandarins, including 11 and 14 plots of each, respectively, they estimated the fixed and variable costs of production for organic as compared with conventional farming, and conducted a sensitivity analysis on price scenarios. They found that the cost savings for organic farmers spending less on pest control products was offset by increases in spending for fertilizers and labour, resulting in higher variable costs for organic producers of both crops. The organic crop yield estimates were lower than conventional, so with no government subsidy or price premium for organic production, the conventional farms were more profitable. They estimated that strong demand in the market and a price premium of between 30 and 40% would be required for the organic farms to be profitable. With the caveat that the price for organic produce fluctuates quite a bit based on supply and demand for a small market volume, they reported prices paid to farmers in 1998–1999 as having a 44 to 46% price premium, plus the European Union supported an organic farming subsidy. Either of these would be enough to tip the scale towards profitability, and if these are present in countries or regions producing limes, a similar outcome, i.e. growth in organic production methods, would be expected.

The USA is the largest purchaser of organic foods with a value of €24.3 billion, followed closely by the EU countries, especially Germany and France. For retail sales, the USA accounts for 43% of all sales, the EU 40% (Germany 13% and France 8%), and China and Canada 4% each (Willer and Lernoud, 2015). Citrus consumption has been estimated to be approximately 5–7% of fresh organic produce sales (Liu, 2003). Though the US imports some organic citrus, the quantity is relatively small because it is also a producer of citrus. Europe, on the other hand, is a major importer of organic citrus, with net imports estimated at around 50,000 t in 2000. Germany, France, Austria, the Netherlands and the UK were each importing between 7000 and 13,000 t. Price premiums for organic citrus averaged 65% overall, with a range of 11% to 144% for various EU countries as compared with conventional citrus crops (Liu, 2003).

Production methods and soil quality

In addition to price premiums, organic farming systems have been adopted by farmers because they provide ecosystem benefits, including enhancement of soil fertility and function. A farm-level field survey was carried out in southern Italy, and documented an increase in soil quality on organically managed citrus orchards (Canali, 2003). The total organic carbon was not statistically different in organically managed soils, though higher, but the carbon mineralization rate was higher. Nitrogen mineralization rate was also higher in the organic soils, but not statistically significant. Nitrogen status of the plants as measured by leaf analysis and the yields were monitored with no significant differences among treatments. However, the soil nitrate levels were lower in the organic treatments, indicating less risk of nitrate leaching. In this study, 27 certified organic farms were paired with conventional farms for the comparisons, and the crops were Navelina and Tarocco orange.

Soil fertility studies specifically on acid lime (*C. aurentifolia* Swingle) (Patel *et al.*, 2012; Rajendra *et al.*, 2013) have also been carried out using organic soil amendments. In these two studies, various rates of N, P and K have been applied in randomized complete block experiments, with a portion of the fertilitizer applied in the chemical form and a portion as animal manure or castor cake. In the first study, the highest rates of manure and vermicompost (8.31 t/ha and 4.16 t/ha, respectively) plus 75% of recommended NPK rates resulted in the highest yield and also the most profitable treatment (Rajendra *et al.*, 2013). In the second experiment, the split application of soluble fertilizer plus castor cake at the highest rate (900-750-500 g/tree of N, P and K, respectively) resulted in the highest yield (Patel *et al.*, 2012). Neither of these treatments would be allowed in certified organic production systems due to the presence of the mineral fertilizer, but these data inform organic growers that they should be aiming for higher rates of fertility than perhaps they would have otherwise.

A study in Portugal conducted on an organic orange plantation tested the application of a model for nutritional diagnosis that had been developed for conventional systems (Domingos *et al.*, 2009). The study found that the nutritional variation in the organically managed trees was consistent with values seen in conventional management. However, the low micronutrient status measured in the experiment was likely due to the calcareous soil, and not the method of production. The nutritional model used proposes to estimate the chlorophyll content from the Mg/Zn ratio in plant tissue. This estimate was not accurate in the organic system as compared with the conventional system, probably as a consequence of the application of organic matter.

A study on potting media for growing citrus rootstocks found that a mix of sand, silt, farmyard manure and compost gave the best results in terms of growth and development for rough lemon rootstock and mandarin saplings (Abbas *et al.*, 2015). Results showed increased mineral content, leaf area index, photosynthetic rate, stomatal conductance and transpiration rate for this treatment. These potting mix components would all be allowed in organic citrus production systems, as long as the source of the manure and compost was documented, not from a concentrated livestock facility and free from added pesticides. Another potting substrate tested in the experiment, coconut husk, would also be allowed in organic production if not contaminated, and all would be superior to mined peat substrates, which are allowed in organic production at this time, but considered unsustainable due to the fact that it is not a renewable resource.

Citrus byproducts can also be used as raw material for compost production. A study in Italy compared citrus industry byproducts (skins and pulp) combined with municipal sludge or composted without the sludge. Both resulted in acceptable composted products, and the one without the sludge could be used by organic farmers, thus providing more sustainability as waste is recycled back into production (Ciaccia *et al.*, 2008).

Pest management: weeds, insects, diseases

Weed control in organic citrus orchards can be achieved by using a variety of methods. In a trial comparing plastic mulches, various depths of applied organic materials (rice straw, oat straw, cattail mulch), mechanical cultivation by machine or by hand and the herbicide glyphosate in Egypt, found that several of the organic alternatives gave superior weed control and yield of the citrus crop as compared with glyphosate (Abouziena *et al.*, 2008). The best treatments in terms of weed control were the 200 μm and 150 μm thick black plastic mulch, the 9 cm deep rice straw mulch and the 12 cm deep cattail mulch. Yields from these treatments were also not statistically different, nor different from the hand hoed treatment and the 8 cm deep cattail mulch. The glyphosate did not give season-long weed control, and yielded higher than the unweeded control plot, but was lower than many of the organic treatments and would not be allowed according to organic standards.

Cover crops would be a less labour intensive way to achieve mulch cover as compared with hand application. Plantings of winter and summer annual cover crops were compared with perennial groundnuts in citrus plantations in Florida (Linares *et al.*, 2008). They found that summer annual cover crops such as sunn hemp (*Crotolaria juncea* L.), hairy indigo (*Indigofera hirsute* L.), cowpea (*Vigna unguiculata* L. Walp.) and alyce clover (*Alsyicarpus vaginalis* L.) all resulted in excellent weed suppression. In general, species mixtures provided better suppression than single species plantings, and the perennial groundnuts took too long to establish to provide much weed control benefit. Though the experiment took place in a certified organic citrus orchard setting (Hamlin and Navel orange varieties), the effect of the cover crop treatments on yield was not measured. When organically approved herbicides such as vinegar were compared with cover crops, none of them provided long-term weed control (Linares *et al.*, 2006). Repeated use of the organically approved herbicides near sprinklers was found to be necessary, however, in combination with cover crops, repeated tillage or other techniques in citrus orchards for weed suppression.

Biological control of both insect and disease pests of lime and other citrus crops is being used in both conventional and organic orchards, and won't be covered in detail here. A successful example of disease suppression includes the use of *Trichoderma viride* isolates to control *Fusarium solani* (Mart.) on acid lime (*C. aurantifolia* Swingle) in pot culture (Kavitha *et al.*, 2006). In a study in Peru, native isolates of bacteria from the pests were used to control the citrus borer worm (*Phyllocnisti citrella*) on acid lime (*C. aurantifolia*) (Sepulveda *et al.*, 2001). Four bacterial strains were combined and field tested, which resulted in 85% control after 24 h, and they were also found to persist in the environment. Mites (*Brevipalpus* spp.) have been controlled on limes in México by using suspensions of mycelia and conidia of *Hirsutella thompsonii* (Roses-Acevedo and Sampedro-Rosas, 2000). A significantly lower amount of fruit damage and a decrease in the population of the mite was recorded in the biological control treatment as compared with traditional farmer management and the untreated control plots. After three foliar applications, the beneficial fungus was found to persist in the treated plots for up to 7 months.

Conclusions and Considerations for Future Research

Limes are important both worldwide, and also regionally. Though much research on organic citrus crops in general can be applied to lime production, there seem to be few studies focused specifically on organic lime production methods. Land in organic production for all crops continues to expand on all continents, reaching as high as 20% of arable land in several European countries (Willer and Lernoud, 2015). Demand for organic fruit crops continues to rise. As more farms are converted to organic production, research needs to take place that is locally and regionally adapted to growers' conditions. Biological control methods and studies are widely applicable to both organic and conventional growers. Soil quality enhancement through the use of cover crops and compost application is also beneficial to all growers. Because organic methods take place within a local context, both experiment station and on-farm research would be useful.

References

Abbas, M., Aftab, M., Zafar-ul-Hye, M., Iqbal, Q., Hussain, M. and Khan, M.M. (2015) Effect of organically amended growing substrates on the growth and physiological attributes of citrus plants. *Communications in Soil Science and Plant Analysis* 46, 1863–1880.

Abdalla, K.M., El-Wakil, A.T. and El-Masiry, H.H. (1979) *Seedbed Treatments with GA in Citrus 1. Morphologic Response of Different Rootstocks*. Ain Shams University, Zagazig University, Zagazig, Egypt. *Research Bulletin of Faculty of Agriculture*, No. 937, p. 29.

Abdelmalek, G.A.M. and Salaheldin, T.A. (2016) Silver nanoparticles as a potent fungicide for citrus phytopathogenic fungi. *Journal of Nanomedicine Research* 3(5), 00065. doi: 10.15406/jnmr.2016.03.00065.

Abouziena, H.F., Hafez, O.M., El-Metwally, I.M., Sharma, S.D. and Singh, M. (2008) Comparison of weed suppression and mandarin fruit yield and quality obtained with organic mulches, synthetic mulches, cultivation and glyphosate. *HortScience* 43, 795–799.

Aggarwal, S. (2008) *Spectral Signature of Earth Objects. Lecture Delivered in Edusat Based Training Programming*. Indian Institute of Remote Sensing, Dehradun, India.

Agusti, M., Garcia-Mari, F. and Guardiola, J.I. (1982) The influence of the flowering intensity on the shedding of reproductive structures in sweet orange. *Scientia Horticulturae* 17, 343–352.

Bak, T. and Jakobsen, H. (2004) Agricultural robotic platform with four wheel steering for weed detection. *Biosystems Engineering* 87, 125–136.

Batchelor, W.D. and Paz, J.O. (1998) Process oriented crop growth models as a tool to evaluate spatial yield variability. In: *Proceedings of the First International Conference on Geospatial Information in Agriculture and Forestry*, Vol. 1, pp. 198–205. ERIM International, Inc., Ann Arbor, Michigan.

Becerra, D.J.A., Lambert, G.F., Flores, J.L.M. and Lasserre, A.A. (2009) *Caracterización de la cadena de suministros de los cítricos en el Estado de Veracruz*. Reporte de Investigación, Universidad Popular Autónoma del Estado de Puebla, Puebla, México.

Belasque, L., Gasparoto, M.C.G. and Marcassa, L.G. (2008) Detection of mechanical and disease stresses in citrus plants by fluorescence spectroscopy. *Applied Optics* 47(11), 1922–1926.

Bertling, J. and Lovatt, C.J. (1996) A possible hormonal basis for vigor/dwarfing in commercially important citrus rootstocks. *Proceedings of the International Society of Citriculture* 2, 745–747.

Boaretto, A.E., Muraoka, T. and de Oliveira, M.W. (1999) Corrective and preventive foliar fertilization in citrus. *Laranja* 20(1), 233–250.

Bondada, B.R., Petracek, P.D. and Syvertsen, J.P. (2006) Cuticular penetration characteristics of urea in citrus leaves. *Journal of Horticultural Sciences and Biotechnology* 81, 219–224.

Boswell, S.B., Lewis, L.N., McCarty, C.D. and Hench, K.W. (1970) Trees spacing of Washington navel orange. *Journal of American Society of Horticultural Sciences* 95, 523–528.

Boswell, S.B., McCarty, C.D., Hench, K.W. and Lewis, L.N. (1975) Effect of tree density on the first ten years of growth and production of Washington navel oranges trees. *Journal of American Society of Horticultural Sciences* 100, 370–373.

Brown, L.R., Gardner, G., Halweil, B. and Malthus, B. (1999) Nineteen dimensions of the population challenge. In: Starke, L. (ed.) *The Worldwatch Environmental Alert Series*. W.W. Norton & Company, New York.

Canali, S. (2003) Soil quality of organically managed citrus orchards in the Mediterranean area. Organic agriculture: sustainability, markets and policies. In: *OECD Workshop on Organic Agriculture*, Washington, DC, 23–26 September 2002. CAB International, Wallingford, UK, pp. 115–125.

Carrasco, E., Bautista, J.A.G. and Mateo, J.V.G. (2007) Automated sequential monitoring of ammonium, phosphate, and nitrite in wastewater by multi-commutated peristaltic and solenoid pumped flow system – a comparative study. *Analytical Chemistry* 52, 757–770.

Ceballos, H., Sanchez, T. and Chavez, A.L. (2006) Variation in crude protein content in cassava (*Manihot esculenta* Crantz) roots. *Journal of Food Composition and Analysis* 19, 589–593.
Chaerle, L., Hagenbeek, D., Vanrobaeys, X. and Van Der Straeten, D. (2007) Jones and Schofield early detection of nutrient and biotic stress in *Phaseolus vulgaris*. *International Journal of Remote Sensing* 28, 3479–3492.
Ciaccia, C., Di Bartolomeo, E., Calabretta, M.L., Intrigliolo, F., Tittarelli, F. and Canali, S. (2008) Quality assessment of citrus-processing industry waste compost for organic and conventional farming. In: Daniel, N. et al., (ed.) *Proceedings of the Second Scientific Conference of the International Society of Organic Agriculture Research (ISOFAR)*. International Society of Organic Agriculture Research (ISOFAR), Modena, Italy, pp. 215–218.
Coggins, C.W. Jr (2000) Plant growth regulators. In: Flint, M.L. (ed.) *UC IPM Pest Management Guidelines, Citrus*. IPM Education and Publications, University of California, Davis, California, pp. E1–E6.
Coggins, C.W. Jr and Hield, H.Z. (1968) Plant growth regulators. In: Reuther, W., Webber, H.J. and Batchelor, I.D. (eds) *The Citrus Industry*, Vol. 1. Division of National Agricultural Sciences, University of California, Berkeley, California, pp. 371–389.
Davenport, T.L. (1983) Daminozide and gibberellin effects on floral induction in *Citrus latifolia*. *HortScience* 18, 947–949.
Davenport, T.L. (1990) Citrus flowering. *Hort. Rev.* 12, 349–408.
de Wit, C.T. (1958) *Transpiration and Crop Yields*. Institute of Biological and Chemical Research on Field Crops and Herbage. *Verse-Landbouwk Onderz* 64(6), 59–84. Wageningen, the Netherlands.
Deng, L. and Zhen, X. (1996) A study of citrus nitrogenous metabolism during flower bud formation. *Programme and Abstracts, VIII Congress of International Society of Citriculture*, Sun City Resort, South Africa, p. 103 (abstract no. P064).
Devi, H.L., Sarkar, S.K., Dhanabati, L. and Majhi, D. (2011) Flushing – flowering behavior and regulation in acid lime – a critical review and research interventions. *Journal of Crop and Weed* 7(2), 87–90.
Dhillon, B.S. and Sidhu, M.S. (1992) Manipulation of fruit crop in Kinnow mandarin. In: Khan, I.A. (ed.) *Proceedings of the 1st International Seminar on Citriculture in Pakistan*, University Press, University of Agriculture, Faisalabad, Pakistan, pp. 229–236.
Diwaker and Katiyar, P.N. (2013) Regeneration of kagzi lime (*Citrus aurantifolia* Swingle) through stem cuttings with the aid of IBA and PHB. *HortFlora Research Spectrum* 2(3), 271–273.
Domingos, I., Candeias, M.F., Correia, P.J. and Pestana, M. (2009) Nutritional evaluation of citrus orchard under organic farming. *Revista de Ciencias Agrarias* 32, 63–69.
El-Otmani, M., Ben-Ismail, M.C., Ait-Obahou, A. and Achouri, M. (1992) Growth regulators use on clementine mandarin to improve fruit set. *Proceedings of the International Society of Citriculture* 1, 500–502.
El-Otmani, M., Kadri, B., Ait-Oubahou, A. and Agusti, A. (1996) Use of auxins to improve fruit size of clementine mandarin–effect on vegetative growth and fruit quality. *Proceedings of the International Society of Citriculture* 2, 1076–1080.
El-Otmani, M., Coggins, C.W., Agusti, M. Jr and Lovatt, C.J. (2000) Plant growth regulators in Citriculture World current uses. *Critical Reviews in Plant Sciences* 19(5), 395–447.
Evans, E.H., Day, J.A. and Price, W.J. (2003) Atomic spectrometry update: advances in atomic emission, absorption and fluorescence spectrometry, and related techniques. *Journal of Atomic Spectrometry* A8, 808–833.
Everitt, J.H., Escobar, D.E., Summy, K.R. and Davis, M.R. (1994) Using airborne video, global positioning system, and geographical information-system technologies for detecting and mapping citrus blackfly infestations. *Southwestern Entomology* 19(2), 129–138.
Fageria, N.K., Barbosa, F.M.P. and Moreira, A. (2009) Foliar fertilization of crop plants. *Journal of Plant Nutrition* 32, 1044–1064. doi: 10.1080/01904160902872826.
FAO (2010) Food and Agriculture Organization. Available at: http://www.fao.org (accessed 29 October 2016).
Fernández Lambert, G., Aguilar Lasserre, A.A., Miranda-Ackerman, M.A., Sánchez, C.G.M., Rivera, B.O.I. and Azzaro-Pantel, C. (2014) An expert system for predicting orchard yield and fruit quality and its impact on the Persian lime supply chain. *Engineering Applications of Artificial Intelligence*. 33, 21–30. doi:10.1016/j.engappai.2014.03.013.
Fernández Lambert, G., Aguilar Lasserre, A.A., Azzaro-Pantel, C., Miranda-Ackerman, M.A., Vázquez, R.P. and Salazar, M.R.P. (2015) Behavior patterns related to the agricultural practices in the production of Persian lime (*Citrus latifolia* tanaka) in the seasonal orchard. *Computers and Electronics in Agriculture* 116, 162–172.

Gaines, T.P. and Gaines, S.T. (1994) Soil texture effect on nitrate leaching in soil percolates. *Communications in Soil Science and Plant Analysis* 25(13 and 14), 2561–2570.

Grift, T.E., Kasten, M. and Nagasaka, Y. (2005) Robotics in agriculture: Asimov meets corn. In: *Proceedings of the Illinois Crop Protection Technology Conference*, Urbana, Illinois, 5–6 January 2005. Available at: http://agronomyday.cropsci.illinois.edu/2004/Tour_A/Robotics/ (accessed 29 October 2016).

Grift, T., Zhang, Q., Kondo, N. and Ting, K.C. (2008) A review of automation and robotics for the bioindustry. *Journal of Biomechatronics Engineering* 1(1), 37–54.

Guardiola, J.L., Agusti, M. and Garcia-Mari, F. (1977) Gibberellic acid and flower bud development in sweet orange. *Proceedings of the International Society of Citriculture* 2, 696–699.

Gupta, G. (2014) Novel therapy of high-priority citrus diseases. *Citrograph Magazine* 2014(Fall), 64–69.

Handson, P.D. and Shelley, B.C. (1993) A review of plant analysis in Australia. *Australian Journal of Experimental Agriculture* 33, 1029–1038.

Hannan, M.W. and Burks, T.F. (2004) Current developments in automated citrus harvesting. *ASAE Paper* No. 04-3087, St. Joseph, Michigan.

Hart, W.G. and Myers, V.I. (1968) Infrared aerial colour photography for detection of populations of brown soft scale in citrus groves. *Journal of Economic Entomology* 61(3), 617–624.

Harty, A.R. and van Staden, J. (1988) Paclobutrazol and temperature effects on lemon. In: *Proceedings of the 6th International Congress*, Tel Aviv, Israel, Vol. 1, pp. 343–353.

He, Z.L., Yao, H. and Chen, G. (1997) Relationship of crop yield to microbial biomass in highly weathered soils of China. In: Ando, T., *et al.* (eds) *Plant Nutrition for Sustainable Food Production and Environment*. Kluwer Academic Publishers, Tokyo, pp. 745–746.

Herbert, R.M. (2009) La inocuidad alimentaria en el mercado Mexicano de Limón Persa (*Citrus latifolia* tanaka), Tesis Doctoral en Socioeconomía, Estadística e Informática Economía. Colegio de Postgraduado. Montecillo, Texcoco, Edo. De México.

Hofstee, J.W., Grift, T.E. and Tian, L. (2004) Development of a tracking and guidance system for a field robot. *Agricultural Engineering Conference*. Leuven, Belgium, 12–16 September 2004.

Huang, W., Lamb, D.W., Niu, Z., Zhang, Y., Liu, L. and Wang, J. (2007) Identification of yellow rust in wheat using in situ spectral reflectance measurements and airborne hyper spectral imaging. *Precision Agriculture* 8(4–5), 187–197.

Husted, S., Persson, D.P. and Laursen, K.H. (2011) Review: the role of atomic spectrometry in plant science. *Journal of Analytical and Atomic Spectrometry* 26, 52–79.

Igual, J.F.J. and Izquierdo, R.J.S. (2001) *Economic and Financial Comparison of Organic and Conventional Citrus-Growing Systems in Spain. A Study Prepared for FAO*. Department of Economics and Social Sciences, University of Valencia, Valencia, Spain. Available at: http://www.fao.org/docrep/003/AC117E/AC117E00.HTM (accessed 27 April 2016).

Jackson, D.I. (2011) Soil, nutrients and water. In: Jackson, D.I., Looney, N.E., Morley-Bunker, M. and Thiele, G.F. (eds) *Temperate and Subtropical Fruit Production*, 3rd edn. CAB International, Wallingford, UK, pp. 101–118.

Jagtap, V.M., Patel, H.C., Nehete, D.S. and Godage, S.S. (2013) Effect of foliar application of plant growth regulators and micronutrients on yield and quality of acid lime cv. KAGZI (*Citrus aurantifolia* Swingle). *Asian Journal of Horticulture* 8(1), 57–59.

Jaskani, M.J., Khan, I.A. and Yakub, M.S. (1996) Regeneration of triploids from citrus endosperm. *Proceedings of International Society of Citriculture* 1, 28–29.

Joa, J.H., Han, C.L. and Seung, G.H. (2006) Microbial activities in dark brown volcanic ash soils was affected by different fertilizer application methods in citrus orchard. In: *Proceedings of the 18th World Congress of Soil Science*, Pennsylvania (Abstract), pp. 138–184.

Kavitha, M., Gopal, K., Ahammed, S.K., Sreenivasulu, Y. and Madhusudhan, P. (2006) Biological control of dry-rot of acid lime (*Citrus aurantiifolia* Swingle) caused by *Fusarium solani* (Mart.) Sacc. *Journal of Biological Control* 20, 105–107.

Kawamura, N., Namikawa, K., Fujiura, T. and Ura, M. (1984) Study on agricultural robot (Part 1). *Journal of the JSAM* 46, 353–358.

Khattab, M.M. and Guindy, L.F. (1996) Effect of gibberellic acid and vapor gard on CO_2 evolved respiration rate and stomatal movement of young sour orange seedlings. *Proceedings of the International Society of Citriculture* 2, 1089–1093.

Knapp, J.L. (1996) *Florida Citrus Pest Management Guide (Florida Cooperative Extension Service Circular; SP 43)*. Florida Cooperative Extension Service, Institute of Food and Agricultural Science, University of Florida, Gainesville, Florida.

Kolthoff, I.M., Sandell, E.B. and Bruckenstein, S. (1969) *Quantitative Chemical Analysis*. Macmillan, London.
Komosa, A. (1990) Influence of selected chemical properties of solutions and nutritional status of plants on the efficiency of foliar nutrition of greenhouse tomato. Dissertation, Scientific Papers of the Agricultural University of Poznań, Poznań, Poland, p. 210 (in Polish).
Kondo, N., Ninomoya, K., Hayashi, S., Tomohiko, O. and Kubota, K. (2005) A new challenge of robot for harvesting strawberry grown on table top culture. *ASAE Paper* No. 053138, St. Joseph, Michigan.
Krezdorn, A.H. and Jernberg, D.C. (1977) Field evaluation of growth regulators for fruit set. *Proceedings of the International Society of Citriculture* 2, 660–663.
Kuepper, G. (2003) Foliar fertilization. ATTRA. Available at: https://attra.ncat.org/attra-pub/summaries/summary.php?pub=286 (accessed 30 October 2016).
Laursen, K.H., Hansen, T.H. and Persson, D.P. (2009) Recent developments in plant tissue analysis. In: *Proceedings of the International Plant Nutrition Colloquium XVI*. University of California, Davis, California.
Linares, J., Scholberg, J., Chase, C., McSorely, R. and Fergusson, J. (2006) Integrative Approaches for weed management in organic citrus orchards. *HortScience* 41, 998.
Linares, J., Scholberg, J., Boote, K., Chase, C.A., Ferguson, J.J. and McSorley, R. (2008) Use of the cover crop weed index to evaluate weed suppression by cover crops in organic citrus orchards. *HortScience* 43, 27–34.
Liu, P. (2003) *World Markets for Organic Citrus and Citrus Juices; Current Market Situation and Medium-Term Prospects*. Food and Agriculture Organization of the United Nations, p. 5. Available at: http://www.fao.org/docs/eims/upload/229991/organic_citrus_markets.pdf (accessed 29 October 2016).
Marschner, H. (1995) *Mineral Nutrition of Higher Plants*, 2nd edn. Academic Press, London.
Menesatti, P., Antonucci, F. and Pallottino, F. (2010) Estimation of plant nutritional status by vis-NIR spectrophotometric analysis on orange leaves. *Biosystems Engineering* 105, 454.
Michałojć, Z. and Szewczuk, C. (2003) Theoretical aspects of foliar nutrition. *Acta Agrophysica* 85, 9–17 (in Polish with English abstract).
Min, M., Lee, W.S. and Burks, T.E. (2008) Design of a hyperspectral nitrogen sensing system for citrus. *Computers and Electronics in Agriculture* 63, 215–226.
Morgan, M., Scholberg, J.M. and Obreza, T. (2006) Citrus nitrogen uptake and nitrification rates in sandy soils. In: *Proceedings of the 18th World of Soil Science*, 9–15 July, 2006 Philadelphia, Pennsylvania.
Moshou, D., Bravo, C., Oberti, R., West, J., Bodria, L., McCartney, A. and Ramon, H. (2005) Plant disease detection based on data fusion of hyper-spectral and multi-spectral fluorescence imaging using Kohonen maps. *Journal of Real Time Image Processing* (Special Issue Spectral Imag. II) 11(2), 75–83.
Moss, G. (1975) The use of growth regulators in citrus culture. In: *Citrus, CIBA-GEIGY Agrochemical Technical Monograph* 4, 61.
Mukunda, L.L., Ramana, K.T.V., Krishna, S., Yuvaraj, K.M., Lakshmi, T.N., Sarada, G., Sankar, T.G., Gopi, V. and Gopal, K. (2014) Effect of growth regulators and chemicals on fruit yield and quality of hasta bahar flowering in acid lime (*Citrus aurantifolia* swingle) cv. balaji. *Research and Reviews: Journal of Agriculture and Allied Sciences* 3(3), 11–13.
Nageswara Rao, P.P., Ravishankar, H.M. and Uday Raj, K.N. (2004) Production estimation of horticultural crops using IRS-LD LISS-III data. *Journal of the Indian Society of Remote Sensing* 32, 393–398.
Nauer, E.M., Boswell, S.B. and Holmes, R.C. (1979) Chemical treatments, greenhouse temperature and supplement day length affect forcing and growth of the newly budded orange trees. *HortScience* 14, 229–231.
Navarro, L. and Juarez, J. (1977) Elimination of citrus pathogens in propagative budwood. II. In vitro propagation. *Proceedings of International Society of Citriculture* 3, 973–987.
Navarro, L., Roistacher, C.N. and Murashige, T. (1975) Improvement of shoot tip grafting *in vitro* for virus free citrus. *Journal of American Society of Horticultural Sciences* 100, 471–479.
Oslund, C.R. and Davenport, T.L. (1987) Seasonal enhancement of flower development in Tahiti limes by marcottage. *HortScience* 22, 498–501.
Patel, K.M., Patel, H.C., Patel, K.A., Chauhan, V.B. and Patel, J.S. (2012) Effect of manures and fertilizers on growth and fruit yield of acid lime cv. 'Kagzi' (*Citrus aurentifolia* Swingle). *The Asian Journal of Horticulture* 7, 481–483.
Rajendra, B.N., Kurubar, A.R., Anasubai, G.H. and Swamy, K.M. (2013) Influence of organic manures and inorganic fertilizers on vegetative development, yield, shelf-life traits and sensory evaluation score of acid lime (*Citrus aurantifolia* Christm.) c.v. Kagzi. *The Asian Journal of Horticulture* 8, 183–187.
Reddy, G.S., Naidu, L.G.K. and Anyaneyulu, K. (2003) Influence of soil nutrient mining on growth, yield and nutritional disorders in sweet orange (*Citrus sinensis* Osbeck) orchards. *Indian Journal of Citriculture* 2, 10–13.

REVIDEC (2003) Necesidades de investigación y transferencia de tecnología de la cadena productiva del limón persa en el Estado de Veracruz, Colegio de Postgraduados, y Fundación Produce de Veracruz, A.C., Tepetates, Veracruz, México.
Ritchie, J.T. (1983) Limitations to efficient water use in crop production. In: Taylor, H.M., Jordan, W.R. and Sinclair, T.R. (eds) *Efficient Water Use on Crop Production: Discussion on the Generality of Relations between Biomass Production and Evapotranspiration*. ASA Spec. Publ. ASA, CSSSA, and SSSA, Madison, Wisconsin, pp. 29–44.
Rivera-Cabrera, F., Ponce-Valadez, M., Sanchez, F.D.L., Villegas-Monter, A. and Perez-Flores, L.J. (2010) Acid limes. A review. *Fresh Produce* 4(Special Issue 1), 116–122.
Robert, P.C., Rust, R.H. and Larson, W.E. (1995) *Site-Specific Management for Agricultural Systems*. ASA-CSSA-SSSA, Madison, Wisconsin, p. 13.
Rojas, E. (1998) Floral response of Tahiti lime (*Citrus latifolia* Tan.) to foliar sprays of hydrogen cyanamide. *Fruits* (Paris) 53(1), 35–40.
Roses-Acevedo, J.L. and Sampedro-Rosas, L. (2000) Biological control of *Brevipalpus* spp. on *Citrus aurantifolia* in Guerrero, Mexico. *Manejo Integrado de Plagas* 55, 56–59.
Rouse, R.E. (1994) Controlling citrus rootstock sprouting in the nursery using Trehold. *Proceedings of the Florida State Horticultural Society* 107, 18–20.
Sanzani, S.M., Schena, L., De Cicco, V. and Ippolito, A. (2012) Detection and quantification of *Botrytis cinerea* in symptomless table grape stamens and berries. *Postharvest Biology and Technology* 68, 64–71.
Schönherr, J. (2001) Cuticular penetration of calcium salts: effects of humidity, anions, and adjuvants. *Journal of Plant Nutrition and Soil Science* 164, 225–231.
Schönherr, J. (2002) Foliar nutrition using inorganic salts: laws of cuticular penetration. *Acta Horticulturae* 594, 77–84.
Schönherr, J. and Luber, M. (2001) Cuticular penetration of potassium salts: effect of humidity, anions and temperature. *Plant Soil* 236(1), 117–122. doi: 10.1023/A:1011976727078.
Schultz, T.W. (1964) *Transforming Traditional Agriculture*. Yale University Press, New Haven, Connecticut.
Schumann, A.W. and Hostler, H.K. (2008) *Yield Monitoring of Citrus Using Color Photography and Ultrasonic Sensors*. ISHS Application of Precision Agriculture for Fruits and Vegetables, Orlando, Florida.
Schumann, A.W. and Zaman, Q.U. (2008) *Quantifying Wild Blueberry Yield with Image Processing*. ASABE, Providence, Rhode Island, p. 1.
Schumann, A.W., Hostler, K.H., Miller, W.M. and Zaman, Q.U. (2004) *Sensor-Based Automatic Yield Monitoring for Manually Harvested Citrus*. ASAE, St. Joseph, Michigan, ASAE Paper No. 041098.
Sepulveda, B., Torres, J., Manrique, G., Tume, P. and Rodriguez-Galvez, E. (2001) Biological control of the citrus borer worm *Phyllocnistis citrella* with native bacteria. *Fitopatologia* 36, 34–42.
Shaaban, M.M. (2009) Injection fertilization: a full nutritional technique for fruit trees saves 90–95% of fertilizers and maintains a clean environment. *Fruit Vegetable and Cereal Science and Biotechnology* 3(1), 22–27.
Sherlock, R.R. and Goh, K.M. (1985) Dynamics of ammonia volatilization from simulated urine patches and aqueous urea applied to pasture. II. Theoretical derivation of a simplified model. *Fertilizer Research* 6, 3–22.
Shrivastava, R.J. and Gebelein, J.L. (2006) Land cover classification and economic assessment of citrus groves using remote sensing. ISPRS *Journal of Photogrammetry* 61, 341–353.
Singh, S., Srivastava, A.K. and Shivankar, V.J. (1999) Citrus based cropping system – a review. *Agricultural Review* 20,123–128.
Skotnikov, A. and Robert, P. (1996) Site-specific crop management – a system approach. In: Robert, P., *et al.* (eds) *Proceedings of the Third International Conference on Precision Agriculture*. American Society of Agron, Madison, Wisconsin, pp. 1145–1152.
Smith, P.F. (1976) Collapse of 'Murcott' tangerine trees. *Journal of American Society of Horticultural Sciences* 101, 23–25.
Smoleń, S. (2012) Foliar nutrition: current state of knowledge and opportunities. In: Srivastava, A.K. (ed.) *Advances in Citrus Nutrition*. Springer, Dordrecht, the Netherlands.
Soon, Y.K. and Kalra, Y.P. (1994) A comparison of plant tissue digestion methods for nitrogen and phosphorus analyses. *Canadian Journal of Soil Science* 75, 243–246.
Srivastava, A.K. (2012) *Advances in Citrus Nutrition*. Springer, Dordrecht, the Netherlands.
Srivastava, A.K. and Singh, S. (2003) Foliar fertilization in citrus – a review. *Agricultural Review* 24(4), 250–264.

Srivastava, A.K. and Singh, S. (2005) Zinc nutrition, a global concern for sustainable citrus production. *Journal of Sustainable Agriculture* 25, 5–42.
Srivastava, A.K. and Singh, S. (2008) Citrus decline: soil fertility and plant nutrition. *Journal of Plant Nutrition* 32(2), 197–245.
Srivastava, A.K., Singh, S. and Marathe, R.A. (2002) Organic citrus: soil fertility and plant nutrition. *Journal of Sustainable Agriculture* 19(3), 5–29.
Srivastava, A.K., Huchche, A.D. and Lallan, R. (2007) Yield prediction in intercropped versus monocropped citrus orchards. *Scientia Horticulturae* 114, 67–70.
Starck, J.R. (1997) *Cultivation and fertilization of horticultural plants*, 3rd edn. Państwowe Wydawnictwo Rolnicze i Leśne, Warszaw.
Stuchi, E.S., Donadio, L.C. and Sempionato, O.R. (2003) Performance of Tahiti lime on *Poncirus trifoliata* var. *monstrosa* Flying Dragon in four densities. *Fruits* 58, 13–17.
Sunding, D. and Zilberman, D. (2001) The agricultural innovation process: research and technology adoption in a changing agricultural sector. In: Gardner, B.L. and Rausser, G.C. (eds) *Handbook of Agricultural and Resource Economics*. Elsevier Science, Amsterdam, pp. 207–261.
Szewczuk, C. and Michałojć, Z. (2003a) Preface – foliar fertilization of plants. *Acta Agrophysica* 85, 7–8 (in Polish with English abstract).
Szewczuk, C. and Michałojć, Z. (2003b) Practical aspects of foliar fertilization. *Acta Agrophysica* 85, 19–29 (in Polish with English abstract).
Tanigaki, K., Fujiura, T., Akase, A. and Imagawa, J. (2008) Cherry-harvesting robot. *Computers and Electronics in Agriculture* 63, 65–72.
Tian, L. (2002) Sensor-based precision chemical application systems. In: *Proceedings of the World Congress of Computers in Agriculture and Natural Resources*, Iguacu Falls, Brazil, 13–15 March 2002, pp. 279–289. American Society of Agricultural and Biological Engineers, St. Joseph, Michigan. doi:10.13031/2013.8342.
Tom, D. (1997) Yield map interpretation. *Crop Insights* 7(25), 1–6.
United States Bureau of the Census (1975) *Historical Statistics of the United States. Colonial Times to 1970*, Parts 1&2. United States Department of Commerce, Washington, DC.
United States Bureau of the Census (1980) *Statistical Abstract of the United States 1980*, 101st edn. United States Department of Commerce, Washington, DC.
United States Bureau of the Census (1998) *Statistical Abstract of the United States 1998*, 118th edn. United States Department of Commerce, Washington, DC.
Van Henten, E.J., Hemming, J., van Tuyl, B.A.J., Kornet, J.G., Meuleman, J., Bontsema, J. and van Os, E.A. (2002) An autonomous robot for harvesting cucumbers in greenhouses. *Autonomous Robots* 13, 241–258.
Vargas, G.A. (1993) Efecto del acido gibberellico y CEPA en el desfasameinto de la floracion en limon Persa (*Citrus latifolia* Tan.) en Martiez de la Torre Ver. *Memorias del II Symposium International Sobre Systemas de produccion en citricos, Programa Interdepartmental de investigacion y Servicio en citricultura*, Chapingo, México, pp. 195–202.
Vargas, G.A. and Espinoza, J.R.E. (1993) Efecto de reguladores del crecimieto en la floracion de los citricos. *Memorias del II Symposium International Sobre Systemas de produccion en citricos, Programa Interdepartmental de investigacion y Servicio en citricultura*, Chapingo, México, pp. 203–221.
Whitney, J.D., Tumbo, S.D., Miller, W.M. and Wheaton, T.A. (2002) *Comparison between Ultrasonic and Manual Measurements of Citrus Tree Canopies*. ASAE, St. Joseph, Michigan, ASAE Paper No. 021052.
Willer, H. and Lernoud, J. (2015) *The World of Organic Agriculture*. Statistics and Emerging Trends. FiBL-IFOAM Report, pp. 88–89.
Youssef, K., Ligorio, A., Nigro, F. and Ippolito, A. (2012) Activity of salts incorporated in wax in controlling postharvest diseases of citrus fruit. *Postharvest Biology and Technology* 65, 39–43.
Zaman, Q., Schumann, A.W. and Hostler, K.H. (2006) Estimation of citrus fruit yield using ultrasonically-sensed tree size. *Applied Engineering and Agriculture* 22(1), 39–44.
Zekri, M. and England, G. (2010) *Foliar Feeding and SAR for Citrus Trees*. University of Florida. Available at: http://citrusagents.ifas.ufl.edu/archived_presentations/fall_prg2010/Citrus%20Agents%20Web%20Site%20-%20-Nutritional.pdf (accessed 29 October 2016).

12 Harvesting and Post-harvest Management

Ahmad Sattar Khan[1]* and Zora Singh[2]

[1]Institute of Horticultural Sciences, University of Agriculture, Faisalabad, Punjab, Pakistan; [2]Curtin Horticulture Research Laboratory, Department of Environment and Agriculture, School of Science, Faculty of Science and Engineering, Curtin University, Perth, Western Australia, Australia

The majority of lime fruit produced in the world are consumed as fresh. Preservation of the natural quality of lime fruit after harvest either for local or international markets is a prerequisite to maintain consumer confidence to buy limes. The literature on the harvesting and post-harvest management of citrus is fairly vast and has been reviewed in the past by various scientists, but information on post-harvest handling of limes is very scant. Similar to other horticultural fresh produce (Mahajan *et al.*, 2014), the major challenge in post-harvest handling of limes is how to reduce post-harvest losses. Harvesting and post-harvest management of citrus fruit are important operations, which subsequently determine the storage and shelf life as well as the quality of lime fruit. Like other non-climacteric citrus fruits, limes are harvested at attainment of full maturity leading to maximum acceptability to consumers. Appropriate harvest maturity and method of harvest ensure good post-harvest handling with higher economic returns. Various post-harvest treatments are applied to limes in order to delay senescence and reduce mechanical injuries, physiological disorders and decay. In this chapter we attempt to summarize briefly all the steps involved in harvesting and post-harvest management including harvesting, post-harvest treatments, packing, storage and transportation of limes, which can positively influence the post-harvest life and fruit quality.

Fruit Maturity and Grade Standards

Limes, being non-climacteric fruit, attain maturity and ripen on the tree. However, the maturity stage at harvest has a significant impact on post-harvest storage life and quality of lime fruit (Singh *et al.*, 2004). After harvest, like other citrus fruits, lime does not undergo any rise in respiratory climacteric. However, variations in fruit maturity depend on species, soil and climatic conditions (Ladaniya, 2008). In subtropical to temperate regions of northern India citrus plants flower in spring, whereas, in the central and southern tropical conditions, trees bloom during spring, summer or monsoon before the onset of winter. Sometime lime trees are forced to flower through regulated water stress, which results in variations in fruit maturity and harvest time. Once lime fruit become mature, they can be stored on the tree beyond their normal harvest time for a few weeks. However, this condition of maturity varies with species, cropping and climatic conditions.

*E-mail: ahmad_khan157@yahoo.com

Fruit maturity indices

Maturity and ripening of most citrus species are judged by biochemical changes such as accumulation of total soluble solids and decrease in titratable acidity. Fruit weight, volume, juice content, acidity and peel colour are important parameters to judge the maturity indices of various lime cultivars. In those areas where lime trees flower year round difficulties arise in estimating harvest maturity. For example, in central India, the fruit colour of acid lime changes from dark green to light green with higher acid and juice contents after 160 days of fruit set (Ladaniya and Singh, 2000). The peel colour acceptability of lime at harvest varies with market requirements; for example, the best harvest colour for the export market is when the peel colour is still dark green or just changing from dark green to light green. Fruit harvested with a light yellow colour are more suitable for domestic markets. Generally at harvest maturity, lime should contain 30% or higher juice content by weight. Random fruit sampling must be performed to analyse juice contents prior to making any decision about harvest. Lime fruit left on the tree for an extended period generally turn yellow and exhibit poor post-harvest life compared with green coloured fruit. In most of the citrus-growing regions of the world, consumers prefer the mature green lime fruit, but in some areas yellow limes also have market value due to the high juice content. Fruit size, smoothness, freedom from bruises, decay and physiological disorder are other important indices considered when judging the quality of lime fruit at maturity. Some non-destructive maturity assessment techniques such as image processing technologies using digital cameras and cell phones have been reported to estimate volume and maturity of sweet lime fruit (Gokul *et al.*, 2015). Harvesting limes too early rather than at optimum harvest maturity makes the fruit more susceptible to chilling injury (CI). Whereas, late harvested limes are more prone to skin breakdown during long-term cold storage than green mature fruit.

Grade standards

The main parameters used to grade lime fruit include shape, size, colour and physical appearance of the fruit skin. Generally shape, size and colour at maturity are varietal dependent characteristics (Saunt, 1990). On the basis of fruit characteristics, limes are broadly divided into two groups: acidic and sweet fruit. In the acid group, there are two important types. One is the small fruited, abundantly seeded sour lime (*Citrus aurantifolia* Swingle). It is also known as Key lime, West Indian lime or Mexican lime. At maturity it is roundish in shape with a small nipple at the apex and a very light neck. It has a smooth texture and extremely thin skin with a very distinctive pungent aroma distinguishing it from the second type (Persian lime). At harvest, the fruit of the sour lime have very tender, light greenish yellow, juicy flesh. Another important quality standard of these fruit is a higher level of acidity, and they contain a higher amount of citric acid than lemons (Saunt, 1990). In comparison, the fruit of the Persian or Tahiti or Bearrs lime (*Citrus latifolia* Tan.) are large in size at full maturity. However, sometimes smaller immature fruit may be mistaken for sour lime fruit. The rind of these fruit is also very thin with the distinctive lime aroma. Like other sweet edible citrus species, Indian sweet lime or Palestine sweet lime has lower sugar content and lower acidity with six times less citric acid than sour lime. The fruit at full maturity is medium in size, with the characteristic nipple at the apex, thin rind and very juicy flesh. It is consumed generally at the domestic level with very limited international trade (Hume, 1957; Saunt, 1990). On the Indo-Pak subcontinent it has captured a high percentage of the domestic market as the earliest available seasonal citrus fruit on the market. In central India and Pakistan, it is available for consumption during August and September when the skin colour is still dark to light green. Along with sour lime, Eustis lime, a hybrid between sour lime and Marumi kumquat, is another small sized roundish/oval shaped, smooth skinned lime fruit. At full maturity, the fruit attains a light yellow colour and is consumed in domestic markets. In some parts of the world, it is also popular for backyard home gardening (Saunt, 1990). The main grade standards for selling lime fruit include fruit weight, ranging from 35–45 g per fruit, high juice content (30–40% of the fruit weight) and skin colour two-thirds or less than 30% of the total surface colour (Ladaniya and Singh, 2000; Kader, 2002; Ladaniya, 2008).

Lime fruit should be firm, with the characteristic shape of the cultivar, free from cuts, injuries, diseases and disorders. In the international trade generally these fruit are graded as extra class, class 1 and class 2. Extra class fruit are superior in quality with no or light superficial defects without any effect on the appearance of the fruit. Class 1 fruit are also very good quality, meeting all the criteria for fruit maturity with slight defects on the skin and in colour. These defects should not affect the internal quality of the fruit (Ladaniya, 2008). Fruit in class 2 should also meet the minimum maturity standards and some defects are allowed as in class 1 fruit. The size of all these fruit should not be smaller than 33 mm.

Post-harvest Physiology

Respiration rate and ethylene production

Lime is a non-climacteric fruit (Kader, 2002), exhibiting a higher respiration rate (150 mg CO_2/kg/h) during the initial growth stage, which declines (80 mg CO_2/kg/h) as the fruit attains full maturity. Even after harvest, the rate of respiration further reduces (40 mg CO_2/kg/h) during 8–10 days of storage at ambient conditions (Ladaniya, 2000). A similar trend in the respiration rate (450, 90 and 40 mg CO_2/kg/h) of lime fruit has been reported after 30, 120 and 180 days of fruit set, respectively (Subramanyam *et al.*, 1965). After harvest these fruit do not exhibit any sharp rise in respiration rate and ethylene production. However, following harvest their respiration rate may accelerate, if these fruit are placed above or below optimal storage temperatures. Fruit stored at 5–7°C exhibited higher respiration rates when exposed to 20°C, as compared with those fruit that were kept at the same temperature without exposure to low temperatures (5–7°C) (Eaks and Masias, 1965). Stimulation of the respiration rate after low temperature storage indicates that some intermediate compounds may accumulate during exposure to low temperatures, which consequently causes irreversible damage to fruit tissues (Eaks, 1960; Lyons, 1973). The respiration rate of lime fruit has shown variation when exposed to chilling temperatures. Singh *et al.* (2004) reported that lime fruit exposed to 0°C exhibited an increase of two to four times in their respiration rate, as compared with those stored at 10°C. Variation in the oxygen levels in the storage environment also influences respiration rate. Fruit stored above or below the normal level of oxygen exhibit higher respiration rates than those with the normal level of oxygen in the storage environment (Sritananan *et al.*, 2006). Limes do not produce ethylene in large quantities. However, ethylene in very low concentrations, less than 0.1 ppm, can be physiologically active in lime fruit. All those actions that reduce accumulation of ethylene around these fruit during storage and marketing help to improve their post-harvest storage life (Wills *et al.*, 1999). Lime fruit need special care during harvesting, storage, transportation and marketing to avoid mechanical injury and physiological disorder through stimulated ethylene production in the injured and wounded tissue (Kader, 2002). Reduction in the respiration rate of citrus fruit during post-harvest handling helps to maintain quality for an extended period of time (Calegario *et al.*, 2001). Therefore, any biotic and abiotic stress such as mechanical injury, disease infestation, CI and exposure to external ethylene can increase the respiration rate of citrus fruit (Burns, 1990; Baldwin, 1993). Generally citrus fruit exhibit low respiration rates and ethylene production (Porat *et al.*, 1999; Artes-Hernandez *et al.*, 2007). Lime fruit stored at 10, 15 and 20°C have been reported to exhibit respiration rates of 3–5, 5–8 and 6–10 ml CO_2/kg/h, respectively; whereas, with the same storage conditions the rate of ethylene production remained lower than 0.1 μl/kg/h (Kader and Arpaia, 2002). In another study, Win *et al.* (2006) found that lime fruit stored at ambient conditions produced 11.3 ml CO_2/kg/h and 0.3 0.1 μl/kg/h ethylene.

Colour metabolism

Limes are harvested when the fruit attain physiological maturity. Fruit harvested at an immature stage cannot ripen off the tree (Baldwin, 1993). Along with other quality parameters, retention of the green colour on the fruit skin increases consumer acceptability. At harvest lime fruit are graded into different classes on the basis of the green colour of the skin. At maturity the green colour of the skin is the main factor affecting its quality. Hence, the colour of lime fruit is the main determinant of its acceptability for consumers. Following harvest limes exhibit rapid degradation of the surface green colour during storage as well as marketing, which reduces their market value. Chlorophyll degradation is the main reason for post-harvest yellowing of lime fruit (Drazkiewice, 1994; Srilaong *et al.*, 2011). Much effort during post-harvest handling of lime fruit revolves around the prevention of chlorophyll degradation, e.g. post-harvest regulation of storage temperature and environment (Salama *et al.*, 1965; Ziena, 2000; Ladaniya, 2004; Sritananan *et al.*, 2006), application of hot water treatment (Obeed and Harhash, 2006), intermittent warming (Kluge *et al.*, 2003a) and surface coating (Bosquez-Molina *et al.*, 2004).

Sugars

After harvest various fruit quality parameters such as sugars and acidity exhibit significant changes depending upon the storage conditions. In the majority of citrus fruits, D-glucose and D-fructose are major monosaccharides present in the juice at maturity (Ting and Deszyck, 1961); whereas, sucrose is a major non-reducing sugar in citrus. The ratios of these sugar components vary in different citrus species. Like oranges, the concentration of glucose is less than that of sucrose, as compared with mandarins and tangerines. In limes and lemons, sucrose is present in much smaller quantities than glucose and fructose. The mature Kaghzi lime has been found to contain 0.84% reducing and 0.82% non-reducing sugars (Selvaraj and Raja, 2000). In Kaghzi lime fruit, the amount of individual sugars changes continuously with stage of maturity; for example, at the dark green mature stage the fruit contains 0.37% glucose and 0.13% fructose; whereas, in light green coloured fruit, the level of glucose and fructose is 0.39% and 0.19%, respectively. The amounts of glucose (0.61%) and fructose (0.23%) increase at the advanced stage of fruit maturity (full yellow) (Ladaniya, 2008). Higher storage temperature coupled with low relative humidity causes rapid water loss, which increases the sugar content in lime fruit (Kohli and Bambota, 1966). Upon ripening, fructose is the major sugar dominating in the lime fruit (Albertini *et al.*, 2006). Crude extract of lime juice has been reported to contain 23.8 g/g reducing sugars, about 2-fold higher than crude extract of grapefruit juice (Guimaraes *et al.*, 2010). In limes, the level of reducing sugars is 4–5-fold higher than non-reducing sugars at harvest maturity (Ladaniya, 2008).

Flavonoids

Polyphenols (tannins) and phenolic contents are widely distributed phytochemicals also known as dietary phenolics (King and Young, 1999). These compounds exhibit strong antioxidant activity and aid in combatting many carcinogenic diseases (Frankel, 1999). Citrus fruit including lime are rich sources of various phytochemicals (Kawaii *et al.*, 1999; Peterson *et al.*, 2006). In acid lime fruit, six different flavonoids, hesperidin, narirutin, diosmin, neoponcirin, isorhoifolin and eriocitrin, have been reported (Nogata *et al.*, 2006). Consumption of lime fruit has been found to promote human health due to their strong anti-cancerous properties (Kawaii *et al.*, 1999; Gharagozloo *et al.*, 2002). Generally diets rich in flavonoids have been reported to reduce the risk of various cardiovascular diseases. The availability of a range of flavonoids in citrus fruit and their health benefits warrant the investigation of how the biosynthesis of the compounds could be upregulated in the fruit. Flavour is one of the main constituents of lime fruit quality. Various flavour compounds have been identified in lime fruit, such as limonene, 1,8-cineole, 4-terpineol, β-pinene, 1,4-cineole, *p*-cymene, α-terpineol, α-bergamotene, β-bisabolene, neryl acetate and germacrene-d (Shaw, 1979; Ranganna *et al.*, 1983; Chamblee and Clark, 1997; Chamblee *et al.*, 1997; Della Porta *et al.*, 1997; Dugo *et al.*, 1998; Venkateswarlu and Selvaraj, 2000; Tincutta and Richard, 2001; Yadav *et al.*, 2004).

Acids

The quality of lime fruit is mainly characterized by their acid content. Acid lime cv. Kaghzi attains over 50% juice content and 9% acidity after 190 days of fruit set in north Karnataka, in Indian climatic conditions (Rao *et al.*, 1983); whereas, in central India the same cultivar takes 160 days to attain 40% juice content and 8% acid content (Ladaniya and Singh, 2000). During the early growth and development of lime fruit, quinic acid is the dominant organic acid; later on, citric acid predominates as a major organic acid (Albertini *et al.*, 2006). Lime fruit also contain considerable amounts of ascorbic acid. The level of ascorbic acid in lime fruit has been found to be between 39 and 62 mg/100 ml juice (Ziena, 2000). Similarly, Kluge *et al.* (2003a) reported that Tahiti lime contains 31 mg/100 ml ascorbic acid. Fresh lime and lemon juice and their concentrates provide higher amounts of citric acid than ready-to-consume grapefruit and orange juice. Fresh lime and lemon juice are rich sources of citric acid (46 and 49 mg/ml, respectively); whereas, juice concentrates of these fruit contain 35 and 39 mg/ml citric acid, respectively (Penniston *et al.*, 2008).

Harvesting Methods and Techniques

Pre-harvest cultural practices, harvesting methods, post-harvest handling, transportation and pack house operations have significant impacts on the post-harvest life and quality of citrus fruit (Hume, 1957; Kader, 2002). Post-harvest loss of citrus fruit occurs due to changes in physiological quality attributes like juice content, weight loss, skin damage, changes in colour or even in certain cases complete loss due to attack of some diseases and physiological disorders. Any delay in the harvest of lime fruit initiates chlorophyll degradation and increases carbohydrate consumption (Gayet and Salvo Filho, 2003). The last operation practised by the grower of citrus fruit in the orchard is the harvesting. The harvesting of the fruit is the beginning of the post-harvest handling system; therefore, the harvesting method, mechanical injuries to fruit during harvesting and weather conditions at the time of harvest greatly influence the losses during subsequent post-harvest handling, storage and supply chain processes (Ladaniya, 2008). Harvesting should preferably be done at a cooler time of the day, i.e. early morning hours or in the afternoon to avoid damage due to high temperatures.

Hand harvest

For fresh consumption, limes are preferably harvested manually either through spot picking or all the crop at once. The latter approach increases the rejection rate in processing, as compared with the spot picking method (Ladaniya, 2008). The harvest method is the key to reduce post-harvest losses and increase profitability of lime fruit. For the best post-harvest potential with minimal losses, fruit should be harvested close to the button by cutting with sharp scissors. Leaving larger stems on the fruit may lead to puncture damage to the other fruit in the harvest basket during post-harvest handling. Pickers at harvest must wear gloves to avoid any injury due to the presence of large numbers of spines on the fruit-bearing twigs. Fruit plugging is not a recommended practice and it should be discouraged as it causes severe post-harvest losses. Traditionally, limes are harvested by just twisting and pulling to break their peduncle at the button end, which leads to rupturing of the button end and increases post-harvest losses. It has been observed that harvesting lime fruit with scissors at the button end increases the percentage of marketable fruit with a reduced level of colour changes and incidence of oleocellosis, as compared to hand pulling and the use of hooks to remove fruit from lime trees (Bassan *et al.*, 2012). Use of bags or containers to hold the harvested fruit is a common practice adopted in advanced countries. However, care must be taken that such bags or containers are free of sharp edges and lined with soft pads or cushions (Ladaniya, 2008). Fruit from these bags or containers are emptied into the big field bins (wooden or plastic), which are directly transported to a pack house or processing unit. Field clipping with sharp scissors is preferred over snap picking to minimize bruises and mechanical injuries. In certain parts of India and Pakistan the lime fruit are harvested with a hook by standing on the ground, which increases losses due to the higher rate of mechanical injuries of these fruit. After harvest, fruit must be carefully transported to the pack house as quickly as possible.

Mechanical harvesting

The high cost of hand harvesting in many countries has encouraged growers to use other alternatives such as mechanical harvesting. Very little information is available about the use of mechanical harvesting for small sized citrus fruit species such as limes and lemons. Mostly these fruits are not produced on a large scale and are harvested for fresh consumption. The use of mechanical harvesting for citrus began in the 1970s (Ladaniya, 2008). Mechanical harvesters work either according to a positioning mechanism (contact removal) or an external force mechanism (mass removal), which shakes the whole tree or limb mechanically. Generally a mass removal system is suitable for processing and contact removal for fresh consumption (Whitney, 1978; Whitney, 1987). Mechanical injuries in the form of splitting, bruising, stem attachment, puncturing and decay have been found to be higher with machine harvesting than with manual picking of fruit (Whitney, 1978). In mass removal mechanical harvesting systems some selective action non-phytotoxic eco-friendly chemicals are used to loosen the contact of the fruit with the twig. These abscission inducing agents increase the efficiency of mass removal mechanical harvesters. Therefore, with various methods of mechanical harvesting (trunk shaker, limb shaker, air blast, robotic arms), use of abscission inducing chemicals reduces the percentage of fruit harvested with stems, which could cause puncture damage to adjacent fruit during handling and transportation (Ladaniya, 2008). Although a lot of research work has been carried out to develop technologies that reduce losses during mechanical harvesting for various citrus fruit such as oranges and mandarins, with varying efficiency, there is further need to standardize the application of these mechanical devices to improve efficiency and reduce post-harvest losses for lime fruit.

Field handling

After harvest, careful field handling is also important to maintain the quality of lime fruit. Poor and delayed handling in the field can significantly increase post-harvest losses (Grierson, 1981). Care must be taken that fruit is harvested during the cool hours of the day, preferably early morning or afternoon. If there is a need to temporarily place harvested fruit in the field, they must be kept in shade. Scratching, pitting, plugging and bruising are common defects found in the lime fruit caused by poor field handling. These defects increase the rejection rate in the pack house operations for fresh consumption and, later on, the rate of deterioration during storage and transportation (Ladaniya, 2008). Clipping citrus fruit with a sharp knife reduces injuries up to 9% as compared with plugging (Sonkar *et al.*, 1999).

Post-harvest Pack House Operations

During processing, transportation and marketing, losses may occur as mechanical damage, physiological disorder and decay. Due to mechanical damage, the lime fruit show both direct (tissue discolouration, deformation) and indirect losses (changes in respiration, ethylene production, colour, texture and aroma of fruit). A detailed outline of pack house operations for lime fruit is given in Fig. 12.1.

Initial inspection and bin dumping

Pack house operations begin with an initial inspection of fruit, which include weighing and physical inspection of field bins for an estimate of the condition of the harvested fruit. This step follows the dumping of these fruit bins into water tanks treated with chlorine (100–200 mg/l) at pH 7–9 to avoid contamination and accumulation of decay causing spores, which can lead to infection at later stages of the pack house operations. Mechanically injured and bruised fruit are more susceptible to these types of decay (Singh and Naqvi, 2001; Singh *et al.*, 2004). The first step in the processing unit is the cleaning of the fruit. The cleaning water is sanitized with 1000 mg/l chlorinated water or 120–150 mg/l household bleach. Lime fruit are sorted on the

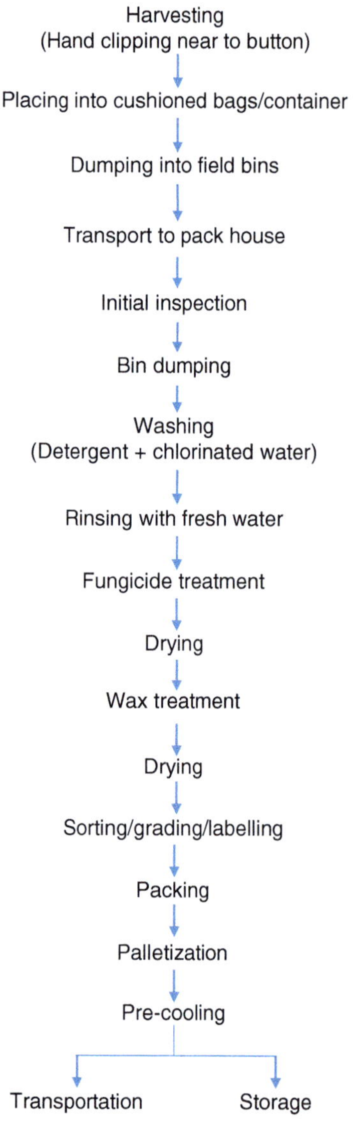

Fig. 12.1. Schematic flow chart for post-harvest handling of lime.

basis of shape, size, colour and diseases. Careful selection of fruit at this step determines the best eating quality and maximum post-harvest storage life during various post-harvest handling procedures. Sorting of citrus fruit is performed through mechanical means rather than manual operation to reduce mechanical injuries and contamination (Kader and Arpaia, 2002). Limes are sorted either on the basis of skin colour with an electronic sorter based on light reflection or on a weight basis using automatic electronic sensors (Wardowski *et al.*, 1986; Singh and Naqvi, 2001).

Trash elimination

Most modern pack houses have provisions for the elimination of trash accompanying the fruit in the field bins. Most of the loose trash such as long stems, leaves and loose sand, which can be injurious to pack house equipment, must be removed before moving the fruit to the next step of the pack house operations. Mostly the roller conveyers have a self-cleaning system that wipes up all sorts of trash from the rollers during processing (Wardowski *et al.*, 1986; Singh and Naqvi, 2001).

Washing

During washing, the fruit is moved between transverse rotating brushes with an exposure of at least 15–20 s for adequate washing. Slow brush speeds give better cleaning than fast speeds. Following washing with detergent, fruit are moved though rinsing nozzles at between 150 and 200 kPa pressure for quick rinsing (Wardowski *et al.*, 1986). Before moving to the next step, the fruit are dried either by passing through hot air or a sponge rubber roller at <75 rpm speed. Some pack house systems are equipped with soft brushes to throw off much of the water from moving fruit (Wardowski *et al.*, 1986; Ladaniya, 2008).

Fungicide treatments

In larger pack houses, cleaning lines are linked with overhead fungicides sprays on the washing lines followed by drying (Wardowski *et al.*, 1986). In various commercial pack houses, lime fruit are treated with 500 mg/l benomyl or 1000 mg/l thiabendazole and imazalil. Similarly, use of 0.5% thiabendazol, 0.05% benlate and 0.05% bavistin effectively controls rots caused by *Penicillium italicum* and *Penicillium digitatum* (Verma and Tikoo, 2003).

Waxing

After fungicide application fruit are treated with wax. Application of wax helps to slow down the drying process and also gives a shine to the fruit. In most advanced pack houses wax is applied as a gentle spray, while fruit move on the roller beds. Waxing is not a necessary step in pack house operations. Application of wax depends upon the market requirements. Among the various methods of wax application, spraying is the most convenient and common method used in most of the citrus-producing countries. The one basic prerequisite is to apply a uniform and controlled coating of wax (Wardowski *et al.*, 1986). To overcome this problem a self-operated automatic control system is installed along the processing line to monitor the flow rate of fruit and wax spray accordingly. Sometimes fungicides are also mixed with waxes. In those units in which wax and fungicides are applied together, care must be taken to clean the roller line at regular intervals to avoid any contamination to moving fruit (Kader, 2002).

Drying

Lime fruit are dried more than once in the processing lines depending upon the steps involved in the processing operations. Normally hot air is delivered over the moving fruit at a constant speed for 1–2 min. The drying unit must have a proper exhaust unit to reduce the saturation of hot air and damage to fruit. Limes are generally passed through hot air at temperatures between 50 and 55°C (Wardowski *et al.*, 1986).

Labelling and packaging

Fruit are labelled before packing. Labels normally contain the trade name or source of origin of the fruit or both. The most common labelling method is attaching paper or plastic labels using an automatic machine onto individual fruit (Wardowski *et al.*, 1986; Kader, 2002). The material used in the labelling must be approved by the regulatory authorities of the consumer markets. Generally, the first criterion for packing of fruit is the size of the fruit. Each packing unit should contain uniform sized fruit. Polyethylene-lined corrugated cardboard boxes are used in various domestic as well as international markets for the packing of lime fruit. Storage of sour lime fruit in punnet bags made of polyethylene terephthalate material at 5°C with 85–90% humidity caused a reduction in fruit weight loss and decay with higher fruit juice acidity and ascorbic acid contents (El-Sheikh *et al.*, 2014). Packaging of fruit is necessary to reduce physical, physiological and pathological deterioration during the different steps in the supply chain and handling. Materials used for the packing of lime fruit also vary with the end user market. Domestic markets have separate packing standards to export markets. For long-term transportation and handling, packing material should be well ventilated and strong enough to minimize bruising injury during transportation and marketing (Ladaniya, 2008). For domestic markets some conventional packing materials such as bamboo baskets, wooden boxes, gunny bags and jute-cloth-lined wooden baskets are used (Fig. 12.2). Sometimes, lime fruit are transported to domestic markets unpacked, which increases post-harvest losses. However, for the international export markets fruit are preferably packed in polyethylene lined corrugated cardboard boxes to minimize these losses (Wardowski *et al.*, 1986; Singh and Naqvi, 2001).

Post-harvest Treatments

Post-harvest handling of lime fruit mainly aims to reduce transpiration, rate of respiration, ethylene production, discolouration, decay and microbial infections. Various post-harvest techniques such as cold storage (Kader, 2002; Ladaniya, 2004), controlled atmosphere (CA) storage (Sritananan *et al.*, 2006), modified atmosphere packaging (Salama *et al.*, 1965), intermittent warming (Kluge *et al.*, 2003a; Kluge *et al.*, 2003b), exogenous application of growth regulators (Gates, 1949; Burns *et al.*, 1964; Win *et al.*, 2006), edible coating (Bisen *et al.*, 2012), oil emulsion (Verma and Dashora, 2000) and radiation (Ladaniya *et al.*, 2003) have been tested for lime fruit with variable success.

Pre-cooling

Pre-cooling is carried out to remove field heat from the fruit after harvest. The shorter the time

Fig. 12.2. Different packing materials used for lime fruit: (A) wooden crate, (B) loose packing of yellow over-mature and (C) green mature lime fruit in jute-cloth-lined wooden baskets, (D) cardboard box, (E) plastic crate, (F) net bag and (G) polyethylene bag.

period between fruit harvest and pre-cooling, the longer will be the impact of pre-cooling on different post-harvest handling practices such as inhibition of decay, fungal growth, respiration rate and water loss (Kader, 2002). Among the various pre-cooling systems, forced air pre-cooling is the most suitable for lime fruit. Proper ventilation levels and staking patterns help to achieve pre-cooling of fruit more effectively (Singh and Naqvi, 2001).

Transportation

Transportation of fruit starts from the field after harvest and continues until it reaches the consumers. Different types of transportation system are used in different citrus-producing countries around the world. In India, for example, limes are transported in gunny bags lined with paddy straw on trucks; whereas, in Pakistan, limes are transported in wooden baskets made of mulberry sticks (Fig. 12.2) and covered with gunny bags to domestic markets on mini trucks. Proper packaging and loading during transportation is necessary to reduce post-harvest losses. Most advanced countries use strong, corrugated cardboard boxes to market lime fruit in local markets, which causes considerable reduction in post-harvest losses (Wardowski *et al.*, 1986).

Cold storage

The temperature of the storage environment plays a vital role in maintaining the quality of citrus fruit. The optimum temperature for prolonged storage of citrus depends upon species and cultivars. Higher temperatures speed up deterioration; whereas, temperatures lower than the optimum cause CI (Kader, 2002). The recommended storage temperature for lime fruit ranges from 7.2–12.2°C with 85–90% relative humidity (Thompson, 2003). Climatic and soil conditions, area of production and pre-harvest management practices affect the storage potential of lime fruit. Acid lime cv. Kaghzi cannot be stored below 8°C due to its susceptibility to CI. At 8°C lime fruit can be stored for 90 days without CI. Whereas, at 10°C fruit have been shown to exhibit higher juice contents compared with those stored at 8°C after 90 days of storage (Ladaniya, 2004). Wills *et al.* (2004) reported that the lowest safe storage temperature for lime fruit is 7°C.

Controlled atmosphere (CA) storage

Lime fruit are best stored between 10 and 12°C with 85–95% relative humidity for up to 8 weeks (Ladaniya, 2004). Storage of the fruit below 8°C results in development of CI in the form of surface pitting. To maintain the green colour of the fruit skin throughout the post-harvest handling chain, application of CA storage has been found to be beneficial (Kader, 2003). Exposure of fresh horticultural produce like fruit and vegetables to low O_2 and higher CO_2 has been used successfully to prolong storage life and maintain quality (Thompson, 1998; Watkins, 2000; Singh *et al.*, 2009). Such conditions help to reduce respiration rate, ethylene production and alleviate low temperature disorder in subtropical fruits (Thompson, 1998; Singh *et al.*, 2009). CA storage containing 5% O_2 and 3% CO_2 has been reported to reduce loss of chlorophyll, changes in peel a* and b* parameter (skin colour) and physiological disorders in lime fruit compared with air storage at 10°C (Sritananan *et al.*, 2006). CA storage can probably delay the degreening of lime fruit; however, such conditions may result in some unfavourable quality changes such as juice loss, rind breakdown and decay. Fruit are stored best at CA comprising 5 kPa O_2 and 0–10 kPa CO_2 at 10–12°C (Kader, 2002; Kader, 2003). In contrast, Spalding and Reeder (1974, 1976) recommended CA containing 5 kPa O_2 and 7 kPa CO_2 to maintain acceptable the green colour without any off-flavour development in lime fruit. Tahiti lime stored in CA comprising 7–10% CO_2 with 21% O_2 at 10°C showed normal rind thickness with acceptable juice contents and skin colour (Spalding and Reeder, 1983). Salama *et al.* (1965) reported that CA storage of Persian lime containing 4% O_2 and 3% CO_2 slowed down the change in green to yellow colour with a higher percentage of decay (up to 80%) (Salama *et al.*, 1965). Sun (1998) stored acid lime for 5 months in CA comprising reduced O_2 (3–6%) and elevated CO_2 (2-5-4%). Low oxygen stimulates accumulation of acetaldehyde and ethanol, and develops off-flavour in citrus fruit.

Evaporative cold storage

The higher cost of energy generation to run refrigerated storage systems has necessitated some cheap alternatives. Developing and underdeveloped countries with limited storage facilities coupled with lack of a regular electricity supply can utilize the advantages of low cold evaporative cold storage systems. These storage chambers can operate without the use of electrical energy (Roy and Khurdiya, 1986). When using such systems, the treatment of citrus fruit with disinfectant and fungicides is necessary due to their higher humidity (Kaushal and Thakur, 1996). Lime fruit treated with 4% wax and 2,4-D (100 mg/l) exhibited storage life 35 days longer in evaporative cooling chambers than control fruit (Thangaraj *et al.*, 1983).

Low pressure storage

Lowering the pressure of the storage atmosphere from normal air pressure has been found to be beneficial to extend the storage and green life of limes with acceptable fruit quality. Low pressure to 30 kPa (50 mm Hg) at 10°C has been reported as optimum to store limes for up to 5 days without skin yellowing, as compared with normal air pressure (Burg and Burg, 1966). Spalding and Reeder (1976) reported that keeping fruit of Tahiti lime at lower storage pressures of up to 176 mm Hg maintained good fruit colour for up to 6 weeks at 10–15°C in contrast to fruit kept under normal air pressure.

Edible coatings

After harvest, lime fruit exhibit an increase in soluble solids and ascorbic acid, and a decrease in acid contents. However, use of different chemicals such as mineral oils and waxes reduces decay during post-harvest storage (Bosquez-Molina *et al.*, 2004). The use of coconut oil emulsion-based edible coating has been reported to reduce physiological weight loss (10%), and increase marketable fruit, soluble solids, ascorbic acid contents and juice contents of Kaghzi lime for up to 78 days of storage with acceptable colour, taste, texture and flavour (Bisen *et al.*, 2012). Kaghzi lime treated with mustard oil and diphenyl exhibited the highest juice and ascorbic acid contents during storage at ambient conditions for 12 days, as compared with neem oil or mustard oil treatment (Verma and Dashora, 2000). Coating of Tahiti acid lime fruit with biofilms (3% cassava starch, 3% corn starch and 3% gelatin) and packing them in burlap sacks maintained their green colour for up to 35 days in ambient conditions (Dotto *et al.*, 2015). Post-harvest application of an edible coating containing non-polluting alternatives such as chitosan can be used effectively to increase the post-harvest life of lime fruit (Palou *et al.*, 2015). Recently, coating of lime fruit with 0.8% (w/w) xanthan gum and 1.5% (w/w) extracted lignin (40% (w/w) NaOH) solution showed reduced fruit weight loss and colour changes and higher anti-fungal activity in contrast to fruit treated with commercial lignin (Jonglertjunya *et al.*, 2014).

Hot water treatment

Being highly nutritious fruit, citrus fruit are a good source of vitamins, minerals, organic acids and various other phytochemicals. However, post-harvest losses limit their storage life. Use of non-chemical treatments such as hot water application has emerged as a safer and non-hazardous approach to eliminate these post-harvest losses (Lurie, 1998a; Fallik, 2004; Sivakumara and Fallik, 2013). To inhibit post-harvest senescence and decay, hot water treatment is gaining popularity to extend the storage life and marketability of fresh fruits including citrus (Lurie, 1998a). However, the hot temperature and duration of exposure during hot water treatment may affect the storage life and quality of lime fruit (Lurie, 1998b). The effect of hot water treatment can be further aggravated by combination with a chemical such as calcium chloride (Martinez-Romero *et al.*, 1999). Mexican lime fruit treated with hot water at 55°C containing 2% calcium chloride had reduced fruit loss with highest juice contents and soluble solids during storage at 12°C (Obeed and Harhash, 2006). Hot water treatment at 40°C for 2 min when fruit were harvested after 6 months of flowering resulted in the lowest loss in fruit weight, chlorophyll degradation and highest juice contents during 20 days of storage, as compared with those fruit harvested after 6 months of flowering and treated with 50°C hot water either for 2 min or 7.5 min. Hot water treatment at 50°C for 5 min delayed the degreening of Tahiti lime with reduced activities of chlorophyllase, chlorophyll-degrading peroxidase and pheophytinase enzymes, respiration rate and ethylene production during storage at 25°C (Kaewsuksaeng *et al.*, 2015).

Other storage options

Tahiti lime can be stored for 4–8 weeks at 10–12°C with 85–95% relative humidity (Kader and Arpaia, 2002). However, long-term storage below 8°C causes CI with increased incidence of decay and superficial rind pitting. Under such conditions some special treatments such as intermittent warming (IW), in which during cold storage fruit are exposed to intervals of high temperatures, help to reduce these losses (Wang, 1993). Tahiti lime treated with IW at 20°C for 48 h every 7–14 days exhibited reduced CI for up to 60 days of cold storage at 5°C, as compared with non-treated fruit at the same temperature (Kluge *et al.*, 2003b). Use of IW in combination with 1-MCP has been reported to improve the maintenance of the green colour in lime fruit during long-term storage with reduced incidence of CI (Kluge *et al.*, 2003a). Various chemicals have also been used previously for lime fruit to delay the undesirable changes in colour and physiological breakdown in peel tissues, such as 2,4-D (Burns *et al.*, 1964), sorbic acid and gibberellic acid (Gates, 1949). Pre-harvest application of GA_3, in combination with post-harvest wax treatment maintained the quality of Mexican lime fruit at 22 ± 2°C for 10 days of storage (Zea-Hernández *et al.*, 2016). Post-harvest treatment of Baramasi lemon with 75 mg/l GA_3 extended the post-harvest life by up to 60 days under ambient conditions (Kaur *et al.*, 2014). Similarly, Salama *et al.* (1965) reported a reduction in the change from green to yellow colour during storage in Persian lime treated with 2,4-D (100 mg/l) and GA_3 (300 mg/l). Lime fruit treated with post-harvest application of 1000 mg/l 2,4,5-trichlorophenoxyacetic acid and wax exhibited delayed senescence with reduced respiration rate, weight loss and changes in peel colour (Ayoub and Abu-Goukh, 2009). Post-harvest application of 1-MCP as an ethylene action inhibitor has the potential to retain the green colour and reduce physiological disorders of non-climacteric fruits including limes (Li *et al.*, 2016). Post-harvest treatment with 250 or 500 nl/l 1-MCP effectively retarded yellowing of acid lime fruit for 21 days in ambient conditions, as compared with higher concentrations of 1-MCP and untreated fruit (Win *et al.*, 2006). Acid lime treated with 250 Gy radiation exhibited reduced stem end breakdown for up to 4 days of low temperature storage, as compared with untreated fruit (Ladaniya *et al.*, 2003).

Post-harvest Disorders and Diseases

Like other tropical fruits, citrus is susceptible to various post-harvest disorders and diseases (Taghipour *et al.*, 2012). These disorders can cause changes in the appearance and flavour of citrus fruit and consequently reduce their quality and acceptability to the consumers. Uncontrolled use of chemicals to overcome these issues in citrus has led to various health and environmental hazards, which consequently prioritized the importance of finding other alternatives such as biological control (Zamani *et al.*, 2008) and application of organic elicitors (Iqbal *et al.*, 2012). Limes are highly perishable in nature and about 25–40% losses occur in the period after harvest up to consumption (Bisen *et al.*, 2012). Limes with bright green coloured skin, free from skin injuries, have high market value. The change in the colour from green to yellow and increase in injuries to the skin of the lime fruit lead to a decline in consumer acceptability.

Chilling injury

Lime fruit stored below 10°C for extended periods of time exhibit CI. In the initial stages CI appears as a small defect in the form of pits on the skin of the fruit. Later on as its severity increases these defects change into brown sunken areas on the rind (Fig. 12.3) (Spalding and Reeder, 1974; Kader, 2002). Limes are sensitive to CI at low storage temperatures. CI in lime is mainly characterized by superficial surface pitting, staining and necrosis of fruit skin (Rivera *et al.*, 2007; Wang, 2010). Because of this low temperature, storage cannot be exploited to its full potential in extending the storage life of lime fruit. Susceptibility to CI in various citrus species mainly depends upon pre-harvest (growing area, season of harvest, species) and post-harvest (storage temperature, storage duration and treatment applications) factors (Sierra *et al.*,

Fig. 12.3. Chilling injury symptoms on Tahiti lime fruit.

1993; Kader, 2002; Kluge *et al.*, 2003b). The recommended storage temperature for lime fruit is 10–12°C; under these conditions fruit can be stored for extended periods of time (8 weeks); whereas, long-term low temperature storage below 8°C induces CI in lime fruit. Low temperature conditioning has been reported to alleviate CI in various fruits including limes (Hatton, 1990). The composition of the fruit also influences the incidence of CI in citrus. Lime fruit with the lowest levels of calcium content in their juice have been shown to have the highest percentage of CI (Slutzky *et al.*, 1981). Harvest maturity and post-harvest treatments also influence the susceptibility of lime to CI. Mexican lime harvested in July and treated with wax + 500 mg/l TBZ + 250 mg/l GA_3 exhibited the best quality and maximum marketability when stored at 10°C with 85–90% relative humidity in contrast to delayed harvested and untreated fruit (Sierra *et al.*, 1993). Storage temperatures of 10–12°C with higher relative humidity (85–90%) are recommended safe temperatures to reduce CI with maximum storage life for lime (Kader and Arpaia, 2002). Lime fruit harvested early in the season and treated with mesquite gum mixed with candelilla wax, mineral oil and stored at 10°C showed better quality with reduced CI and higher retention of skin surface green colour in contrast to untreated and late harvested fruit (Bosquez-Molina *et al.*, 2004). Conditioning of lime fruit with low temperature (13°C) for 48 h has been found to be effective to induce tolerance against CI by maintaining higher activities of peroxidase and superoxide dismutase enzymes (Rivera *et al.*, 2007). Sour limes retain a good green colour when stored at 5–7°C but exhibit severe rind pitting (Ladaniya, 2004). Lime fruit have been found to maintain better rind colour when kept at 4°C; however, such fruit are more subject to CI in the form of pitting, with limited storage life (Murata, 1997).

Degreening

Degradation of green colour is a key issue in post-harvest handling of lime. A uniform green surface colour of lime fruit should be maintained throughout the supply chain for maximum consumer acceptability. Many techniques are being used commercially to reduce the degradation of the green colour such as hot water dipping at 50–55°C for 2–4 min and post-harvest application of growth regulators (Obeed and Harhash, 2006; Win *et al.*, 2006). In certain cases, use of an ethylene absorbent compound such as potassium permanganate has been found to be useful to inhibit the ethylene induced yellowing of fruit skin (Kader, 2002). Application of electrostatic atomized water particles significantly activated the hydrogen peroxide scavenging activity and reduced the degreening of Nagato-yuzukichi and Yuzu citrus fruit stored at 20°C (Yamauchi *et al.*, 2014). Care must be taken to avoid storage of lime fruit with higher ethylene producing fruit even for short periods of time.

Oil spotting

In lime, the most common skin disorder is oleocellosis, which occurs when disruption of the oil glands with release of oil contents from skin cells leads to the damaging of epidermal tissues. The main symptoms of this disorder are breaking of oil cells in the flavedo with release of oil that damages the surrounding tissues. This disorder is also known as oil spotting, which is mainly caused by the toxic action of peel oil through wounded, injured or bruised skin tissues. Mishandling, mechanical injury during harvesting and pack house operations are the main reasons for this problem. Infected fruit also show oddly shaped brown or yellow spots and discolouration. Symptoms normally appear several days after harvest, which makes it difficult to manage this problem at that time. Preventive measures give better control for the management of oleocellosis (Singh *et al.*, 2004).

Stylar end breakdown

Harvesting of fruit during hot hours of the day and exposure of fruit to higher temperatures during transportation increase the susceptibility of lime fruit to stem end breakdown disorder. Water soaked greyish spots appears near the blossom end with dark coloured lesions, which become sunken and dried out with further progression of severity of this issue (Singh *et al.*, 2004; Ladaniya, 2008). Keeping fruit at the optimum temperature (10–12°C) during the whole supply chain helps to minimize stylar end breakdown. The rupturing of the juice vesicle located at the periphery of the locules causes stylar end breakdown (Petracek *et al.*, 2006).

Blue and green mould

Post-harvest diseases caused by blue and green mould lead to significant losses in citrus. Blue mould produces soft, watery spots on the skin of the fruit followed by development of blue mould. Injured, bruised or mechanically damaged fruit are also susceptible to green mould. Disease appears as olive green colour growth surrounded by white mycelium. Infected fruit in the packing line, packing containers or storage can spread spores of this disease to healthy fruit. Biological methods for control of blue and green moulds are becoming popular due to the effects of fungicides on fruit. Post-harvest application of two bacterial antagonists have shown effective control of green mould in Mexican lime, as compared with chemical and heat treatment (Ketabchi *et al.*, 2012). Native isolates *Debaryomyces hansenii* obtained from the marine environment (*D. hansenii* isolates DhhBCS06, LL1 and LL2) exhibited 80% reduction in the incidence of blue mould in Mexican lime under *in vitro* and in simulated industrial pack house conditions for 2 weeks (Hernandez-Montiel *et al.*, 2010). Similarly, *Saccharomyces cerevisiae* isolates caused 100% control of green mould in Tahiti lime fruit when treated with a quarter dose of imazalil and stored at 27°C and 70% relative humidity for 2 weeks (Moretto *et al.*, 2014).

Stem end rot

Lime fruit are also attacked by stem end rot, which begins near the stem end of the fruit as light or dark coloured spots, followed by uneven downward finger like markings. Healthy tissues show a clear line of separation from the diseased tissues in lime fruit affected with stem end rot (Singh *et al.*, 2004; Ladaniya, 2008).

Conclusions and Future Prospects

Limes are important citrus fruit produced in all tropical and subtropical countries throughout the world. Both pre- and post-harvest practices influence their keeping quality after harvest during handling. Minimizing post-harvest losses in horticultural fresh produce including lime fruit is a notable challenge to post-harvest physiologists and biologists (Mahajan *et al.*, 2014). Harvesting of fruit at the appropriate stage of maturity leading to maximum consumer acceptability is the key to ensure maximum storability and market price. To ensure maximum uniformity of fruit maturity, growers should adopt the spot picking technique for harvesting lime fruit instead of bulk harvest. Practices such as mechanical fruit processing (washing, fungicide and wash treatment, grading, labelling), use of vented and strong packing material and pre-cooling should be encouraged for the handling of lime fruit both for the domestic and international markets. Limes are highly sensitive to CI; therefore, these fruit must be kept at optimum temperatures (10–12°C with 85–90% relative humidity) throughout the supply chain to avoid development of CI. To reduce other post-harvest physiological disorders such as oil spotting and stylar end breakdown, fruit should be harvested at the correct maturity and kept at optimum temperature and relative humidity. Like other citrus fruits, lime fruit should be packed in corrugated cardboard boxes on a commercial scale to replace traditional methods of fruit packing. Limited information is available regarding commercial applications of edible fruit coating materials and MAP in lime fruit, which warrant further investigation. Pre- and post-harvest management of diseases of lime fruit with appropriate use of fungicides and organic elicitors, followed by proper harvest and careful post-harvest handing to reduce bruises, injuries and decay are all prerequisites. Sanitation of both orchards and pack houses are necessary to further reduce these problems. There is a need to minimize or replace the use of synthetic chemicals with plant-based product or organic elicitors (1-MCP, aminoethoxyvinylglycine, polyamines, salicylic acid, etc.) for post-harvest treatment of lime fruit to reduce hazardous health effects on consumers.

References

Albertini, M.-V., Carcouet, E., Pailly, O., Gambotti, C., Luro, F. and Berti, L. (2006) Changes in organic acids and sugars during early stages of development of acidic and acidless citrus fruit. *Journal of Agricultural and Food Chemistry* 54, 8335–8339.

Artes-Hernandez, F., Rivera-Cabrera, F. and Kader, A.A. (2007) Quality retention and potential shelf-life of fresh-cut lemons as affected by cut type and temperature. *Postharvest Biology and Technology* 43, 245–254.

Ayoub, S.O. and Abu-Goukh, A.A. (2009) Effect of 2,4,5-trichlorophenoxy acetic acid and waxing on quality and storability of lime fruits. *University of Khartoum Journal of Agricultural Sciences* 17, 183–197.

Baldwin, E.A. (1993) Citrus fruit. In: Seymour, G.B., Taylor, J.E. and Tucker, G.A. (eds) *Biochemistry of Fruit Ripening*. Chapman & Hall, London, pp. 107–149.

Bassan, M.M., Moura Filho, F.d.A.A., Caron, V.C., Couto, H.T.Z.d. and Jacomino, A.P. (2012) The harvesting system affects the quality and conservation of the 'Tahiti' acid lime. *Scientia Horticulturae* 155, 72–77.

Bisen, A., Pandey, S.K. and Patel, N. (2012) Effect of skin coatings on prolonging shelf life of kagzi lime fruits (*Citrus aurantifolia* Swingle). *Journal of Food Science and Technology* 49, 753–759.

Bosquez-Molina, E., Domínguez-Soberanes, J., Pérez-Flores, L.J., Díaz-de-Léon-Sánchez, F. and Vernon-Carter, J. (2004) Effect of edible coatings on storage life of Mexican limes (*Citrus aurantifolia* swingle) harvested in two different periods. *Acta Horticulturae* 632, 329–335.

Burg, S.P. and Burg, E.A. (1966) Fruit storage and atmospheric pressure. *Science* 153, 314–315.

Burns, J.K. (1990) Respiratory rates and glycosidase activities of juice vesicles associated with section-drying in citrus. *HortScience* 25, 544–546.

Burns, R., Rosedale, D.O., Pehrson, J.E. and Coggins, C.W. (1964) Gibberellins sprays delay lime maturity. *California Agriculture* 17, 14–15.

Calegario, F.F., Cosso, R.G., Almeida, F.V., Vercesi, A.E. and Jardim, W.F. (2001) Determination of respiration rate of tomato fruit using flow analysis. *Postharvest Biology and Technology* 22, 249–256.

Chamblee, T.S. and Clark, B.C. (1997) Analysis and chemistry of distilled lime oil (*Citrus aurantifolia* Swingle.). *Journal of Essential Oil Research* 9, 267–274.

Chamblee, T.S., Karelitz, T., Radford, T. and Clark, B.C. (1997) Identification of sesquiterpenes in citrus essential oils by cryofocussing GC/FT-IR. *Journal of Essential Oil Research* 9, 127–132.

Della Porta, E., Reverehon, D. and Chouchi Barth, D. (1997) Mandarin and lime peel oil processing by supercritical CO_2 desorption; deterpenation and high molecular weight compound elimination. *Journal of Essential Oil Research* 9, 515–522.

Dotto, M., Pirola, K., Wagner Júnior, A., Radaelli, J.C. and Danner, M.A. (2015) Biofilms and packaging on post-harvest Tahiti acid lime. *Revista Brasileira de Ciências Agrárias* 10, 365–369.

Drazkiewice, M. (1994) Chloropyllase: occurrence, function, mechanism of action, effects of external factors. *Photosyntheica* 30, 301–309.

Dugo, P., Cotroneo, A., Bonarcorsi, I. and Mondello, L. (1998) On the genuineness of citrus essential oils Part L VII. The composition of distilled lime oil. *Flavour Fragrances Journal* 13, 93–97.

Eaks, I.L. (1960) Physiological studies on chilling injury in citrus fruit. *Plant Physiology* 35, 632–636.

Eaks, I.L. and Masias, E. (1965) Chemical and physiological changes in lime fruits during and after storage. *Journal of Food Science* 30, 509–515.

El-Sheikh, M.H., Abd El-Migeed, M.M.M. and Saleh, M.M.S. (2014) Effect of bagging types under different conditions of cold temperature on storability of lime fruits. *World Applied Sciences Journal* 30, 395–401.

Fallik, E. (2004) Prestorage hot water treatments (immersion, rinsing and brushing). *Postharvest Biology and Technology* 32, 125–134.

Frankel, E.N. (1999) Food antioxidants and phytochemicals: present and future perspectives. *European Journal of Lipid Science and Technology* 101, 450–455.

Gates, C.M. (1949) The possibilities of 2,4-D for the control of post-harvest decay in Florida limes. *Proceedings of the Florida State Horticulture Society* 62, 220–222.

Gayet, J.P. and Salvo Filho, A.D. (2003) Colheita e beneficiamento. In: Mattos Junior, D.D., De Negri, J.D. and Figueiredo, J.O.D. (eds) *Lima ácida 'Tahiti'*. Instituto Agronômico, Campinas.

Gharagozloo, M., Doroudchi, M. and Ghaderi, A. (2002) Effects of *Citrus aurantifolia* concentrated extract on the spontaneous proliferation of MDA-MB-453 and RPMI-8866 tumor cell lines. *Phytomedicine* 9, 475–477.

Gokul, P.R., Raj, S. and Suriyamoorthi, P. (2015) Estimation of volume and maturity of sweet lime fruit using image processing algorithm. *Proceedings of the 2015 International Conference on Communications and Signal Processing (ICCSP)*, pp. 1227–1229.

Grierson, W. (1981) Harvesting Florida citrus for overseas export. *Proceedings of the Florida State Horticultural Society* 94, 252–254.

Guimaraes, R., Barros, L., Barreira, J.C.M., Sousa, M.J., Carvalho, A.M. and Ferreira, I.C.F.R. (2010) Targeting excessive free radicals with peels and juices of citrus fruits: grapefruit, lemon, lime and orange. *Food and Chemical Toxicology* 48, 99–106.

Hatton, T.T. (1990) Reduction of chilling injury with temperature manipulation. In: Wang, C.Y. (ed.) *Chilling Injury of Horticultural Crops*. CRC Press, Boca Raton, Florida, pp. 269–280.

Hernandez-Montiel, L.G., Ochoa, J.L., Troyo-Diaguez, E. and Larralde-Corona, C.P. (2010) Biocontrol of postharvest blue mold (*Penicillium italicum* Wehmer) on Mexican lime by marine and citrus *Debaryomyces hansenii* isolates. *Postharvest Biology and Technology* 56, 181–187.

Hume, H.H. (1957) *Citrus Fruits: Revised Edition of the Cultivation of Citrus Fruits.* Macmillan, New York.

Iqbal, Z., Singh, Z., Khangura, R. and Ahmad, S. (2012) Management of citrus blue and green moulds through application of organic elicitors. *Australasian Plant Pathology* 41, 69–77.

Jonglertjunya, W., Juntong, T., Pakkang, N., Srimarut, N. and Sakdaronnarong, C. (2014) Properties of lignin extracted from sugarcane bagasse and its efficacy in maintaining postharvest quality of limes during storage. *LWT-Food Science and Technology* 57, 116–125.

Kader, A.A. (2002) *Postharvest Technology of Horticultural Crops*. University of California, Agriculture and Natural Resources, Oakland, California, Publication 3311.

Kader, A.A. (2003) A summary of CA requirements and recommendations for fruits other than apples and pears. *Acta Horticulturae* 600, 737–740.

Kader, A.A. and Arpaia, M.L. (2002) Postharvest handling systems: subtropical fruit. In: Kader, A.A. (ed.) *Postharvest Technology of Horticultural Crops*. University of California, Division of Agricultural and Natural Resources, Oakland, California, pp. 375–384.

Kaewsuksaeng, S., Tatmala, N., Srilaong, V. and Pongprasert, N. (2015) Postharvest heat treatment delays chlorophyll degradation and maintains quality in Thai lime (*Citrus aurantifolia* Swingle cv. Paan) fruit. *Postharvest Biology and Technology* 100, 1–7.

Kaur, S., Jawandha, S.K. and Singh, H. (2014) Response of Baramasi lemon to various post-harvest treatments. *International Journal of Agriculture, Environment and Biotechnology* 7, 895–902.

Kaushal, B.B.L. and Thakur, K.S. (1996) Influence of ambient and evaporative cool chamber storage condition on the quality of polyethene packed Kinnow fruit. *Advances in Horticulture Science* 10, 179–184.

Kawaii, S., Tomono, Y., Katase, E., Ogawa, K. and Yano, M. (1999) Quantitation of flavonoid compound in citrus fruit. *Journal of Agricultural and Food Chemistry* 47, 3565–3571.

Ketabchi, S., Taghipour, M.A. and Sharzei, A. (2012) Identification of lime fruit surface colonizing bacteria antagonistic against the green mold (*Penicillium digitatum*) and comparison of biological control with heat treatment and chemical control. *Asian Journal of Experimental Biological Science* 3, 287–292.

King, A. and Young, G. (1999) Characteristics and occurrence of phenolic phytochemicals. *Journal of the American Dietetics Association* 99, 213–218.

Kluge, R.A., Jomori, M.L.L., Jacomino, A.P., Vitti, M.C.D. and Padula, M. (2003a) Intermittent warming in 'Tahiti' lime treated with an ethylene inhibitor. *Postharvest Biology and Technology* 29, 195–203.

Kluge, R.A., Jomori, M.L.L., Jacomino, A.P., Vitti, M.C.D. and Vitti, D.C.C. (2003b) Intermittent warming of 'Tahiti' lime to prevent chilling injury during cold storage. *Scientia Agricola* 60, 729–734.

Kohli, R.R. and Bambota, J.R. (1966) Storage of lime (*C. aurantifolia* Swingle). *Indian Journal of Horticulture* 23, 140–146.

Ladaniya, M.S. (2000) Response of 'Nagpur' mandarin to degreening by ethylene trickle method, mechanized wax coating and storage at ambient condition. *Journal of Food Science and Technology-Mysore* 37, 103–110.

Ladaniya, M.S. (2004) Response of 'Kagzi' acid lime to low temperature regimes during storage. *Journal of Food Science and Technology* 41, 284–288.

Ladaniya, M.S. (2008) *Citrus Fruit Biology, Technology and Evaluation.* Academic Press, San Diego, California.

Ladaniya, M.S. and Singh, S. (2000) Maturity indices for acid lime (*Citrus aurantifolia*) cultivar 'Kagzi' grown in central India. *Indian Journal of Agriculture Science* 70, 292–295.

Ladaniya, M.S., Singh, S. and Wadhawan, A.K. (2003) Response of 'Nagpur' mandarin, 'Mosambi' sweet orange and 'Kagzi' acid lime to gamma radiation. *Radiation Physics and Chemistry* 67, 665–675.

Li, L., Lichter, A., Chalupowicz, D., Gamrasni, D., Goldberg, T., Nerya, O., Ben-Arie, R. and Porat, R. (2016) Effects of the ethylene-action inhibitor 1-methylcyclopropene on postharvest quality of non-climacteric fruit crops. *Postharvest Biology and Technology* 111, 322–329.

Lurie, S. (1998a) Postharvest heat treatments. *Postharvest Biology and Technology* 14, 257–269.
Lurie, S. (1998b) Postharvest heat treatments of horticultural crops. *Horticultural Review* 22, 91–121.
Lyons, J.M. (1973) Chilling injury in plants. *Annual Review of Plant Physiology* 24, 445–466.
Mahajan, P.V., Caleb, O.J., Singh, Z., Watkins, C.B. and Geyer, M. (2014) Postharvest treatments of fresh produce. *Philosophical Transactions of the Royal Society A: Mathematical, Physical and Engineering Sciences* 2014, 5 May, 372 (2017), 20130309. doi: 10.1098/rsta.2013.0309.
Martinez-Romero, D., Serrano, M., Carbonell, A., Brugos, L. and Riquelme, F. (1999) Effects of post-harvest putresciene and calcium treatments on reducing mechanical damage. *Journal of the Science of Food and Agriculture* 79, 1589–1595.
Moretto, C., Cervantes, A.L.L., Batista Filho, A. and Kupper, K.C. (2014) Integrated control of green mold to reduce chemical treatment in post-harvest citrus fruits. *Scientia Horticulturae* 165, 433–438.
Murata, T. (1997) Citrus. In: Mitra, S.K. (ed.) *Postharvest Physiology and Storage of Tropical and Subtropical Fruits.* CAB International, Wallingford, pp. 21–46.
Nogata, Y., Sakamoto, K., Shiratsuchi, H., Ishii, T., Yano, M. and Ohta, H. (2006) Flavonoid composition of fruit tissues of citrus species. *Bioscience, Biotechnology, and Biochemistry* 70, 178–192.
Obeed, R.S. and Harhash, M.M. (2006) Impact of postharvest treatment on storage life and quality of 'Mexican' lime. *International Journal of Advance Agricultural Research* 11(3), 189–193.
Palou, L., Valencia-Chamorro, S.A. and Perez-Gago, M.B. (2015) Antifungal edible coating of fresh fruit: a review. *Coatings* 5, 962–986.
Penniston, K.L., Nakada, S.Y., Holmes, R.P. and Assimos, D.G. (2008) Quantitative assessment of citric acid in lemon juice, lime juice, and commercially-available fruit juice products. *Journal of Endourology* 22, 567–570.
Peterson, J.J., Beecher, G.R., Bhagwat, S.A., Dwyer, J.T., Gebhardt, S.E., Haytowitz, D.B. and Holden, J.M. (2006) Flavanones in grapefruit, lemons, and limes: a compilation and review of the data from the analytical literature. *Journal of Food Composition and Analysis* 19, Supplement, S74–S80.
Petracek, P.D., Kelsey, D.F. and Grierson, W. (2006) Physiological peel disorders. In: Wardowski, W.F., Miller, W.M., Hall, D.J. and Grierson, W. (eds) *Fresh Citrus Fruits*, 2nd edn. Florida Science Source. Inc., Longboat Key, Florida, pp. 397–419.
Porat, R., Weiss, B., Cohen, L., Daus, A., Goren, R. and Droby, S. (1999) Effects of ethylene and 1-methylcyclopropene on the postharvest qualities of "Shamouti" oranges. *Postharvest Biology and Technology* 15, 155–163.
Ranganna, S., Govindrajan, V.S. and Ramana, K.V.R. (1983) Citrus fruits – varieties, chemistry, technology, and quality evaluation. Part II. Chemistry, technology, and quality evaluation. A. Chemistry. *Critical Review in Food Science and Nutrition* 18, 313–386.
Rao, M.M., Hitallmani, S.V. and Bojappa, K.M. (1983) A comparative study on the developmental physiology on Kagzi (*C. aurantifolia* Swingle) and Tahiti (*C. latifolia* Tanaka) lime fruits. *Proceedings of International Citrus Symposium, Bangalore, Horticultural Society of India*, pp. 211–216.
Rivera, F., Pelayo-Zaldivar, C., De León, F.D., Buentello, B., Castillo-Rivera, M. and Pérez-Flores, L.J. (2007) Cold-conditioning treatment reduces chilling injury in Mexican limes (*Citrus aurantifolia* S.) stored at different temperatures. *Journal of Food Quality* 30, 121–134.
Roy, S.K. and Khurdiya, D.S. (1986) Studies on the evaporatively cooled zero energy input chamber for the storage of horticulture produce. *Indian Horticulture* 40, 26–31.
Salama, S.B., Grierson, W. and Oberbacher, M.F. (1965) Storage trials with limes, avocados and lemons in modified atmospheres. *Proceedings of the Florida State Horticultural Society* 78, 353–358.
Saunt, J. (1990) *Citrus Varieties of the World.* Sinclair International Limited, Norwich, UK.
Selvaraj, Y. and Raja, M.E. (2000) Biochemistry of ripening of Kagzi lime (*Citrus aurantifolia* Swingle) fruit. *Indian Journal of Horticulture* 57, 1–8.
Shaw, P.E. (1979) Review of quantitative analysis of citrus essential oils. *Journal of Agricultural and Food Chemistry* 27, 246–257.
Sierra, C.C., Molina, E.B., Zaldivar, C.P., Flores, L.P. and Garcia, L.P.D.L. (1993) Effect of harvesting season and postharvest treatments on storage life of Mexican limes (*Citrus aurantifolia* Swingle). *Journal of Food Quality* 16, 339–354.
Singh, S. and Naqvi, S.A.M.H. (2001) *Citrus.* International Book Distribution Company, Lucknow, UP, India.
Singh, S., Shivankar, V.J., Srivastava, A.K. and Singh, I.P. (2004) *Advances in the Citriculture*. Jagmander Book Company, Delhi, India.

Singh, Z., Singh, S.P. and Yahia, E.M. (2009) Subtropical Fruits. In: Yahia, E.M. (ed.) *Modified and Controlled Atmospheres for the Storage, Transportation, and Packaging of Horticultural Commodities*. CRC Press, Boca Raton, Florida, pp. 317–361.
Sivakumara, H. and Fallik, E. (2013) Influence of neat treatments on quality retention of fresh and fresh-cut produce. *Food Reviews International* 29, 294–320.
Slutzky, B., Gonzalez-Abreu, A. and Berdam, I. (1981) Chilling injury related to mineral composition of grapefruit and limes during cold storage. *Proceedings of the International Society of Citriculture* 1981, 779–782.
Sonkar, R.K., Ladaniya, M.S. and Singh, S. (1999) Effect of harvesting methods and postharvest treatments on storage behaviour of 'Nagpur' mandarin (*Citrus reticulata* Blanco) fruit. *Indian Journal of Agriculture Science* 69, 434–437.
Spalding, D.H. and Reeder, W.F. (1974) Quality of Tahiti limes stored in a controlled atmosphere or under low pressure. *Proceedings of the Tropical Region, American Society for Horticultural Science* 18, 128–135.
Spalding, D.H. and Reeder, W.F. (1976) Low pressure (hypobaric) storage of limes. *Journal of the American Society for Horticultural Science* 101, 367–370.
Spalding, D.H. and Reeder, W.F. (1983) Conditioning Tahiti lime to reduce chilling injury. *Proceedings of the Florida State Horticultural Society* 96, 231–232.
Srilaong, V., Aimla-or, S., Soontornwat, A., Shigyo, M. and Yamauchi, N. (2011) UV-B irradiation retards chlorophyll degradation in lime (*Citrus latifolia* Tan.) fruit. *Postharvest Biology and Technology* 59, 110–112.
Sritananan, S., Uthairatanakij, A., Srilaong, V., Kanlayanarat, S. and Wongs-Aree, C. (2006) Efficacy of controlled atmosphere storage on physiological changes of lime fruit. *Acta Horticulturae* 712, 591–598.
Subramanyam, H., Narsimhan, P. and Srivastava, H.C. (1965) Studies on the physical and biochemical changes in lime (*C. aurantifolia* Swingle) during growth and development. *Journal of Indian Botanical Society* 44, 105–109.
Sun, Y.T. (1998) Citrus fruit respiration and CA storage. *South China Fruits* 27, 16–17.
Taghipour, M.A., Ketabchi, S. and Sharzei, A. (2012) Identification of lime fruit surface colonizing bacteria antagonistic against the green mold (*Penicillium digitatum*) and comparison of biological control with heat treatment and chemical control. *Asian Journal of Experimental Biological Science* 3, 287–292.
Thangaraj, T., Trulappan, I. and Rangasawamy, P. (1983) Pre-storage treatment of acid lime fruits under cool chamber storage. *National Seminar on Optimization of Production and Productivity of Acid Lime.* The Horticultural Society of India and Horticultural College and Research Institute, Periyakulam, Tamilnadu, India.
Thompson, A.K. (1998) *Controlled Atmosphere Storage of Fruits and Vegetables*. CAB International, Wallingford, UK.
Thompson, A.K. (2003) *Fruits and Vegetables: Harvesting, Handling and Storage*. Blackwell Publishing, Oxford, UK.
Tincutta, V. and Richard, S. (2001) High-speed GC and GC/Time-of-Flight MS of lemon and lime oil samples. *Analytical Chemistry* 73, 4395–4402.
Ting, S.V. and Deszyck, E.J. (1961) The carbohydrates in the peel of oranges and grapefruits. *Journal of Food Science* 26, 146–152.
Venkateswarlu, G. and Selvaraj, Y. (2000) Changes in peel oil composition of kagzi lime during ripening. *Journal of Essential Oil Research* 12, 50–52.
Verma, P. and Dashora, L.K. (2000) Post harvest physiconutritional changes in Kagzi limes (*Citrus aurantifolia* Swingle) treated with selected oil emulsions and diphenyl. *Plant Foods for Human Nutrition* 55, 279–284.
Verma, V.S. and Tikoo, M.L. (2003) Seasonal disease profile and chemical management of fungal rot loss of citrus fruits. *Annals of Plant Protection Sciences* 11, 79–82.
Wang, C.Y. (1993) Approaches to reduce chilling injury of fruits and vegetables. *Horticultural Review* 15, 63–95.
Wang, C.Y. (2010) Alleviation of chilling injury in tropical and subtropical fruits. *Acta Horticulturae* 864, 276–273.
Wardowski, W.F., Nagy, S. and Grierson, W. (1986) *Fresh Citrus Fruits*. Van Nostrand Reinhold Company Inc., New York.
Watkins, C.B. (2000) Responses of horticultural commodities to high carbon dioxides as related to modified atmosphere packaging. *HortTechnology* 6, 65–72.

Whitney, J.D. (1978) Air-shakers for removal of oranges in Florida. *Proceedings of the Society of Citriculture, Australia*, 1, 91–93.
Whitney, J.D. (1987) Shakers affect Florida orange fruit yields and harvesting efficiency. *Applied Engineering Agriculture* 3, 20–24.
Wills, R.B.H., Shohot, D. and Kim, G.H. (1999) Importance of low ethylene levels to delay senescence of non-climacteric fruit and vegetables. *Australian Journal of Experimental Agriculture* 39, 221–224.
Wills, R., McGlasson, B., Graham, D. and Joyce, D. (2004) *Postharvest: An Introduction to the Physiology and Handling of Fruit, Vegetables and Ornamentals.* UNSW Press Sydney, Australia and CAB International, Wallingford, UK.
Win, T.O., Srilaong, V., Heyes, J., Kyu, K.L. and Kanlayanarat, S. (2006) Effects of different concentrations of 1-MCP on the yellowing of West Indian lime (*Citrus aurantifolia*, Swingle) fruit. *Postharvest Biology and Technology* 42, 23–30.
Yadav, A.R., Chauhan, A.S., Rekha, M.N., Rao, L.J.M. and Ramteke, R.S. (2004) Flavour quality of dehydrated lime [*Citrus aurantifolia* (Christm.) Swingle]. *Food Chemistry* 85, 59–62.
Yamauchi, N., Takamura, K., Shigyo, M., Migita, C.T., Masuda, Y. and Maekawa, T. (2014) Control of degreening in postharvest green sour citrus fruit by electrostatic atomized water particles. *Food Chemistry* 156, 160–164.
Zamani, M., Tehrani, A.S., Ahmadzadeh, M., Behboodi, K. and Hosseininaveh, V. (2008) Biological control of *Penicillium digitatum* on oranges using *Pseudomonas* spp. either alone or in combination with hot sodium bicarbonate dipping. *Australasian Plant Pathology* 37, 605–608.
Zea-Hernández, L.O., Saucedo-Veloz, C., Cruz-Huerta, N., Ramírez-Guzmán, M.E. and Robles-González, M.M. (2016) Evaluation of post-harvest applications of gibberellic acid on the quality and shelf life of three varieties of Mexican lime. *Revista Chapingo Serie Horticultura* 22, 17–26.
Ziena, H.M.S. (2000) Quality attributes of Bears seedless lime (*Citrus latifolia* Tan) juice during storage. *Food Chemistry* 71, 161–172.

13 Traditional/Commercial Uses and Future Dynamics

M. Asif Hanif[1]*, Smitha Padmanabhan[2], Mostafa Waly[2] and Ahmed Al-Maskari[3]
[1]Department of Chemistry, University of Agriculture, Faisalabad, Pakistan; [2]Department of Food Science and Nutrition, College of Agricultural and Marine Sciences, Sultan Qaboos University, Al Khod, Oman; [3]Department of Crop Sciences, College of Agricultural and Marine Sciences, Sultan Qaboos University, Al Khod, Oman

Introduction

Citrus aurantifolia (lime) belongs to the Rutaceae family (orange family) and is a genus of flowering plants. This tree is extensively cultivated in subtropical and tropical areas because of its edible fruits (FAO, 2009). The tree rarely reaches more than 5 m in height and 7.5 m in width; if not pruned it becomes shrub-like. Citrus is probably the most extensively planted fruit for direct human use in the world (Bakare *et al*., 2012). The lime fruit and its juice are key ingredients in many drinks, confections, pickles and sauces. The flavour and bouquet of limeade and other lime-flavoured drinks are quite different from those prepared from lemons. Lime juice is largely used for citric acid production and is often used extensively for medicinal purposes with lemon juice. With the advent of technology, many valuable products are being manufactured for domestic as well as industrial applications. The most important quality of lime is its preservation ability. Products like lime oil, pectin, citric acid, lime juice, etc. are currently in trend – there is a growing market for all these products year-round. Lime essential oil is utilized for flavouring purposes, particularly in artificial lime juice cordials, which consist of mineral water and a sweetened solution of tartaric acid (Morton, 1987a; Ashurst, 2016).

Chemistry

Lime fruit are rich in vitamins, phenolic compounds, minerals, essential oils, dietary fibre and carotenoids, and lime is regarded as the third most significant citrus species (González-Molina *et al*., 2009). Lime fruit flavedo is mainly composed of cellulosic material and other components such as fatty acids, paraffin waxes, essential oils, bitter principles, enzymes, steroids and carotenoids. The mesocarp or albedo contains celluloses and soluble carbohydrates, flavonoids, amino acids, pectin and vitamins, while the endocarp is the edible portion with seeds which comprise juice sacs or vesicles (Ho *et al*., 2013). Lime essential oil is extensively used for medicinal purposes, such as for fungicidal, virucidal, bactericidal, insecticidal, anti-parasitical applications, and is also utilized in different industries, especially in the cosmetic, pharmaceutical, food, agricultural and sanitary industries.

*E-mail: muhammadasifhanif@ymail.com

Chemical Composition

Lime is a medicinal fruit, and it contains phosphorus, potassium, iron, calcium, riboflavin, B2, thiamin, B1, Vitamin C, niacin, Vitamin A, carbohydrates, fat, crude fibre and protein (Vand and Abdullah, 2012). The chief commercial products of lime are the juice and essential oil (Spadaro *et al.*, 2012). The total soluble solid in lime ranges from 5.5–8.5% and total acid content is varied from 5–7.5%. The juices of Mexican lime contain approximately 40–60 mg of ascorbic acid/100 g of fresh weight. The primary organic acid constituent of the juice of lime is citric acid, which is present in the soluble solid of the juice along with sugar. The maturity indices of the fruit and the quality of a variety are determined by the acid contents in the juice. The supply of vitamins, especially ascorbic acid (Vitamin C), is certainly the chief role of citrus fruits in human nutrition, as Vitamin C is among the most important vitamins (Shrestha *et al.*, 2012). Citrus peel is a valuable source of essential oils and other important secondary metabolites of plant, though it is usually taken as agro-industrial waste (Kamal *et al.*, 2011). Lime peel and leaf essential oils are used for flavouring beverages and manufactured foods, as perfume ingredients and also in pharmaceutical products (Sandoval-Montemayor *et al.*, 2012).

Phytochemistry

Lime extracts are known to have anti-bacterial, anti-oxidant, anti-tuberculosis, anti-cholinesterase and cytotoxic properties. The phytochemical constituents of lime extracts are palmitic acid, 5-geranyloxy-7-methoxycoumarin, 5-geranyloxypsoralen, 5,8-dimethoxypsoralen, 5,7-dimethoxycoumarin, 1-methoxy-cyclohexene, 5-methoxypsoralen, 3-methyl-1,2-cyclopentanedione, corylone and umbelliferone. Some other constituents are apigenin, quercetin, kaempferol, rutin and nobiletin. Furthermore, the limonoids reported are limonin, isolimonexic acid and limonexic acid (Lawal *et al.*, 2014).

Citrus essential oils show a great variation in chemical composition, which may be a result of numerous factors. These factors include seasons as well as environmental factors, such as the nature of the soil and climate, the type of varieties considered, genetic factors, the geographical location, the plant part used for the oil extraction and processing and the extraction method (Lawal *et al.*, 2014). The main compounds of volatile oils of the leaves of *C. aurantifolia* are limonene, geranial and neral (Dongmo *et al.*, 2013). The major constituents of lime peel oil are β-pinene, γ-terpinene, limonene, citral, terpinolene and α-terpineol (Spadaro *et al.*, 2012; Costa *et al.*, 2014).

Given the economic, dietary and medical values of citrus seed oil, there has been a recent surge of studies on the chemical composition (fatty acid content in particular) of the oil of seeds of different species of citrus. Many works have measured the oil content of citrus seeds: Brazilian Rangpur lime seeds (32.0–38.3%), Tunisian citrus seeds (26.1–36.1%), Tunisian sweet orange (50–55%), Egyptian citrus seeds (40.2–45.5%), lemon seeds (70–80%) and Pakistani citrus seeds (27.0–36.5%) (Reazai *et al.*, 2014).

The seeds of *C. aurantifolia* fruits, the waste of the fruit processing industry, are used to produce seed oil by the cold physical expression of these seeds. Lime seed oil is not an essential oil but a fixed triglyceride oil. The crude oil has a remarkable citrus odour, pale green colour, and needs additional processing for cosmetic and nutritional use. After refining, lime seed oil is liquid at 25°C, has a pale yellow colour and when cooled to below 5°C it becomes a waxy paste. The essential fatty acids are present in large amounts in lime seed oil (Table 13.1). The major lipid profile (70%) is made up of the key Ω3, 6 and 9 fatty acids. In addition, this oil mainly comprises palmitic acid (<20%). Principle fatty acids are palmitic (C16:0), oleic (Ω9) (C18:1), linoleic (Ω6) (C18:2) and alpha linolenic (Ω3) (C18:3) acid (Matthaus and Özcan, 2012).

Lime Processing and Value Addition

C. aurantifolia (Christm) Swingle fruit are used for non-culinary and culinary purposes throughout the world, primarily for their juice, which has both cleaning and culinary uses (Morton, 1987b). Each and every part of the lime fruit is utilized by the lime processing industry for

Table 13.1. Fatty acid composition of *Citrus aurantifolia*. (From: Matthaus and Özcan, 2012.)

Sample	Palmitic	Stearic	Oleic	Vaccenic	Linoleic	Linolenic	Arachidic	Total
Citrus aurantifolia	22.6	4	17.7	1.6	44.3	8.1	0.4	98.7

production of highly marketable products (Ho *et al.*, 2013).There are usually two commercial products of citrus fruit processing: the juice and the peel essential oil. Clarified lime juice is used in many food products and is also utilized as a health drink. The citrus species are a possible source of volatile oil, which is used in food and other industries (Njoku and Evbuomwan, 2014). A lime juice production set-up often contains a peel oil extractor as a vital part. The essential oil is either used in some renowned beverages or the taste of lime juice is enhanced by mixing it back into the juice. In the flavour industry, *C. aurantifolia* oil is certainly one of the most worldwide traded citrus oils. Lime essential oil is extracted by hydrodistillation (Costa *et al.*, 2014) and steam distillation (Khan *et al.*, 2012; Njoku and Evbuomwan, 2014). In Indonesia, lime juice is frequently used as a medicine for acute respiratory tract infections by mixing it with honey or sweet soy sauce (Eveline and Yuliana, 2014). Lime essential oils are value-added products and are broadly used as nutritional supplements, in pharmaceutical components, and for the aromatherapy and cosmetic industries (Colecio-Juárez *et al.*, 2012).

Uses

Human beings have been using lime for many centuries but initially its applications were limited. Citrus fruit or juice consumption is established to be inversely related to the occurrence of a number of diseases. Though the citrus fruits are largely used for desserts and salads, due to their aromatic compounds they are also sources of essential oils. Anti-oxidant and radical-scavenging activities of these essential oils have been reported by many researchers. Anti-tumour, anti-inflammatory, anti-fungal and blood clot inhibition activities are found to be associated with citrus fruits (Guimarães *et al.*, 2010; Jabri Karoui and Marzouk, 2013). The essential oils of lime are not only used in pharmaceutical forms and as constituents in fragrances and perfumes, but also as flavouring agents in manufactured foods and beverages (Dongmo *et al.*, 2013). Citron (*Citrus medica*) is a hybrid citrus related to both lemons and limes and is the source of the note citron, which is used in perfumery, and its twigs and the leaves are used to distill cedrat petitgrain.

Culinary uses

Fresh limes are an excellent food seasoning choice as they are very high in Vitamin C, fat-free and contain almost no sodium. The fruit is also used for mixed drinks, lemonade and is squeezed onto seafood or other foods to bring out the flavour. It is also used in carbonated beverages and bottled juice. The principal byproduct of lime is lime oil, which is mainly used for flavouring and cosmetics (Khan, 2007; Vand and Abdullah, 2012). Fresh lime or dry powdered lime is an essential ingredient for Omani culinary delights as well as Irani, Iraqi and Indian dishes. It is a good appetizer and digestive (Abrahimy and Haji Vand, 2004; Vand and Abdullah, 2012).

Traditional medical uses

In traditional medicines, lime (*C. aurantifolia*) is utilized as an astringent, antiseptic, an anti-helminthic, mosquito repellent, digestive and appetite stimulant, for stomach diseases, as an anti-scorbutic, as a tonic, diuretic, for arthritis, and for headache, sore throats, coughs and colds. Previous studies have found terpenoids, coumarins and flavonoids to be present in lime. Furthermore, lime leaves show defensive activity against induced platelet aggregation and osteoporosis (Sandoval-Montemayor *et al.*, 2012). *C. aurantifolia* alleviates nervousness and anxiety. Lime also relieves stress-related disorders such as nervous originated digestive disorders or insomnia. It also possesses anti-inflammatory potential for the digestive system (Sandoval-Montemayor *et al.*, 2012). The peel essential oil of lime has been shown to have anti-helminthic, anti-microbial, anti-cholinesterase, radical scavenging and anti-cancer activities. Lime essence has anti-spasmodic virtues that can be employed during spasms of the digestive system (diarrhoea, distension). It has an anti-coagulant property, which renders it very valuable for people with cardiovascular risks. It is also used against headaches, fever and colds (Dongmo *et al.*, 2013). Extracts of *C. aurantifolia* are known to have anti-oxidant, anti-tuberculosis, anti-bacterial, anti-cholinesterase and cytotoxic activities (Sandoval-Montemayor *et al.*, 2012).

Anti-microbial activity

Anti-microbial activity of *C. aurantifolia* Swingle essential oil has been studied on a representative American Type Culture Collection (ATCC)) and clinical strains of yeasts and Gram-positive as well as Gram-negative bacteria. The methods used were disc diffusion and the broth microdilution method. The lime essential oil was found to consist of γ-terpinene, β-pinene, limonene, α-terpineol, citral and terpinolene as major components. These components of lime oil are considered to be responsible for good anti-microbial activity, in particular on *Bacillus subtilis*, *Staphylococcus aureus* and *Staphylococcus epidermidis* (Gram-positive bacteria) (Costa *et al.*, 2014). Lime solutions have been reported to rapidly kill toxogenic *Vibrio cholerae* O1 (Rowe *et al.*, 1998). Controlled case series have reported that lime may have anti-bacterial and anti-cholera effects (Rodrigues *et al.*, 1997; Rodrigues *et al.*, 2000).

Anti-*Mycobacterium tuberculosis* activity

The hexane extract of lime (*C. aurantifolia*) fruit peels was tested for its anti-*Mycobacterium tuberculosis* activity. This extract showed significant effect against isoniazid, streptomycin or ethambutol mono-resistant *M. tuberculosis* strains with minimum inhibitory concentrations (MICs) in the range of 25–50 μg/ml (Camacho-Corona *et al.*, 2008).

Anti-oxidant activity

Limes contain several putative bioactive compounds and certain compounds are known to possess anti-oxidant activity. It is a vital resource of ascorbic acid (Vitamin C) for human nutrition (Shrestha *et al.*, 2012). The fruit peel, seeds and juice have been found to have anti-oxidative activity. Both peel and juice have anti-oxidant properties, but it is not a good idea to consume excessive amounts of lime juice. Lime juice and peel may have different anti-oxidant effects due to differences in the concentration and types of flavonoids (Boshtam *et al.*, 2011). The anti-oxidant potential of different solvent extracts of lime seeds has also been studied. Defatted lime seeds were extracted with ethyl acetate, methanol and methanol/water, and tested using ABTS (2,2′-azinobis-(3-ethylbenzothiazoline-6-sulfonic acid)) and DPPH (1,1-diphenyl-2-picryl-hydrazyl) methods, for their radical scavenging activity. The methanol extract showed maximum radical scavenging activity in both ABTS and DPPH assays at four concentrations. The phenolic content was the highest in extracts of ethyl acetate, followed by methanol and methanol/water. The results suggest that lime seeds have potential for scavenging free radicals, which can be further utilized for health maintaining properties (Patil *et al.*, 2014).

Anti-proliferative effects

In vitro study showed that proliferation of phytohaemagglutinin (PHA) stimulated mononuclear cells were appreciably controlled and reduced by 250 and 500 mcg/ml of concentrated lime juice extract, while just 500 mcg/ml of the extract could inhibit proliferation of staphylococcal protein A stimulated mononuclear cells (Gharagozloo and Ghaderi, 2001).

Pharmacological Activities

Lime has health benefits including enhanced digestion, weight loss, prevention of scurvy, skincare benefits, relief from constipation and respiratory disorders, benefits for eye care, treatment of peptic ulcers, gout, bleeding gums, piles and urinary disorders, etc.

Scurvy

Scurvy has frequent infection symptoms similar to the common cold, such as cracked lip corners and lips, ulcers in the mouth and on the tongue. It occurs due to Vitamin C deficiency and is cured by lime. In old times, lime was given to sailors and soldiers to protect them from scurvy, which was a potentially fatal and horrible disease for them. Lime is still distributed among workers working in polluted environments like mines, cement factories, heat treatments, furnaces, painting shops and other unsafe work environments to protect them from this disease (Chick *et al.*, 1918; Smith, 1919; Carpenter, 1988).

Skin care

The juice and essential and fatty oils of lime are very valuable for skin and can be applied externally or consumed orally. It refreshes and shines the skin and lessens body odour due to the occurrence of a great deal of flavonoids and Vitamin C. It protects the skin from infection as lime contains highly beneficial anti-oxidants, and has disinfectant and antibiotic effects. On external skin application, the acids present in lime scrub out bruises, dead cells, rashes and clear out dandruff. Lime oil or juice can be added to water for a refreshing bath due to these properties (Harsha *et al.*, 2003; Rajendran *et al.*, 2008).

Digestion

The irresistible aroma of lime causes the mouth to water (the digestive saliva) even before you taste it and this actually helps in primary digestion. The rest of the job is done by the natural acidity of the lime. Natural acids present in lime promote digestion of food macro molecules. The flavonoids found in lime essential oils activate the digestive system and boost the secretion of acids, digestive juices and bile. The flavonoids present in the lime also stimulate peristaltic motion. These properties are the motive behind using lime and lemon pickle with lunch and dinner as a traditional practice in Pakistan, India and various other South Asian countries (Guimarães *et al.*, 2010; Ryan and Prescott, 2010; Rodríguez-Roque *et al.*, 2013).

Constipation

Lime washes out and clears the excretory system due to the presence of acids. The fibre in lime also eases constipation, but high acidity is the most helpful component. An overdose of salted lime juice is an outstanding purgative with no side effects and gives excellent relief from constipation (Suryawanshi, 2011).

Diabetes

Researchers working on diabetes found citrus fruits including lime to be a diabetes super food because of the high soluble fibre contents. The presence of high soluble fibre contents makes lime an ideal nutritional support as it not only helps to control the sugar absorption into the bloodstream but also reduces the occurrence of blood sugar spikes. Limes have a low glycaemic index (Owa *et al.*, 2016).

Heart disease

Lime also helps to lower blood pressure and remove low-density lipoprotein (LDL) cholesterol due to the presence of soluble fibre. Furthermore, soluble fibre reduces inflammation of the blood vessels and acts as a preventative agent against heart attacks, strokes and heart disease (Suryawanshi, 2011).

Peptic ulcers

Lime stimulates the healing process of peptic and oral ulcers due to the antibiotic, anti-carcinogenic, anti-oxidant and detoxifying properties of limonoids such as limonin glucoside and ascorbic acid (Rozza *et al.*, 2011).

Respiratory disorders

Flavonoid rich lime oil is used widely in inhalers, balms, vaporizers and in other anti-congestive medicines due to the presence of kaempferol. An immediate relief from nausea and congestion is obtained by inhaling vapours of lime peel just by scratching it.

Arthritis

One of the many causes of arthritis is the excessive accumulation of uric acid in the body. Uric acid is one of the excretory products that is eliminated from the body during normal urination. If too much uric acid accumulates in the body, it can make the arthritis pain and inflammation even worse. Lime juice containing citric acid acts a solvent for uric acid and increasing amounts of uric acid are excreted in the urine (Kikuchi *et al.*, 2009).

Eye care

Vitamin C present in lime can protect the eyes from macular degeneration and ageing. Flavonoids of lime help to protect the eyes from infections (Znaiden *et al.*, 2000).

Fever

Citrus fruits including lime are particularly important in reducing fever. In the case of very high fever, the diet should be restricted to lime juice and water. Vitamin C naturally lowers the body temperature (Adegoke *et al.*, 2011).

Gout

The two main causes of gout are the build up of free radicals and toxins in the body (particularly uric acid). Lime is a brilliant source of detoxifiers and anti-oxidants (flavonoids and Vitamin C) and thus helps to prevent both causes of gout by reducing the number of free radicals as well as detoxifying the body (Okwu and Emenike, 2006).

Piles

Lime eliminates all the reasons for piles as it provides relief from constipation and helps heal wounds and ulcers in the excretory system and the digestive tracts. Piles (haemorrhoids) occur in the anal region and can result in discomfort and bleeding both during general activity and excretion. Certain types of cancer can also be caused by piles. Limes can assist in avoiding their formation or reoccurrence.

Cholera

Lime juice acts as an effective disinfectant when added to possibly infected water and helps to reduce cholera-caused fatalities (Dalsgaard *et al.*, 1997; Rodrigues *et al.*, 2000).

Weight loss

The citric acid occurring in lime is an excellent fat reducer and excellent agent for weight loss. Lime mixed with warm water acts as a brilliant refresher and anti-oxidant drink (Bent *et al.*, 2004).

Urinary disorders

Limes contain high concentrations of potassium, which is very efficient in the elimination of toxic substances and the precipitates accumulated in the urinary bladder and kidneys. Furthermore, it stops the growth of the prostate gland, which is very common in males above 40 years and can remove calcium precipitates from the urinary tract to clear a blockage of urine (Rahmatullah *et al.*, 2010).

Cytochrome P450 inhibitory effects

Bergamottin (a furanocoumarin) plays a vital role in drug interactions by competitively inhibiting CYP 3A4 monooxygenase activities (Bailey *et al.*, 2003; Paine *et al.*, 2005).

Tumour promoting effects

Some evidence suggests that tumours might be promoted if distilled lime oils are used in the presence of carcinogenic compounds (Leung, 1980).

Industrial Products of Lime

Lime fruit juice

Lime fruit juice is an important industrial product of lime. The lime fruit is used primarily for its juice all over the world for culinary and non-culinary purposes (Morton, 1987b). Clarified lime juice is utilized as a health drink and is added to many food items. The juice of lime contains ascorbic acid, citric acid and sugar as the main components. The supply of ascorbic acid (Vitamin C) is a major contribution of lime fruits to human food (Shrestha *et al.*, 2012). Fresh limes may be used for extraction of juice, or unsweetened and sweetened juice can be purchased in bottles.

Lime essential oils

Among the citrus oils, lime oil is undoubtedly one of the most traded oils all over the world in the flavour industry. In many industrial set-ups, the extraction of lime peel oil is frequently an essential step after juice production. The essential oil of lime is extracted by cold compression of fresh lime peels, by steam distillation (Njoku and Evbuomwan, 2014) or hydrodistillation (Costa *et al.*, 2014). The lime oil so obtained is either mixed into the juice for taste improvement, or used in other well known beverages. This valuable oil is widely used in sorbets, pickles, squash, jams, marmalades, beverages, sauces, desserts, cosmetics and numerous other industrial products.

Limeade

Limeade is the perfect cool, refreshing and tasty drink with a light summer meal eaten outdoors. Limes are also much cheaper than lemons, which saves money and makes an equally delicious drink. This amazing drink is actually quite simple to make.

Lime juice is used to make limeade and also used in many cocktails. Limeade is made with lime flavouring or juice. It is available in cloudy and clear varieties. India, Guyana, Trinidad, Thailand, Pakistan and all other South-east Asian countries are the nations where limeade is especially popular due to the common availability of limes. Thai-style limeade often does not contain any sugar and tastes salty. Most major beverage companies now present their own brand of limeade, for example, A.G. Cherry Limeade drink has been introduced by Barr of Glasgow and Newman's Own and Sonic Drive-In in response to the popularity of limeade (Wilbur, 2002).

Lemon-lime drinks

Lemon-lime drinks are carbonated soft drinks with lemon and lime flavouring. Usually these drinks are colourless, though cloudy varieties such as Limca also exist. Lemon-lime soft drinks are often available in green bottles to better differentiate them from soda water. These beverages are also known as lemonade in Australia, the UK and New Zealand. The trendy brands of lemon-lime drinks consist of 7Up, Sprite and Sierra Mist.

Jam, jelly and pickle of lime

Limes are often used to make jam, jelly and marmalade. In Malaya, syrup is used to preserve lime. Limes are also pickled by first making four slices of the fruits, covering them with salt and then preserving in vinegar. These pickled fruits are eaten as appetizers after frying in coconut oil and sugar before serving.

In India, pickling of lime is done in the same way. The difference is, in India the lime pickle jar is placed in the sun for 3–4 days and the contents are mixed daily. Spices like turmeric, green chilli peppers and ginger may also be added. The preserving quality is enhanced by lastly adding coconut or another edible oil. Another method of pickling is steeping the scraped fruits in lime juice and then salting and placing in the sun (Morton, 1987a).

References

Abrahimy, Y. and Haji Vand, S. (2004) *Comparison of Minab lime, Rodan lime and Persian lime cultivars on different citrus rootstocks in Minab, Hormozgan, Iran*. Iran Citrus Research Institute – ICRI, Bandar Abbas, Iran.

Adegoke, S., Oyelami, O., Olatunya, O. and Adeyemi, L. (2011) Effects of lime juice on malaria parasite clearance. *Phytotherapy Research* 25, 1547–1550.

Ashurst, P.R. (2016) *Chemistry and Technology of Soft Drinks and Fruit Juices*. Wiley, Oxford, UK.

Bailey, D.G., Dresser, G.K. and Bend, J.R. (2003) Bergamottin, lime juice, and red wine as inhibitors of cytochrome P450 3A4 activity: comparison with grapefruit juice. *Clinical Pharmacology and Therapeutics* 73, 529–537.
Bakare, A.A., Bassey, R.B., Okoko, I.-I.E., Sanyaolu, A.O., Ashamu, A.E. and Ademola, A.(2012) Effect of lime juice (*Citrus aurantifolia*) on histomorphological alterations of the ovaries and uterus of cyclic Sprague–Dawley rats. *European Journal of Scientific Research* 67, 607–616.
Bent, S., Padula, A. and Neuhaus, J. (2004) Safety and efficacy of *Citrus aurantium* for weight loss. *The American Journal of Cardiology* 94, 1359–1361.
Boshtam, M., Moshtaghian, J., Naderi, G., Asgary, S. and Nayeri, H. (2011) Antioxidant effects of *Citrus aurantifolia* (Christm) juice and peel extract on LDL oxidation. *Journal of Research in Medical Sciences: The Official Journal of Isfahan University of Medical Sciences* 16, 951.
Camacho-Corona, M.D.R., Ramirez-Cabrera, M.A., Santiago, O.G., Garza-Gonzalez, E., Palacios, I.D.P. and Luna-Herrera, J. (2008) Activity against drug resistant-tuberculosis strains of plants used in Mexican traditional medicine to treat tuberculosis and other respiratory diseases. *Phytotherapy Research* 22, 82–85.
Carpenter, K.J. (1988) *The History of Scurvy and Vitamin C*. Cambridge University Press, Cambridge, UK.
Chick, H., Hume, E.M., Skelton, R.F. and Smith, A.H. (1918) The relative content of antiscorbutic principle in limes and lemons, together with some new facts and some old observations concerning the value of "lime juice" in the prevention of scurvy. *Lancet* 30, 735–738.
Colecio-Juárez, M.C., Rubio-Nunez, R.E., Botello-Álvarez, J.E., Martinez-Gonzalez, G.M., Navarrete-Bolanos, J.L. and Jimenez-Islas, H. (2012) Characterization of volatile compounds in the essential oil of sweet lime (*Citrus limetta* Risso). *Chilean Journal of Agricultural Research* 72, 275.
Costa, R., Bisignano, C., Filocamo, A., Grasso, E., Occhiuto, F. and Spadaro, F. (2014) Antimicrobial activity and chemical composition of *Citrus aurantifolia* (Christm.) Swingle essential oil from Italian organic crops. *Journal of Essential Oil Research* 26, 400–408.
Dalsgaard, A., Reichert, P., Mortensen, H.F., Sandstrom, A., Kofoed, P.E., Larsen, J.L. and Molbak, K. (1997) Application of lime (*Citrus aurantifolia*) juice to drinking water and food as a cholera-preventive measure. *Journal of Food Protection* 60, 1329–1333.
Dongmo, P.M.J., Tchoumbougnang, F., Boyom, F.F., Sonwa, E.T., Zollo, P.H.A. and Menut, C. (2013) Antiradical, antioxidant activities and anti-inflammatory potential of the essential oils of the varieties of *Citrus limon* and *Citrus aurantifolia* growing in Cameroon. *Journal of Asian Scientific Research* 3, 1046–1057.
Eveline, H. and Yuliana, F. (2014) The influence of sweet soy sauce or honey addition toward antibacterial activity of lime (*Citrus aurantifolia* Swingle) Juice. *International Journal of Tropical Medicine* 9, 15–20.
Food and Agricultural Organization of the United Nations (FAO) (2009) FAOSTAT. Available at: http://faostat.fao.org/site/567/DesktopDefault.aspx?PageID=567#ancor (accessed 30 August 2016).
Gharagozloo, M. and Ghaderi, A. (2001) Immunomodulatory effect of concentrated lime juice extract on activated human mononuclear cells. *Journal of Ethnopharmacology* 77, 85–90.
González-Molina, E., Moreno, D.A. and Garcia-Viguera, C. (2009) A new drink rich in healthy bioactives combining lemon and pomegranate juices. *Food Chemistry* 115, 1364–1372.
Guimarães, R., Barros, L., Barreira, J.C., Sousa, M.J., Carvalho, A.M. and Ferreira, I.C. (2010) Targeting excessive free radicals with peels and juices of citrus fruits: grapefruit, lemon, lime and orange. *Food and Chemical Toxicology* 48, 99–106.
Harsha, V., Hebbar, S., Shripathi, V. and Hegde, G. (2003) Ethnomedicobotany of Uttara Kannada District in Karnataka, India—plants in treatment of skin diseases. *Journal of Ethnopharmacology* 84, 37–40.
Ho, C.T., Mussinan, C., Shahidi, F. and Contis, E.T. (2013) *Nutrition, Functional and Sensory Properties of Foods*. Royal Society of Chemistry, London.
Jabri Karoui, I. and Marzouk, B. (2013) Characterization of bioactive compounds in Tunisian bitter orange (*Citrus aurantium* L.) peel and juice and determination of their antioxidant activities. *BioMed Research International*, 345–415.
Kamal, G., Anwar, F., Hussain, A., Sarri, N. and Ashraf, M. (2011) Yield and chemical composition of Citrus essential oils as affected by drying pretreatment of peels. *International Food Research Journal* 18(4), 1275–1282.
Khan, I.A. (2007) *Citrus Genetics, Breeding and Biotechnology*. CAB International, Cambridge, Massachusetts.
Khan, M.M., Iqbal, M., Hanif, M.A., Mahmood, M.S., Naqvi, S.A., Shahid, M. and Jaskani, M.J. (2012) Antioxidant and antipathogenic activities of citrus peel oils. *Journal of Essential Oil Bearing Plants* 15, 972–979.

Kikuchi, M., Matsuura, K., Matsumoto, Y., Inagaki, T. and Ueda, R. (2009) Bibliographical investigation of complementary alternative medicines for osteoarthritis and rheumatoid arthritis. *Geriatrics and Gerontology International* 9, 29–40.

Lawal, O.A., Ogunwande, I.A., Owolabi, M.S., Giwa-Ajeniya, A.O., Kasali, A.A., Abudu, F.A., Sanni, A.A. and Opoku, A.R. (2014) Comparative analysis of essential oils of *Citrus aurantifolia* Swingle and *Citrus reticulata* Blanco, from two different localities of Lagos State, Nigeria. *American Journal of Essential Oils and Natural Products* 2, 8–12.

Leung, A.Y. (1980) *Encyclopedia of Common Natural Ingredients Used in Food, Drugs, and Cosmetics*. Wiley, Oxford, UK.

Matthaus, B. and Özcan, M. (2012) Chemical evaluation of citrus seeds, an agro-industrial waste, as a new potential source of vegetable oils. *Grasas y Aceites* 63, 313–320.

Morton, J. (1987a) Mexican lime. In: Morton, J.F. (ed.) *Fruits of Warm Climates*, 1st edn. JF Morton, Miami, Florida.

Morton, J.F. (1987b) *Fruits of Warm Climates*. JF Morton, Miami, Florida.

Njoku, V. and Evbuomwan, B. (2014) Quantitative and qualitative analysis, and comparative study of essential oil extracted from Nigerian orange, lemon and lime peels. *International Journal of Applied Science and Engineering Research* 3, 519–531.

Okwu, D. and Emenike, I. (2006) Evaluation of the phytonutrients and vitamins content of citrus fruits. *International Journal of Molecular Medicine and Advanced Sciences* 2, 1–6.

Owa, S.O., Taiwo, A.A., Okosun, J.A., Otohinoyi, D.A., Akujobi, Y.O., Oyewalw, D.G., Ibraheem, O., Edewor-Ikupoyi, T.I. and Adeyemi, O.S. (2016) The biochemical effects of lime concentrate 'Aporo' and Mucuna pruriens seeds extract on Alloxan-induced diabetic rats. *Journal of Taibah University Medical Sciences* 11(3), 260–267.

Paine, M.F., Criss, A.B. and Watkins, P.B. (2005) Two major grapefruit juice components differ in time to onset of intestinal CYP3A4 inhibition. *Journal of Pharmacology and Experimental Therapeutics* 312, 1151–1160.

Patil, J.R., Jayaprakasha, G.K., Patil, B.S. and Chetti, M.B. (2014) Antioxidant activities of kagzi lime (*Citrus aurantifolia* swingle). In: Carlin, M.J.A. and Fayard, G (eds) *III International Symposium on Human Health Effects of Fruits and Vegetables – FAVHEALTH 2009*. International Society for Horticultural Science (ISHS), Leuven, Belgium, pp. 381–384.

Rahmatullah, M., Khatun, A., Morshed, N., Neogi, P.K., Khan, S.U.A., Hossan, S., Mahal, M.J. and Jahan, R. (2010) A randomized survey of medicinal plants used by folk medicinal healers of Sylhet Division, Bangladesh. *Advances in Natural and Applied Sciences* 4, 52–63.

Rajendran, K., Balaji, P. and Basu, M.J. (2008) Medicinal plants and their utilization by villagers in southern districts of Tamil Nadu. *Indian Journal of Traditional Knowledge* 7, 417–420.

Reazai, M., Mohammadpourfard, I., Nazmara, S., Jahanbaksh, M. and Shiri, L. (2014) Physicochemical characteristics of citrus seed oils from Kerman, Iran. *Journal of Lipids* 2014, 1–3.

Rodrigues, A., Brun, H. and Sandstrom, A. (1997) Risk factors for cholera infection in the initial phase of an epidemic in Guinea-Bissau: protection by lime juice. *The American Journal of Tropical Medicine and Hygiene* 57, 601–604.

Rodrigues, A., Sandstrom, A., Ca, T., Steinsland, H., Jensen, H. and Aaby, P. (2000) Protection from cholera by adding lime juice to food–results from community and laboratory studies in Guinea-Bissau, West Africa. *Tropical Medicine and International Health* 5, 418–422.

Rodríguez-Roque, M.J., Rojas-Graü, M.A.A., Elez-Martínez, P. and Martín-Belloso, O. (2013) Changes in vitamin C, phenolic, and carotenoid profiles throughout in vitro gastrointestinal digestion of a blended fruit juice. *Journal of Agricultural and Food Chemistry* 61, 1859–1867.

Rowe, A.K., Angulo, F.J. and Tauxe, R.V. (1998) A lime in a litre rapidly kills toxogenic *Vibrio cholerae* O1. *Tropical Doctor* 28, 247.

Rozza, A.L., De Mello Moraes, T., Kushima, H., Tanimoto, A., Marques, M.O.M., Bauab, T.M., Hiruma-Lima, C.A. and Pellizzon, C.H. (2011) Gastroprotective mechanisms of *Citrus lemon* (Rutaceae) essential oil and its majority compounds limonene and β-pinene: involvement of heat-shock protein-70, vasoactive intestinal peptide, glutathione, sulfhydryl compounds, nitric oxide and prostaglandin E 2. *Chemico-biological Interactions* 189, 82–89.

Ryan, L. and Prescott, S.L. (2010) Stability of the antioxidant capacity of twenty-five commercially available fruit juices subjected to an in vitro digestion. *International Journal of Food Science and Technology* 45, 1191–1197.

Sandoval-Montemayor, N.E., Garcia, A., Elizondo-Trevino, E., Garza-Gonzalez, E., Alvarez, L. and Del Rayo Camacho-Corona, M. (2012) Chemical composition of hexane extract of *Citrus aurantifolia* and anti-*Mycobacterium tuberculosis* activity of some of its constituents. *Molecules* 17, 11173–11184.

Shrestha, R.L., Dhakal, D.D., Gautum, D.M., Paudyal, K.P. and Shrestha, S. (2012) Variation of physiochemical components of acid lime (*Citrus aurantifolia* Swingle) fruits at different sides of the tree in Nepal. *American Journal of Plant Sciences* 3, 1688–1692.
Smith, A.H. (1919) A historical inquiry into the efficacy of lime-juice for the prevention and cure of scurvy. *Journal of the Royal Army Medical Corps* 32, 93–116.
Spadaro, F., Costa, R., Circosta, C. and Occhiuto, F. (2012) Volatile composition and biological activity of key lime *Citrus aurantifolia* essential oil. *Natural Product Communications* 7, 1523–1526.
Suryawanshi, J.A.S. (2011) An overview of *Citrus aurantium* used in treatment of various diseases. *African Journal of Plant Science* 5, 390–395.
Vand, S.H. and Abdullah, T.L. (2012) Identification and introduction of thornless lime (*Citrus aurantifolia*) in Hormozgan, Iran. *Indian Journal of Science and Technology* 5, 3670–3673.
Wilbur, T. (2002) *Top Secret Recipes – Sodas, Smoothies, Spirits, and Shakes: Creating Cool Kitchen Clones of America's Favorite Brand-Name Drinks*. Penguin, London.
Znaiden, A.P., Crotty, B.A. and Gural, P.D. (2000) Moistened cosmetic eye treatment fads. Google Patents. Available at: www.google.com/patents/US6159487 (accessed 2 March 2017).

Index

Page numbers in *italic* refer to Figures.